MW01505658

The U.S. Department of Energy's Building America
Program is reengineering the American home for
energy efficiency and affordability. Building America
works with the residential building industry to develop
and implement innovative building processes and
technologies innovations that save builders and
homeowners millions of dollars in construction and
energy costs. This industry-led, cost-shared partnership program uses a systems engineering approach to
reduce energy use, utility bills, construction time, and
construction waste.

The research conducted by Building America teams
improves the energy efficiency, sustainability, quality,
and performance of today's homes and provides
valuable information for homes of the future. By
supporting the development of innovative building
methods and technologies that achieve significant
energy and cost savings, the Building America Program is helping to shape the future of American
Homes.

FOR MORE INFORMATION VISIT OUR WEB SITE AT:
www.buildingamerica.gov

RESEARCH AND DEVELOPMENT OF BUILDINGS

Our nation's buildings consume more energy than any other sector of the U.S. economy, including transportation and industry. Fortunately, the opportunity to reduce building energy use—and the associated environmental impacts—are significant.

DOE's Building Technologies Program works to improve the energy efficiency of our nation's buildings through innovative new technologies and better building practices. The program focuses on two key areas:

- **Emerging Technologies**
  Research and development of the next generation of energy-efficient components, materials and equipment

- **Technology Integration**
  Integration of new technologies with innovative building methods to optimize building performance and savings

TO LEARN MORE ABOUT BUILDING AMERICA, CONTACT:

www.buildingamerica.gov   www.pathnet.org   www.energystar.gov

George James
DOE's Building America Program
(202) 586-9472
fax: (202) 586-8134
email: George.James@hq.doe.gov

Brad Oberg
IBACOS Consortium
(412) 765-3664
fax: (412) 765-3738
email: boberg@ibacos.com

Steven Winter
Consortium for Advanced Residential Buildings
(203) 857-0200
fax: (203) 852-0741
email: swinter@swinter.com

Tom Kenney
NAHB Research Center
(800) 638-8556
fax: (301) 430-6180
www.nahbrc.org

Ren Anderson
National Renewable Energy Laboratory
(303) 384-7433
fax: (303) 384-7540
email: ren_anderson@nrel.gov

Betsy Pettit
Building Science Consortium
(978) 589-5100
fax: (978) 589-5103
email: betsy@buildingscience.com

Robert Hammon
Building Industry Research Alliance
(209) 473-5000
fax: (209) 474-0817
email: Rob@consol.ws

Subrato Chandra
Industrialized Housing Partnership
(321) 638-1412
fax: (321) 638-1439
email: subrato@fsec.ucf.edu

David Springer
Davis Energy Group
(530) 753-110
(503) 753-4125
email: springer@davisenergy.com

Pat M. Love
Oak Ridge National Laboratory
(865) 574-4346
fax: (865) 574-9331
lovepm@ornl.gov

A small contingent of building professionals representing the United States, Canada, and Sweden, gathered in Pine Island, Minnesota in early 1982 to develop criteria for the construction of buildings that were more energy efficient. These forward thinking pioneers created the Energy Efficient Building Association.

Today, EEBA's mission reflects the vision of those pioneers in its goal to provide education and resources to transform the residential design, development and construction industries to profitability deliver energy efficiency and environmentally responsible buildings and communities. The strength of the organization lies in the diversity and talent of its partners, which includes: architects, builders, developers, manufacturers, engineers, utilities, code officials, researchers, educators, and environmentalists.

As a leading provider of educational resources EEBA offers the industry the following:

The EEBA **Institute of Building Construction Technology** (**www.eeba.org/institute**) is the organization's primary educational outreach initiative. The institute is based on principles of building science and the "System's Approach" to construction for improved building performance and durability. The curriculum consists of a 12-module/100 credit hour format and optional MasterBuilder Project with both a class-room and online option.

Based on research from the U.S. Department of Energy's Building America Program, EEBA's highly acclaimed, **Houses That Work**™ (**www.eeba.org/events/buildingamerica**) offers climate specific building science education using regional case studies of high performance homes and the dramatic results that can be achieved. Each 8-hour session provides attendees with resource materials, and a modified selection of the EEBA Bookstore and support materials as well as live demonstrations and a mini-expo.

For over 23 years the **Building Solutions Conference** has provided the industry with the most comprehensive building performance educational opportunity through classroom sessions, live demonstrations, regionally specific tours and expo. Today EEBA has expanded its conference capabilities by providing the management to the educational components of other industry shows such as the Sunbelt Builders Show, regional training sessions and colleges (**www.eeba.org/conference**).

The EEBA organization has long served as a broker for the gathering and dissemination of best practice literature and information. Nationally recognized books such as this **Climate Specific Builder's Guide** and EEBA **White Papers**, such as the **Water Management Guide** provide highly pictorial guidance for the proper selection and installation of building materials for better building performance. These and other leading industry resources are offered through the organization's online **Bookstore** (**www.eeba.org/bookstore**) and web resource sights as a one-stop shop for the most comprehensive collection of industry materials offered today.

The EEBA organization continues to develop alliances within the industry to bring builders, contractors, manufacturers and industry organizations together to better serve the needs of the housing market. Together we are building the future today.

For more information on the Energy & Environmental Building Association visit **www.eeba.org** or contact:

Energy & Environmental Building Association
10740 Lyndale Avenue South
Suite 10 West
Minneapolis, MN 55420
952.881.1098 fax 952.881.3048

Builder's Guide

Hot-Humid Climates

A systems approach to designing and building homes that are safe, healthy, durable, comfortable, energy efficient and environmentally responsible.

Written by: **Joseph Lstiburek, Ph.D. , P. Eng.**
Building Science Corporation
70 Main Street
Westford, MA 01886
(978) 589-5100
www.buildingscience.com

Comments and constructive criticism of this publication are welcomed by the author and all such comments will be considered in future revisions. Please contact the author directly at joe@buildingscience.com.

Every effort has been made to provide current, reliable information on best practices associated with the construction of building enclosures and building subsystems. However, in every complex undertaking, such as the creation of this guide, errors occur. As new information becomes available a change or modification to a detail or recommended practice may be warranted. This information is posted and updated regularly on the "*Builders Guide Errata*" webpage that can be found on the Building Science Corporation website at www.buildingscience.com.

This guide was significantly enhanced by the thoughtful review of the following individuals whose input is greatly appreciated:

Terry Brennan, Comroden Associates

Dennis Creech, Southface Energy Institute

Stephanie Finegan, Building Science Corporation

Don Gatley, Gatley & Associates

Neil Moyer, Florida Solar Energy Center

Betsy Pettit, Building Science Corporation

Armin Rudd, Building Science Corporation

Kohta Ueno, University of Waterloo

Book Design and   Stephanie Finegan
Illustrations by: Building Science Corporation

*Rob ;
Draw the Rain
on the Pluy*

Acknowledgments

The building science information presented in this guide and the companion guides for other climates has evolved over the past 100 years. Many individuals and numerous institutions, organizations and agencies have contributed significantly, often anonymously, to the wealth of information and experience contained here. Fundamental research from the research establishments of four nations — Canada, Norway, Sweden and the United States — provides the foundation for the construction details and approaches presented. More significantly, the lessons learned from the construction of the Arkansas House, Saskatchewan Conservation House, Leger House, Canada's R-2000 program and the U.S. Building America Program were applied, altered, improved, discarded, rediscovered and massaged by builders, contractors, architects and engineers throughout North America in thousands of field tests and experiments, sometimes referred to as home-building. The experience from these lessons can be found in the following pages.

Joseph Lstiburek,
Westford, MA
July, 2000

*Revised
Second printing
June, 2005*

About the Author

Joseph Lstiburek is principal of Building Science Corporation in Westford, Massachusetts. He is an ASHRAE Fellow and has been a licensed professional engineer since 1982.

He received an undergraduate degree in Mechanical Engineering from the University of Toronto, a master's degree in Civil Engineering from the University of Toronto and a doctorate in Building Science from the University of Toronto.

He used to be a contractor—and then he crossed to the "dark side."

When we build, let us think that we build forever. Let it not be for present delight nor for present use alone. Let it be such work as our descendants will thank us for; and let us think, as we lay stone on stone, that a time is to come when those stones will be held sacred because our hands have touched them, and that people will say, as they look upon the labor and wrought substance of them, "See! This our parents did for us."

John Ruskin

Contents

About This Guide xii
Hygro-Thermal xiv
Approach xviii

Introduction 1
Changes in Construction 3
Thermal Insulation 3
Tighter Building Enclosures and New Materials 3
Heating and Cooling Systems 4
Integration 5

The House System 7
Functional Relationships 7
Prioritization 12
People Priorities 13
Building Priorities 13
Environmental Priorities 13

PART I DESIGN 15
Design 17
Site Planning and Building Form 17
House Layout 21
Thermal Mass 25
Basic Structure and Dimensions 32
Building Enclosure 32
Vapor Barriers 36
Perimeter Slab Edge Insulation 36
Roofs 39
Unvented Crawlspaces 41
Mechanical Systems 41
Controlled Ventilation and Equipment Selection 43
Fireplaces 57
Material Selection 57
Appliances 59
Solar Water Heating 61
Commissioning 61

Rain, Drainage Planes and Flashings 63
Physics of Rain 64
Water Management Approaches 65

Drainage Planes/Water Resistant Barriers 67
Rainwater Control 68

3 **Air Barriers** 101
Performance Requirements 102
Approaches 103

4 **Insulations, Sheathings and Vapor Retarders** 113
Vapor Diffusion and Air Transport of Vapor 113
Vapor Retarders 115
Water Vapor Permeability 115
Air Barriers.. 118
Magnitude of Vapor Diffusion and
Air Transport of Vapor 119
Combining Approaches 120
Overall Strategy 120
Cold Climates 121
Hot Climates.. 121
Mixed Climates 122
Control of Condensing Surface Temperatures.............. 123
Sheathings and Cavity Insulations 126

5 **Wall Design** 139
Moisture Balance 139

6 **Roof Design** 155
Approach 155
Vented Roof Design 156
Unvented Roof Design 157

7 **Foundation and Wall Assemblies** 175
Water Managed Foundations 175
Soil Gas 176
Moisture 176
Slab Construction 177
Slab Moisture Problems 177
Carpets 184
Crawlspaces 185
Insects and Termites 188

8 **Creatures and Dust Control** 223
To Know the Critter is to Control the Critter 223
Keeping Them Out 224
Reducing Food and Water 224
Pesticides 224
Dust 225
Enrty Control 225

Cleanable Surfaces 225
Filtration 225

PART II CONSTRUCTION 227

9   **General Contractor** 229
Concerns 229
Materials on the Building Site 231

10   **Concrete and Excavation** 233
Polyethylene Under Slabs 239

11   **Wood Frame Construction** 243
Concerns 243
Frame Movement 244
Rain 245
Air Barrier 245
Moisture 245
Paint and Trim 245

12   **Masonry Construction** 303
Concerns 303

13   **HVAC** 321
Concerns 321
Choices 322
Ventilation Requirements 323
Indoor Humidity and Airborne Pollutants 325
Dehumidification 327
Combustion Appliances 328
Garages 329
Smoke 329
Recirculating Fans 329
Whole House Fans 329
Ceiling Fans 340
System Sizing 340
Air Handlers and Ductwork 342
Sheet Metal Duct Sealing 343
Flex Duct Sealing 343
Sealing Boots and Grilles 343
Plenums 346
Return Path 347
Cooling Coils and Drain Pans 347
Air Filtration 348
Large Exhaust Fans 349
Clothes Dryers 349
Central Vacuum Systems 349

14 **Plumbing** 371
Concerns 371
Tubs and Shower Stalls 371
Water Consumption 372

15 **Electrical** 375
Concerns 375

16 **Insulation** 381
Concerns 381
Fiberglass 382
Cellulose 387
Roofs 387
Crawlspaces 388
Spray Foam 388

17 **Drywall** 389
Concerns 389
Truss Uplift 390
Air Barriers 390
Ceramic Tile Tub and Shower Enclosures 391
Winter Construction 391

18 **Painting** 399
Concerns 399
Paint vs. Stains 399
Primers 400
Wood Substrates 400
Stucco Substrates 401
Deck Substrates 402
Peeling Paint on Siding and Trim 403
Stucco and Housewrap 404
Interior Surfaces 404

**PART III ALTERNATIVES TO WOOD
FRAME CONSTRUCTION** 407

19 **Insulating Concrete Forms (ICF)** 409
Concerns 409

20 **Structural Insulated Panel Systems (SIPS)** 433
Concerns 433

21 **Precast Autoclaved Aerated Concrete (PAAC)** 461
Concerns 461

X **APPENDICES** 477
Appendix I: Design Data 477
Appendix II: Air Leakage Testing, Pressure Balancing,
               and Combustion Safety 503
Appendix III: Best Practices for High Performance
               Homes in Hot-Humid Climates 511
Appendix IV: Transfer Grille Sizing Charts 517
Appendix V: Additional Resources 521
Appendix VI: Glossary 525

A-Z **Index** 531

Firmness **C**ommodity **D**elight

"These are properly designed, when due regard is had to the country and climate in which they are erected. For the method of building which is suited to Egypt would be very improper in Spain, and that in use in Pontus would be absurd at Rome: so in other parts of the world a style suitable to one climate, would be very unsuitable to another: for one part of the world is under the sun's course, another is distant from it, and another, between the two is temperate."

Marcus Vitruvious Pollio

About this Guide

Buildings should be suited to their environment. It is irrational to expect to construct the same manner of building in Montreal, Memphis, Mojave and Miami. It's cold in Montreal, it's humid in Memphis, it's hot and dry in Mojave and it's hot and wet in Miami. And that's just the outside environment. It is equally irrational to expect to construct the same manner of building to enclose a warehouse, a house or a health club with a swimming pool. The interior environment also clearly matters.

We have accepted that design and construction must be responsive to varying seismic regions, wind loads and snow loads. We also consider soil conditions and frost depth, orientation and solar radiation. Yet we typically ignore the variances in temperature, humidity, rain of the exterior climate and the variances in the interior climate.

Building enclosures and mechanical systems should be designed for a specific hygro-thermal region, rain exposure zone and interior climate class in addition to the previously mentioned external environmental loads. The following hygro-thermal regions, rain exposure zones and interior climate classes influence design:

Hygro-Thermal Regions (see pages xiv and xv)
    Arctic/subarctic
    Very cold
    Cold
    Marine
    Mixed-humid
    Hot-humid
    Hot-dry/Mixed-dry

Rain Exposure Zones (see page 70)
    Extreme (above 60 inches annual precipitation)
    High (40 to 60 inches annual precipitation)
    Moderate (20 to 40 inches annual precipitation)
    Low (less than 20 inches annual precipitation)

Interior Climate Classes
    I   Warehouses, Garages, Storage Rooms
        Temperature moderated
        Vapor pressure uncontrolled
        Air pressure uncontrolled

II   Houses, Apartments, Condominiums, Offices, Schools
     Temperature controlled
     Vapor pressure moderated
     Air pressure moderated

III Hospitals, Museums, Swimming Pool Enclosures and Computer Facilities
     Temperature controlled
     Vapor pressure controlled
     Air pressure controlled

This builder's guide focuses on construction in hot-humid climate hygro-thermal regions with extreme to moderate rain exposure zones for building enclosures and mechanical systems suited for a Class II interior climate — that is an interior climate that is temperature controlled, vapor pressure moderated and air pressure moderated. In other words houses, apartment, condominiums, townhouses, and manufactured housing.

Hygro-Thermal Region for this Guide

This guide contains information that is applicable to hot-humid climates. A hot-humid climate is defined as a region that receives more than 20 inches (50 cm) of annual precipitation and where one or both of the following occur:*

- a 67°F (19.5°C) or higher wet bulb temperature for 3,000 or more hours during the warmest six consecutive months of the year; or

- 73°F (23°C) or higher wet bulb temperature for 1,500 or more hours during the warmest six consecutive months of the year.

Figure A, Climate Zones, illustrates the seven major climate zones in North America used to distinguish the range of applicability of this guide and the companion guides for cold, mixed-humid and hot-dry/mixed-dry climates. Each climate zone specified is broad and general for simplicity. The climate zones are generally based on Herbertson's Thermal Regions, a modified Koppen classification (see Goode's World Atlas, 19th Edition, Rand McNally & Company, New York, NY, 1990), the ASHRAE definition of warm-humid climates (see ASHRAE Fundamentals, ASHRAE, Atlanta, GA, 1997), the International Energy Conservation Code (IECC) Climate Zones, and average annual precipitation obtained from the U.S. Department of Agriculture. For a specific climate zone, designers and builders should consider weather records, local experience, and the micro-climate around a building. Elevation, incident solar radiation, nearby water and wetlands, vegetation, and undergrowth can all affect the micro-climate.

Although this guide provides general recommendations with applicability based on Figure A, local experience and local building codes should also be considered. Where a conflict between local code and regulatory requirements and the recommendations in this guide occur, authorities having jurisdiction should be consulted or the local code and regulatory requirements should govern.

The recommendations on construction relating to wind loading and seismic zones are extremely general and are not intended to substitute for professional design by a professional engineer or licensed architect and are not intended to conflict with specific code requirements. The recommendations are intended to alert the reader that conditions vary

---

* These last two criteria are identical to those used in the ASHRAE definition of warm-humid climates and are very closely aligned with a region where monthly average outdoor temperature remains above 45°F (7°C) throughout the year.

significantly across the hot-dry and mixed-dry climate zones defined in Figure A and that due diligence should occur.

Illustrations depicting wood framing are shown with exterior walls framed using 2x6 framing techniques. The colored lines or colored shading on all illustrations represent materials that form the air barrier system.

Precise specification of materials and products is not typically provided on the illustrations or in the text to provide maximum flexibility. It is the responsibility of the designer, builder, supplier and manufacturer to determine specific material compatibility and appropriateness of use. For example, there are a wide range of performance and cost issues dealing with sealants, adhesives, tapes and gaskets. Hot weather or cold weather construction and oily, damp or dusty surfaces affect performance along with substrate compatibility issues. Tapes must be matched to substrates. Similarly, sealants and adhesives must be matched to materials and joint geometry. In hot-humid and hot-dry climates, intense solar exposure typically results in damage to materials from ultraviolet radiation. Paints, sealants and coatings have greatly reduced service lives. Roofing materials and wood products are particularly sensitive.

Generally, several different tapes, sealants, adhesives or gaskets can be found to provide satisfactory performance when installed in the locations illustrated in this guide. Premium tapes, sealants, adhesives or gaskets typically (but not always) out perform budget tapes, sealants, adhesives or gaskets. It is always advisable to obtain samples and test compatibility and performance on actual material substrates prior to construction and over an extended period of time.

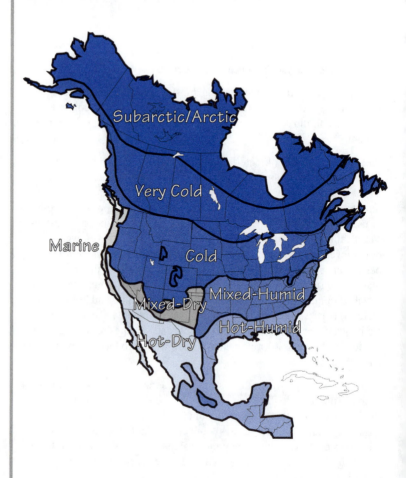

Figure A
Hygro-Thermal Regions
- Based on Herbertson's Thermal Regions, a modified Koppen Classification, the ASHRAE definition of hot-humid climates, the International Energy Conservation Code (IECC) Climate Zones, and average annual precipitation from the U.S. Department of Agriculture and Environment Canada

1 Celsius: 7,000 heating degree days (18°C basis)
2 Celsius: 5,000 heating degree days (18°C basis)
3 Celsius: 3,000 heating degree days (18°C basis)
4 Celsius: 5,000 heating degree days (18°C basis)
5 Celsius: 3,000 heating degree days (18°C basis)

Legend

Subarctic/Arctic

 A subarctic and arctic climate is defined as a region with approximately 12, 600 heating degree days (65°F basis)[1] or greater

Very Cold

 A very cold climate is defined as a region with approximately 9,000 heating degree days (65°F basis)[2] or greater and less than approximately 12,600 heating degree days (65°F basis)

Cold

 A cold climate is defined as a region with approximately 5,400 heating degree days (65°F basis)[3] or greater and less than approximately 9,000 heating degree days (65°F basis)[4]

Mixed-Humid

 A mixed-humid climate is defined as a region that receives more than 20 inches (50 cm) of annual precipitation, has approximately 5,400 heating degree days (65°F basis)[5] or less, and where the monthly average outdoor temperature drops below 45°F (7°C) during the winter months

Marine

 A marine climate meets all of the following criteria:
- A mean temperature of the coldest month between 27°F (-3°C) and 65°F (18°C)
- A warmest month mean of less than 72°F (22°C)
- At least four months with mean temperatures over 50°F (10°C)
- A dry season in summer. The month with the heaviest precipitation in the cold season has at least three times as much precipitation as the month with the least precipitation in the rest of the year. The cold season is October through March in the Northern Hemisphere and April through September in the Southern Hemisphere.

Hot-Humid

 A hot-humid climate is defined as a region that receives more than 20 inches (50 cm) of annual precipitation and where one or both of the following occur:*
- a 67°F (19.5°C) or higher wet bulb temperature for 3,000 or more hours during the warmest six consecutive months of the year; or

- a 73°F (23°C) or higher wet bulb temperature for 1,500 or more hours during the warmest six consecutive months of the year.

* These last two criteria are identical to those used in the ASHRAE definition of warm-humid climates and are very closely aligned with a region where monthly average outdoor temperature remains above 45°F (7°C) throughout the year.

Mixed-Dry

 A mixed-dry climate is defined as a region that receives less than 20 inches (50 cm) of annual precipitation, has approximately 5,400 heating degree days (65°F basis) or less, and where the average monthly outdoor temperature drops below 45°F (7°C) during the winter months

Hot-Dry

A hot-dry climate is defined as a region that receives less than 20 inches (50 cm) of annual precipitation and where the monthly average outdoor temperature remains above 45°F (7° C) throughout the year

Approach

Reducing a builder's number one headache: warranty and callback expenses, should be considered a "payback" concept. Reducing warranty and callback expenses often involves increasing initial construction costs. The good news to the builder is that the up front cost to the builder results in subsequent cost savings also to the builder. It is logical to pay a little more up front to prevent warranty and callback expenses, rather than a great deal more down the road to deal with warranty and callback expenses.

Building energy-efficient homes can also be considered a "payback" concept. Energy-efficient construction usually involves upgrading materials or equipment to increase the energy efficiency of a home. These changes typically add to the initial cost of a home. Economically, the increased cost is often justified to the home owner based on subsequent cost savings from reduced energy bills. This is also a "pay more up front" and get savings "in the long term" approach. Unfortunately, for the builder, the up front cost is to the builder, but the subsequent cost savings is to the home owner. Builders have to pass on the up front costs to the home owner resulting in a more expensive home. Not all home owners recognize the long term savings from this approach and typically object to the higher up front costs.

The payback concept is not the only approach to reducing warranty and callback expenses or the only approach to energy-efficient construction. There are "break points" where the cost of the warranty and callback reduction strategies as well as the energy-efficient features are balanced by the reductions of other construction costs. These "break points" involve construction strategies or levels of energy efficiency that allow a specific component of a building to be downsized or deleted. For example, construction costs can be increased by changes and improvements to the building enclosure that reduce warranty and callback expenses as well as reduce heat gain and heat loss. The improved building enclosure performance allows the mechanical equipment to be downsized. The initial construction cost increases are offset by the reduced costs associated with the downsized mechanical system.

The construction cost savings that occur from applying a systems engineering approach to warranty and callback reduction, and energy conservation, are typically able to pay for the increased costs associated with "healthy housing" and resource efficiency. The end result be-comes a home that is healthier, safer, more comfortable, durable and affordable, with no increased cost to the builder or home buyer.

This builder's guide addresses both warranty and callback expenses as well as energy-efficient construction. Strategies are presented to reduce drywall cracking, nail pops, paint and trim problems, dust marking of carpets, comfort complaints and "ghost-marking" of studs and other framing.

Introduction

The basic requirement for buildings is to create an indoor environment different from the outdoors. In this regard, buildings are environmental separators. They allow the regulation of temperature, air movement, humidity, rain, snow, light, dust, odors, noise, vibrations, insects and vermin. They must accomplish this in a safe, healthy and durable manner.

In the past, the design and construction of buildings had been based on building practices developed through an evolutionary process of trial and error. Since the end of the Second World War, the technology of construction has become exceedingly complex and there have been major developments in materials, products and systems. The evolutionary process of trial and error is no longer adequate.

Specialization and an abundance of specialists has been one of the consequences of the complexity of current construction. These specialists have tended to focus on their own disciplines. The list of specialists involved in house construction can be endless, from the financial consultant and interior designer to the soils engineer, truss designer, lighting consultant and environmental advisor. However, all of these disciplines are interrelated and have an effect on the durability and performance of a building.

Building design and construction have become fragmented as a consequence of the increasing specialization and sheer quantity of available information. Prior to the post-war building boom, the architectural profession and the "master builder" provided the understanding of the construction process. Architects and master builders were the generalists who knew all aspects of the construction process; they were the original "building scientists." Architects and master builders applied knowledge of traditional materials combined with aesthetic judgment to create magnificent and effective buildings. Tradition and past experience assisted in improving the understanding of a slowly evolving construction process. Architects and master builders provided the "systems view"; tradition and past experience provided a "system model."

During the post-war building boom, the emphasis on educating architects shifted to aesthetics and design theory relating to aesthetics and away from the fundamental aspects of construction and an understanding of materials, assemblies, building systems and subsystems. For builders, the focus became finance, land development and marketing. Accountants became kings. As in the case of architects, builders began to lose their understanding of the fundamental aspects of construction. Tradition and past experience remained dominant, but fundamental understanding waned.

At the same time, service conditions changed as well as materials. Slow, evolutionary changes in the construction process gave way to rapid changes. Buildings became more heavily insulated, building enclosures became tighter, forced air heating and cooling systems were introduced, countless new construction materials were introduced, and the post-war consumer revolution resulted in a plethora of interior surface finishes, furnishings and consumer amenities. Larger and more sophisticated kitchens and more and larger bathrooms became the norm. Rooflines became more complex, and overhangs became narrower or non-existent. Lifestyles changed resulting in more time spent within conditioned spaces and occupant comfort needs became preeminent. However, the construction process remained based on tradition.

As architects and builders focused more on aesthetics and finance and less on the construction process, gaps in the process began to be filled by specialists. Some specialists such as subcontractors have extensive training and experience in putting the pieces together efficiently. As a consequence, we typically get buildings constructed on time and on budget. However, the understanding of performance and predictability has become lost, so, although buildings are constructed on time and on budget, they often do not work.

The current process remains based on tradition and past experience without understanding. Tradition has a great weakness in that it deals only with a way of doing something but not with an understanding of why the traditional methods work.

In order to impose rationality on the current fragmented design/build process, a systems view and a system model is necessary. A systems view of buildings is necessary to provide predictability and understanding.

Changes in Construction

In the last fifty years there have been three important changes to the way we build homes:

- the introduction of thermal insulation
- development of tighter building enclosures
- the advent of forced air heating and cooling systems

Each of these changes has made homes more comfortable, but has also made the same houses less durable and more unpredictable in terms of performance.

Thermal Insulation

Thermal insulation was added to wall cavities and ceilings to keep the heat in the home during the winter, keep it out in the summer and make the home more comfortable. Wall cavities and attics became colder because the insulation did its job. As a result, two things happened; the walls got wetter because as they got colder the relative humidities went up and therefore so did moisture contents, and the ability of these assemblies to dry when they get wet from either interior or exterior sources was reduced because there was less energy available to dry them. How do you dry a material? You heat it. No heat, no drying. The addition of thermal insulation increased the "wetting potential" of the building enclosures while at the same time it reduced their "drying potential."

Tighter Building Enclosures and New Materials

Homes built today are much tighter than the homes of yesterday. We use plywood, gypsum board and other sheet sheathings in place of board sheathing and plaster. We platform frame instead of balloon frame. We use prefabricated windows instead of site glazing. House-wraps come in 10-foot-wide rolls unlike building papers which come in 3-foot-wide rolls. We put more caulk, glue and sealants on our houses than ever before, and we can buy material that actually sticks and holds. The results are fewer holes and a lower air change. The lower the air change, the less the dilution of interior pollutants such as moisture (from people, soil and appliances), formaldehyde (from particle board, insulation, furniture and kitchen and bathroom cabinets), volatile organic compounds (from carpets, paints, cleaners and adhesives), radon (from basements, slabs, crawlspaces and water supplies) and carbon dioxide (from people).

This trend to lower air change occurred simultaneously with the introduction of hundreds of thousands of new chemical compounds, materials and products which were developed to satisfy the growing consumer demand for household goods and furnishings. Interior pollutant sources have increased while the dilution of these pollutants has decreased. As a result, indoor air pollutant concentrations have increased.

Additionally, chimneys don't work well in tight homes. In tight homes, other exhaust fans compete with chimneys and flues for available air. The chimneys and flues typically lose in the competition for available air, resulting in spillage of combustion products, and backdrafting of furnaces and fireplaces.

As air change goes down, interior moisture levels rise causing condensation problems on windows where the surface is cold, mold on walls, dust mites in carpets and decay in wall cavities and attic spaces even in traditionally forgiving climates such as hot-dry and mixed-dry climates. Even though interior moisture levels are rising, builders continue to install central humidifiers rather than installing dilution ventilation or dehumidification and neglect to warm potential condensing surfaces such as window frames and glazing by installing thermally broken frames and insulating glass.

Traditional chimneys in many new homes have been replaced with power vented, sealed combustion furnaces. Many new homes have no chimneys or flues and rely on heat pumps or electric heating. Traditional chimneys ("active chimneys") acted as exhaust fans. They extracted great quantities of air from the conditioned space that resulted in frequent air changes and the subsequent dilution of interior pollutants. Eliminating the "chimney fan" has led to an increase in interior pollutant levels such as moisture.

Active chimneys also tended to depressurize conditioned spaces during heating periods. Depressurization in cold climates led to a reduced wetting of building assemblies from interior air-transported moisture and therefore a more forgiving building enclosure.

Heating and Cooling Systems

Today, forced air systems (heating and air conditioning) move large quantities of air within building enclosures of increasing tightness. The tighter the building enclosure the easier it is to pressurize or depressurize. Improperly installed forced air systems can lead to serious health, safety, durability, and operating cost issues.

Supply duct systems are typically more extensive than return duct systems. There are usually supply registers in each room with common centralized returns. Centralized returns replaced distributed returns to reduce costs. Pressurization of rooms and depressurization of common areas is created by the combination of more extensive supply systems, leaky supply and return ductwork combined with interior door closure.

Typical ductwork leaks. Air handlers leak. When leaky supply ducts are run outside the building enclosure in garages, vented attics, roof and crawlspaces, depressurization of the building enclosure occurs. Similarly, locating air handlers outside the building enclosure leads to the same result. Depressurization can cause infiltration of radon, moisture, pesticides and soil gas into foundations as well as the probable spillage and backdrafting of combustion appliances and potential flame roll-out resulting in fire.

Leaky supply ducts and chases connected to exterior spaces can lead to depressurization of the building enclosure. Depressurization in hot-humid climates can lead to the infiltration of hot moisture-laden air into wall and roof cavities which are at lower drying potentials because of higher levels of insulation.

Integration

The three important changes in the way we build homes today interact with each other. This is further complicated by the effects of climate and occupant lifestyle. The interrelationship of all of these factors has led to major warranty problems that include health, safety, durability, comfort and affordability concerns. Problems are occurring despite the use of good materials and good workmanship.

In hot-humid climates the use of higher levels of thermal insulation and better window systems has led to a reduction of heat gain. As this "sensible" load is reduced the ability of the air conditioning systems to remove moisture is reduced. At the same time, leaky ductwork leads to an increase in infiltration increasing "latent" or moisture loads. This unbalance of the latent to sensible load ratios assures moisture problems. Even "correctly" sized air conditioning systems are no longer up to the task of simultaneously providing temperature and humidity control.

We cannot return to constructing drafty building enclosures without thermal insulation, without consumer amenities, and with less efficient heating and air conditioning systems. The marketplace demands sophisticated, high performance buildings operated and maintained intelligently. As such, buildings must be treated as integrated systems that address health, safety, durability, comfort and affordability.

Quality construction consists of more than good materials and good workmanship. If you do the wrong thing with good materials and good workmanship, it is still wrong. You must do the right thing with good materials and good workmanship. The purpose of this guide is to promote the use of good materials and good workmanship in a systematic way, so that all the parts work together and promote good performance, durability, comfort, health and safety in an environmentally responsible manner.

The House System

Functional Relationships

Residential construction is a complex operation including thousands of processes by dozens of industries, bringing together hundreds of components and sub-systems into a house. A house is a complex, interrelated system of people, the building itself and the environment (Figure HS.1).

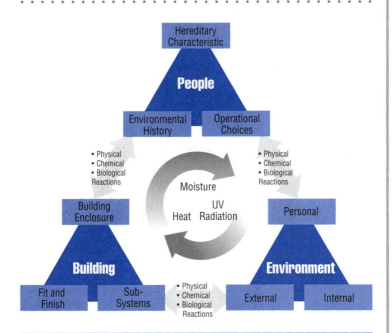

Figure HS.1
Analytical Model of the House System – Functional Relationships

A house consists of the building enclosure, the sub-systems contained within it and the fit and finish. The building enclosure is composed of assemblies. Assemblies are composed of elements. Sub-systems are composed of components. The fit and finish is composed of surfaces, appliances, trim, fixtures and the furnishings.

The building enclosure, assemblies, elements, sub-systems, components and the fit and finish are all interrelated. A change in an element can change the performance of an assembly, affect the building enclosure and subsequently change the characteristics of the house. Similarly, a change in a sub-system can influence the house, an assembly or an element of an assembly.

The house, in turn, interacts with the people who live in the house and with the local environment where the house is located. The functional relationships between the parameters are driven by physical, chemical and biological reactions. The basic factors controlling the physical, chemical and biological reactions are:

- heat flow
- moisture flow
- ultra-violet radiation

These basic factors are referred to as "damage functions."* Damage functions are responsible for the breakdown of materials via decay, corrosion, mold, spalling, efflorescence, etc.

Because large quantities of moisture can be transported by air — as well as the particulate and gaseous breakdown products of damage function action on building materials — air flow is also a significant factor affecting the functional relationships.

Controlling heat flow, moisture flow and ultra-violet radiation (the "damage functions") along with air flow will control the interactions among the physical elements of the house, its occupants and the environment.

Building houses is really about the durability of people (health, safety and well being of people), the durability of buildings (the useful service life of a building is typically limited by its durability) and the durability of the planet (the well being of the local and global environment).

---

* Ozone is also a key damage function and has been implicated in numerous indoor contaminant and odor problems, however, it is not considered to be at the same level of significance as the primary damage functions listed.

The relationships that define major elements of the residential construction process as well as continuing home operation are represented in Figures HS.2, HS.3, and HS.4.

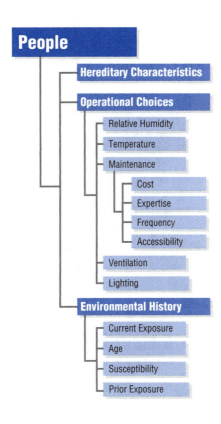

Figure HS.2
Hierarchical Relationships – People

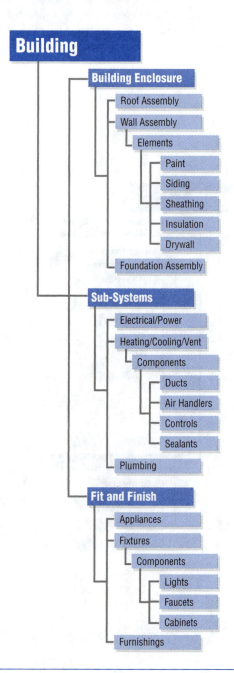

Figure HS.3
Hierarchical Relationships – Building

The House System

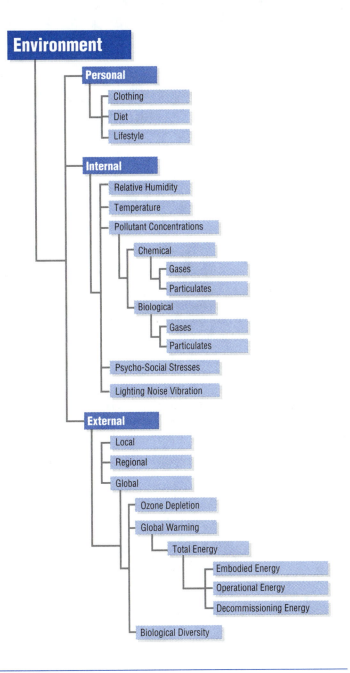

Environment

- **Personal**
  - Clothing
  - Diet
  - Lifestyle
- **Internal**
  - Relative Humidity
  - Temperature
  - Pollutant Concentrations
    - Chemical
      - Gases
      - Particulates
    - Biological
      - Gases
      - Particulates
  - Psycho-Social Stresses
  - Lighting Noise Vibration
- **External**
  - Local
  - Regional
  - Global
    - Ozone Depletion
    - Global Warming
      - Total Energy
        - Embodied Energy
        - Operational Energy
        - Decommissioning Energy
    - Biological Diversity

Figure HS.4
Hierarchical Relationships – Environment

Prioritization

The construction process should minimize needs for energy, water and materials and satisfy these needs in the least disruptive manner possible (Figure HS.5).

The interior environment, or conditioned space, should be safe, healthy and comfortable. The building should be both durable and affordable in terms of purchase price and operating costs. And the building should be built in a manner that does the least harm to the local environment, including the construction site and the land around it. The impact of home production on the global environment should also be considered. What resources will be used to build and operate the building? Are they renewable? What is the effect on the environment of extracting them?

> **Minimize Need for Energy, Water and Materials**
>
> **Satisfy Need with Least Disruption. Reduce, Reuse, Recycle Managed Resource Extraction and Processing**

**Figure HS.5
Minimization of Needs**

The sometimes conflicting needs among people, buildings and the environment should be prioritized. For example, the needs of people should be considered before the needs of a building. The internal environment created by a building should be considered before the planetary environment. Short-term concerns should be considered before long-term concerns (Figure HS.6). But all should be considered.

Applying prioritization to the construction process would identify the immediate health risk from carbon monoxide poisoning as a result of improper installation of combustion appliances as more significant than long-term health concerns from the infiltration of radon gas.

People (personal environment) → Buildings (indoor environment) → Environment (exterior environment)

Short-Term Concerns → Mid-Term Concerns → Long-Term Concerns

Local Concerns → Regional Concerns → Global Concerns

**Figure HS.6
Priorities**

Extending the prioritization process further shows that job-site recycling and reducing construction waste should have precedence over a global concern such as ozone depletion in the upper atmosphere. Finally, ozone depletion — a global short-term risk — should have precedence over global warming — a global long-term risk.

People Priorities

Houses should be safe, healthy, comfortable and affordable. A safe, healthy home is one in which concerns about clean water, fire and smoke spread, structural adequacy, indoor air quality, and security have been addressed. Comfort involves satisfying people's sensory perception. Comfort is addressed by dealing with thermal comfort, interior relative humidity, odors, natural light, sound and vibrations. Affordability means the designer, the builder and the various subcontractors and suppliers should be able to make a profit, and the occupant, for whom the home is built, should be able to purchase it and afford the operating and maintenance costs (Figure HS.7).

Building Priorities

Houses should be durable and capable of being maintained. The single most important factor affecting durability is deterioration of materials by moisture. Houses should be protected from wetting during construction and operation, and be designed to dry should they get wet.

Homeowners and occupants should be instructed on how to maintain and operate their buildings. Houses should be able to be renewed and renovated as new technologies, materials and products emerge. The house built today will likely be renovated at some point in the future. Houses should be able to be adapted as families and occupancy change. Finally, houses should be designed and constructed with decommissioning at the end of the useful service life in mind. This means taking into consideration how the materials that go into the house will ultimately be disposed of (Figure HS.7).

Environmental Priorities

Houses should be constructed in a manner that reduces construction waste and operated in a manner that reduces occupancy waste. Recycling of construction and operating waste should be encouraged. Use of construction water, domestic water and irrigation water should be minimized. Erosion of soil during site preparation and the construction process should be controlled. Storm water should be infiltrated back into the site.

Activities that contribute to air pollution during construction, such as construction dust, painting and burning of trash, should be minimized. Materials and systems that contribute to ozone depletion by releasing chlorofluorocarbons (CFCs) into the air should be avoided.

Biological diversity of plant and animal species should be protected by using materials from managed forests and managed mineral extraction processes. Finally, the use of energy to operate the building and to make and transport building products (embodied energy) should be minimized to reduce the production of greenhouse gases (e.g., carbon dioxide) that contribute to global warming (Figure HS.7).

People Priorities	Building Priorities	Environmental Priorities
Health & Safety	**Durability**	**Local Environment**
• potable water and sewage • fire and smoke spread • structure • indoor environment (air quality) • security • accessibility	• deterioration due to physical, chemical and biological reactions • operation, housekeeping and maintenance	• construction waste • operating waste • construction water • operating water • rain water run-off and local hydrology • erosion of soil
Comfort	**Renewal, Reuse and Renovation**	**Regional Environment**
• temperature • moisture (relative humidity) • odors • sound/vibrations • light • aesthetics	• future sub-system upgrading, such as communications, space conditioning and power • adaptability	• contamination of groundwater, streams and lakes (acid rainfall and acidification of lakes) • regional air pollution • regional recycled materials and waste disposal
Affordability	**Decommissioning/ Disassembly**	**Global Environment**
• capital cost, financing • operating cost from energy, water and maintenance	• benign materials • disposable and further recyclable materials	• ozone depletion affected by CFC-containing materials and systems • global warming affected by operating and embodied energy • biological diversity affected by utilizing materials from non-sustainable sources

Figure HS.7
People, Building and Environmental Priorities

Part I

1
2
3
4
5
6
7
8

Design

During the design phase, the designer makes fundamental decisions with respect to the siting, massing, layout and design of the house. The designer is often the general contractor but can be an architect or the home buyer. In many cases, design decisions are shared among the designer, general contractor, and home buyer. The designer must understand and be aware of the limitations of the project budget, the needs of the home buyer and the requirements of the general contractor. The designer must understand the local climate and the specific limitations of the proposed site. Furthermore, the designer must understand the project time, labor, construction sequence, material characteristics, the process of construction and the intended durability or expected lifespan of the structure (Figure 1.1).

During the design phase, fundamental decisions relating to foundation design, wall design, roof design and the design of the heating, ventilation and air conditioning (HVAC) system also occur. In order to make these decisions, the designer must be familiar with site and climate related issues such as the thermal and humidity conditions, annual rainfall, wind intensity and seismic zone.

Site Planning and Building Form

The ideal site is seemingly never available and there never is enough money. However, where flexibility exists, sites facing southeast, south, or southwest provide the best opportunities for optimizing a building's orientation with respect to daylighting and passive solar gain (Figure 1.2 and Figure 1.3). Sites sheltered from winter winds and open to summer breezes are warmer in winter and cooler in summer. Bodies of water and areas of vegetation moderate air temperature. Sites shaded by deciduous trees are cooler in summer. Sites that are well drained reduce the stress on drainage systems and water management. Building in a swamp is always more difficult than building on the top of a hill.

The site access, site clearing, excavations, site manipulation and shap-

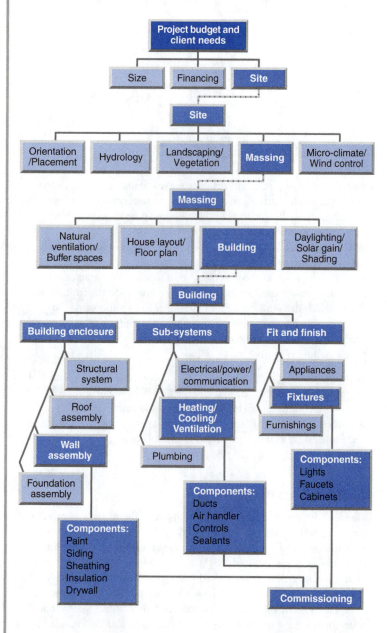

Figure 1.1
Process of House Construction

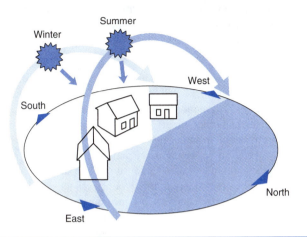

Figure 1.2
Summer vs. Winter Sun Angle

ing, construction process, site development and landscaping all need to be considered with respect to soil erosion, and the existing hydrology. Sculpting the ground to permit a slab-on-grade rather than a walkout basement or elevated crawlspace can save thousands of dollars, but may also create severe environmental stress from the loss of topsoil and damage to vegetation.

Landscaping should be used to buffer the house from winter winds, allow winter solar gain and daylighting, and provide summer shading and cooling (Figure 1.4). Vegetation, walls, fences and other buildings can be utilized as wind breaks. Wind breaks and the building form can be used to channel breezes into buildings and outdoor spaces (Figure 1.5). Overhead structures can be used to provide shade for exterior walls and outdoor use areas (Figure 1.6). West and southwest facades, that provide the greatest potential for summer overheating, should be shaded from low-angled sun. Light colored walls or fences can be used to facilitate daylighting by reflecting sunlight into north windows. Paving should be minimized and shaded from the sun. Understory vegetation should be cleared and maintained to admit summer breezes. Air can be cooled by directing it through and under vegetation (Figure 1.7).

Site hydrology and the management of storm water have a major environmental impact. Under natural conditions most rainfall percolates into the ground upon which it falls, whereas almost all the rain that falls on a built-up area contributes to surface run-off. The principle advantage of vegetation is that it controls soil erosion from run-off water. Grass encourages percolation into the water table and is an effective vegetation for erosion control. In fact, many grasses work to reduce the impact of pollutants by breaking them down before they reach the water table. Turf grass requires considerably more water than other ground covers. Xeriscaping (vegetation or amenities which require little or no water) should be considered.

An ideal situation would be one in which no increase in run-off occurs as a result of development, so that problems are localized rather than passed on to others. This concept can be promoted by reducing the amount of paving and other impervious surfaces or using porous pavement in order to permit a greater portion of the storm water to seep into the ground. Vegetation can be increased and/or retained in order to maximize the amount of storm water consumed and stored by plants. Storm water can be drained to temporary or permanent storage areas by means of surface collection and low volume underground systems. It's okay to have puddles of water after a rain. Just don't have those puddles right next to the house. The most important area is the first ten feet from the house. If the puddles are kept at least ten feet from the house it is much less likely that water ends up in your house.

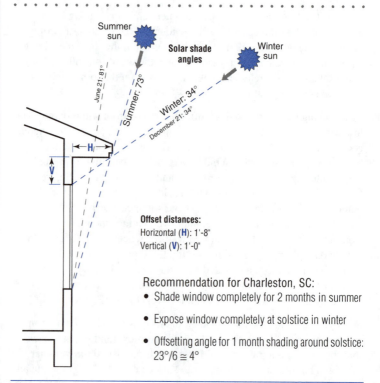

Figure 1.3
Overhang Sizing for Winter Gain and Summer Shade

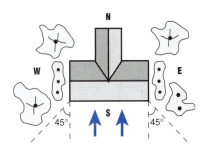

Figure 1.4
Landscaping
- Deciduous trees, trellises or other features placed on the east and west sides of a house may be used to shade windows from summer solar gain but also permit solar gain in winter
- Ground cover can be used to decrease heat absorption
- Evergreens and other plantings may be used as wind breaks or be used to direct wind towards or away from buildings

House Layout

The building layout is a major factor influencing cost and performance. Heat gain and heat loss occur across building surfaces. Maximizing volume while minimizing surface area will increase operating efficiencies and will minimize the use of materials. Building layouts that are compact are typically more resource and energy efficient than those which are spread out. The trend to move into finished basements and finished attics is an example of this logic. Unused attics, crawlspaces and basements are a waste of resources.

In terms of floor plans, locating the most actively used spaces where they will most benefit from daylighting makes the most sense. Kitchens, living rooms and family rooms should be located to the south side. Buffer spaces should be utilized both within a building (vestibules) and external to a building (porches, sunspaces, sheltered patios) to temper weather extremes.

Outdoor use areas used primarily during winter months should be located adjacent to south side of buildings and southern exposures should be free from obstructions except with respect to shading from the summer sun. The opposite is true for outdoor use areas used primarily during summer months. Large paved surfaces should be avoided on windward sides of buildings. Such surfaces, if necessary, should be located to the lee side of use areas with respect to summer breezes so as to minimize summer thermal mass effects.

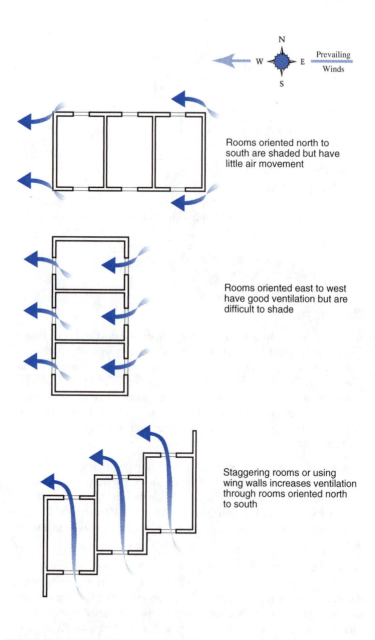

Rooms oriented north to south are shaded but have little air movement

Rooms oriented east to west have good ventilation but are difficult to shade

Staggering rooms or using wing walls increases ventilation through rooms oriented north to south

Figure 1.5
Solar and Wind Orientation

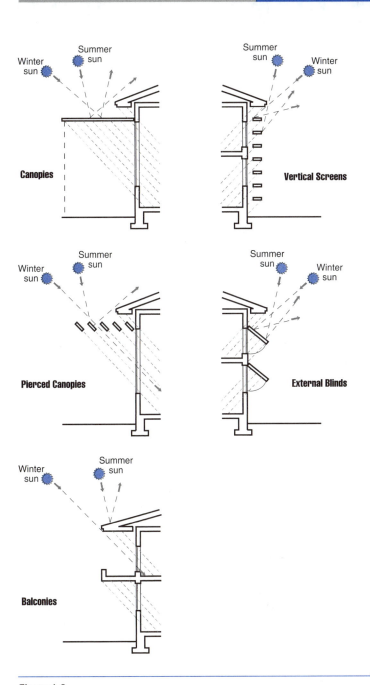

Figure 1.6
Shade for Exterior Walls for Both Summer and Winter

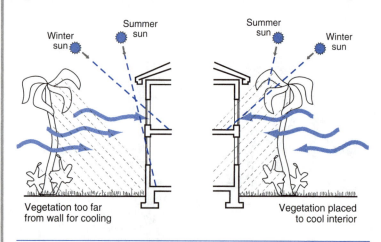

Vegetation too far
from wall for cooling

Vegetation placed
to cool interior

Figure 1.7
Shading Air Using Vegetation

Open floor plans allow for air flow and ventilation efficiencies, increased daylighting, and summer cross ventilation. During heating, heat is more evenly distributed.

Windows should be sized and positioned so as to decrease heat loss during the winter and decrease heat gain in the summer while providing daylighting year round. In hot-humid climates, north-facing glazing is preferred. East/west glazing should be minimized — particularly west-facing glazing due to potential overheating. Moderate amounts of southern glazing with appropriate shading can provide improved comfort during heating (sun tempering) while providing views without an energy penalty. Substantial south- and west-facing glass can lead to overheating in summer as well as winter and should be avoided.

In designing for cooling load, east and west facing windows typically cause the most summertime overheating. Spectrally selective glazing should be used throughout, but are essential for extensive east- and west-facing glazing. These types of window systems reduce solar gain, while still providing daylighting. They are sometimes referred to as "southern Low E glass." These types of windows have low solar heat gain coefficients, high visible light transmittance and low U-values.

Shade screens can provide a low cost and flexible method of solar control for windows. Shade screens reflect, absorb and dissipate a large portion of the sun's heat, between 60 and 70 percent depending on the shading coefficient, before it reaches the windows. The main drawback of shade screens is that the low shading coefficient also corresponds to a low light transmittance. They are best coupled with spectrally selective glazing ("southern Low E glass").

Large areas of glazing overhead (large skylights) should be avoided. Vegetation such as deciduous trees should be considered to shade windows during the summertime while leaving windows unshaded in winter time. Fencing or a trellis can also be used to shade glazing, particularly on east and west exposures.

Light colored walls, floors and ceilings should be used to reflect incoming light deep into rooms. A small skylight can be used to get light to the back of the rooms. Light ground colors should be avoided in front of south facing windows to minimize reflected summer heat gain, but should be considered for walls and fences on north exposures to reflect sunlight into north windows for improved daylighting.

The design challenge has always been to distribute operable windows and glazing in such a manner as to provide flow through ventilation without the penalties associated with excessive solar thermal gains during the summer while still providing sun tempering during the winter. Local practice and historical experience can often provide examples of elegant climatic optimization (Figure 1.8) in this regard.

With the advent of spectrally selective glazing, the introduction of new construction materials, coupled with traditional experience and the application of building science analysis it is possible to displace significant cooling loads and allow glazing where traditionally it would have contributed to excessive solar gain.

Thermal Mass

In hot, dry climates, mass construction out of adobe, rammed earth or rendered masonry were historically the materials and approaches of choice. This was before the advent of thermal insulation and air conditioning. Thermal mass walls and heat absorbing clay tile roofs served to moderate temperature extremes between cool evenings and hot days. Small windows, sheltered from direct solar exposure were sited to maximize cross ventilation, and when coupled with heat sink towers and the venturi effect, the resultant enhanced air flow could provide significant air change during cool evenings.

In regions with high diurnal temperature savings, air change during the evenings was relied upon to cool the interior thermal mass. By maximizing air change during the evenings and reducing solar gain and air change during the day, interior conditions could be maintained significantly cooler than exterior daily peaks (Figure 1.9). This approach is not as applicable in hot-humid climates because of warmer nights and high humidity.

In some middle eastern countries with dry desert climates, notably Iran and Iraq, this approach was historically supplemented by evaporative cooling from interior ponds and fountains, significantly extending the time interior conditions could be maintained within the comfort range. Again, this strategy has limited applications in hot-humid climates.

Until the twentieth century, these were the only strategies available. The major limitation of these approaches is the inability to provide acceptable indoor comfort, for extended periods, by the standards of today — especially with respect to humidity control.

The optimal thickness of an uninsulated mass wall assembly was historically determined by many factors: climatic exposure, emissivity, thermal conductivity, thermal diffusivity, air flow coupling, structural strength, available tools, available equipment and the experience of the designer and crew.

Ideally, the rate of heat gain through the exterior had to be balanced by the rate of heat removal to the interior by ventilation under a diurnal phase shift - maximum heat gain occurs approximately 12 hours out of phase from maximum cooling. Historically, 1-to-2-foot thick walls were common.

The introduction of thermal insulation significantly changed hot climate mass construction. By insulating the mass externally, thereby protecting it from the heat gain during the day, performance of mass buildings could be significantly enhanced, in effect making uninsulated mass walls such as adobe and rammed earth obsolete. Temperature moderation became easier and less air flow was required to provide comparable performance. Additionally, the hot climate mass design problem became much less complicated and was reduced to two questions: how much insulation do you need on the outside, and how much mass do you need on the inside? Coincidentally, this ultimately became the same design problem in passive solar heated mass structures in mixed and cold climates. The answers to both design problems, not surprisingly, are remarkably similar.

Experience has shown that of more importance than the quantity of mass, is the thermal coupling of the mass to the cooling source. In practice this leads to maximizing the surface area of mass exposed to moving air. In this regard, distributed mass is typically more beneficial than the same quantity of mass in one location. Additionally, the ability to pull heat out of mass easily is also significant. Gypsum board installed over furring strips on a masonry wall is not as effective a mass wall assembly as plaster installed directly on masonry. The furring strips create an air space that interferes with heat transfer.

Experience has also shown that more than 2 inches of mass on the interior of a well insulated exterior wall provides negligible benefit. It proves difficult to get much heat into and out of more than 2 inches of mass without relying on active means. This limitation has lead to the argument, with some justification, that 2 layers of gypsum board on the interior of an insulated wood frame wall may be as effective as an assembly comprising 2 inches of rigid insulation over a masonry wall rendered on the interior with plaster (Figure 1.10).

Figure 1.11 illustrates an effective use of externally insulated thermal mass and whole house ventilation to moderate interior temperatures. Note that under this approach, whole house ventilation is only effective as a cooling strategy when exterior temperatures are below the interior temperature. When exterior temperatures are above the desired or acceptable range of interior temperature and when the exterior moisture levels are above the interior moisture levels, whole house ventilation cannot be used to cool the thermal mass. Under these conditions, mechanical cooling is necessary to provide acceptable interior conditions from both a temperature and humidity perspective.

If a house with substantial thermal mass is allowed to heat up, it may be a thermal liability if mechanical cooling is now used to provide cooling to cool it to the desired interior temperature - unless the mechanical cooling is limited to evenings when exterior temperatures are lower than daytime exterior temperatures. Mechanical cooling equipment experiences a significant decrease in EER value as exterior temperatures rise. By limiting mechanical cooling to evenings, this EER value reduction can be substantially moderated.

The optimum approach to utilizing externally insulated thermal mass walls, whole house ventilation and mechanical cooling relies on whole house ventilation when exterior temperatures and moisture levels are below interior temperatures and moisture levels and relies on mechanical cooling when they are not. However, in both cases, the interior thermal mass is used to shift the cooling periods towards the evenings, maximizing the periods when whole house ventilation can be used and minimizing the EER value erosion when whole house ventilation cannot be used and mechanical cooling must be used to maintain acceptable interior conditions.

One of the best uses of thermal mass in hot-humid climates appears to be the shifting of mechanical cooling to off-peak hours to take advantage of off-peak utility rates if they are available.

1

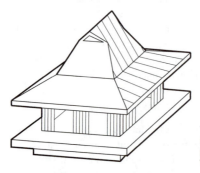

Traditional Hawaiian architecture incorporates gable vents, open floor plans and high ceilings for ventilation with broad overhangs for shading

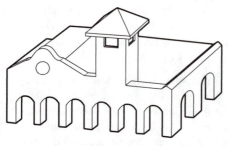

Traditional Spanish mission architecture incorporates thick thermal mass walls, small openings, exterior arch sheltered verrandas acting as heat screens, heat absorbing, self-vented clay tile roofs and massive heat sink towers acting as thermal chimneys providing vertical ventilation in the absence of wind

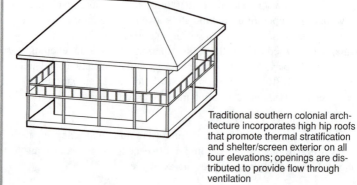

Traditional southern colonial architecture incorporates high hip roofs that promote thermal stratification and shelter/screen exterior on all four elevations; openings are distributed to provide flow through ventilation

Figure 1.8
Traditional House Styles

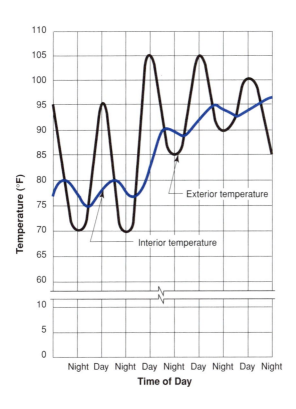

Figure 1.9
Interior Temperatures Compared to Exterior Temperatures

- Thermal mass coupled with air change during evenings has traditionally been effective in maintaining the interior of desert buildings significantly cooler than exterior daily peaks
- Note that this strategy has limitations in that the interior can never be cooler than the coldest evening temperature of the previous day and, in practice, only approaches the coldest evening temperature of the previous day
- For most individuals, relying on this strategy alone will provide unacceptable interior temperatures for extended periods according to today's standards, especially with respect to humidity control.

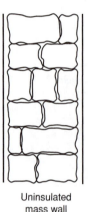
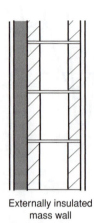
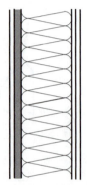

| Uninsulated mass wall | Externally insulated mass wall | Light frame wall, heavily insulated with supplemental interior thermal mass |

Figure 1.10
Thermal Mass Assemblies

- A heavily insulated "light" frame wall (interior gypsum board over a 2x6 wood frame, R-19 cavity insulation and 1" of exterior rigid insulation) may provide comparable overall thermal performance to a moderately insulated "mass" exterior wall (interior plaster installed directly over 8" masonry with 2" of exterior rigid insulation)
- An externally insulated mass wall and an insulated frame wall with interior mass significantly outperforms an uninsulated mass wall with respect to temperature moderation and thermal phase shifting

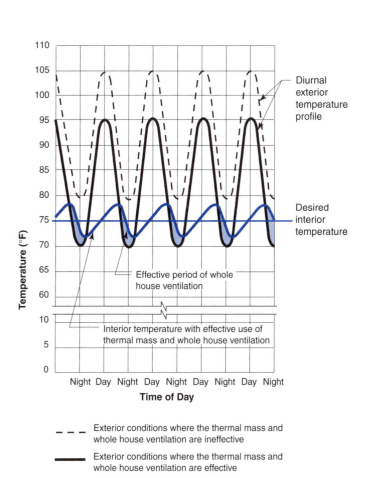

- - - Exterior conditions where the thermal mass and whole house ventilation are ineffective

━━━ Exterior conditions where the thermal mass and whole house ventilation are effective

Figure 1.11
Externally Insulated Thermal Mass and Whole House Ventilation
- Thermal mass and whole house ventilation are only effective as a cooling strategy when exterior temperatures drop below the interior temperature
- When the lowest exterior temperatures are above the desired or acceptable interior temperature, whole house ventilation to cool the thermal mass is of no benefit; if a house with substantial thermal mass is allowed to heat up, it may be a thermal liability if mechanical cooling is now used to provide cooling to cool it to the desired interior temperature, unless the mechanical cooling is limited to evenings when exterior temperatures are lower than exterior daily peaks, thereby limiting mechanical cooling to periods when equipment can operate at high EER values or when off-peak utility rates are available

Basic Structure and Dimensions

The house layout and massing define the basic structure. The type of foundation system (crawlspace, slab or basement), roof system (attic, cathedral ceiling or flat), floor system (joist, truss, or slab) and structural system (wood frame, steel frame, concrete, masonry, insulated concrete forms (ICF), structural insulated panels (SIPS), etc.) are selected by the designer based on costs, availability of materials, regional practices and preferences, site conditions and micro-climate, environmental impact and availability of experienced trades.

The type of structural system chosen affects building dimensions. For example, plywood, oriented strand board (OSB), and insulating sheathings typically come in 4-foot-by-8-foot sheets. It makes sense to design out-to-out dimensions on 2-foot increments to reduce sheet good waste and to maximize the efficiency of the structural frame (Figure 1.12). Similarly, concrete block, ICF and SIPS material dimensions should be considered when dimensioning floor plans to minimize waste and labor.

Roof slopes and overhangs should also be selected and dimensioned to take advantage of 2-foot increments. We often specify 4:12, 6:12 or 8:12 roof pitches when we can just as easily specify a 4.217: 12 roof pitch or a roof pitch by degrees from horizontal (i.e. 19.4 degrees) and minimize sheet good waste (Figure 1.12).

When using wood framing, studs should be spaced on 24-inch centers rather than 16-inch centers. Using this approach, coupled with single plates, stack framing, two stud corners, and elimination of cripples, it is possible to frame with 2x6's less expensively than with 2x4's. The volume (board footage) of lumber is about the same, but there are 30 percent fewer pieces, so the building goes together faster (less labor/ less cost). In this way the wall is thicker and therefore can accommodate more insulation (Figure 1.13).

In frame systems, roof framing components should line up with wall framing and floor framing. Windows and doors should be located on 2-foot grid increments to eliminate extra studs on the sides of openings (Figure 1.13).

Building Enclosure

The building enclosure should exclude rain, control rain water absorption, control air flow and vapor diffusion. Additionally, the building enclosure should be forgiving so that if it does get wet it can dry to the interior or exterior. Interior vapor barriers such as vinyl wall coverings and polyethylene should be avoided except in very cold and arctic/subarctic climates.

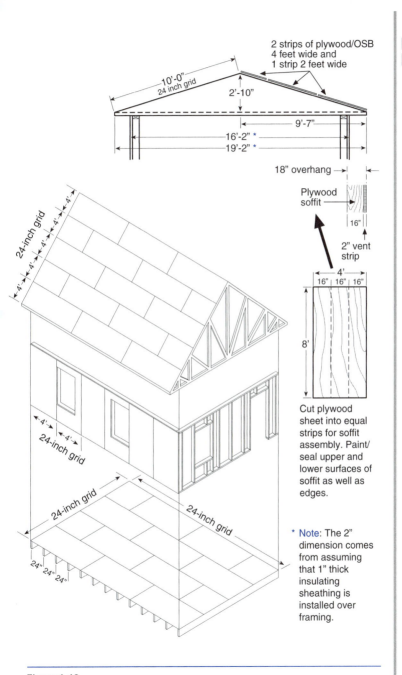

2 strips of plywood/OSB
4 feet wide and
1 strip 2 feet wide

10'-0"
24 inch grid

2'-10"

9'-7"

16'-2" *
19'-2" *

18" overhang

Plywood
soffit

16"

2" vent
strip

4'
16" | 16" | 16"

8'

Cut plywood
sheet into equal
strips for soffit
assembly. Paint/
seal upper and
lower surfaces of
soffit as well as
edges.

24-inch grid

4'

24-inch grid

24-inch grid

24-inch grid

24" 24" 24"

* Note: The 2"
dimension comes
from assuming
that 1" thick
insulating
sheathing is
installed over
framing.

Figure 1.12
Efficient Material Use by Design

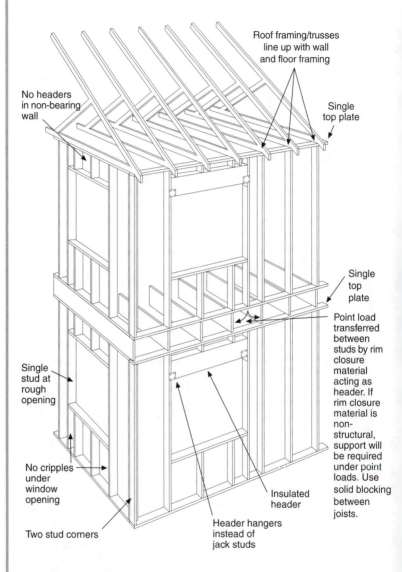

Roof framing/trusses
line up with wall
and floor framing

No headers
in non-bearing
wall

Single
top plate

Single
top
plate

Point load
transferred
between
studs by rim
closure
material
acting as
header. If
rim closure
material is
non-
structural,
support will
be required
under point
loads. Use
solid blocking
between
joists.

Single
stud at
rough
opening

No cripples
under
window
opening

Insulated
header

Two stud corners

Header hangers
instead of
jack studs

Figure 1.13
Stack Framing
- Eliminate headers in non-bearing interior walls
- Headers not needed for openings less than four feet wide in non-load bearing exterior walls

Specifically, the building enclosure must:
- hold the building up
- keep the rain water out
- keep the groundwater out
- keep the wind out
- keep the water vapor out
- let the water and vapor out if they get in
- keep the soil gas out
- keep the heat in during the winter
- keep the heat out during the summer
- keep the noise out

The designer has to choose materials, equipment and systems to make all this work. Will exterior sheathings be permeable or impermeable? Will building papers or housewraps be used? What type of air barrier system will be used? What will the thermal resistance of the building enclosure be? Chapter 3 discusses air barriers and Chapter 4 discusses exterior sheathings, permeability and building papers.

A key concept should be used at this point in the design approach - the concept of "break points." A break point denotes the situation where an increase in cost in one area is balanced by a reduction in cost in another area.

For example, increasing thermal resistance of the building enclosure by using insulating sheathing and high performance glazing will result in an increased cost. However, the heating and cooling system can be made much smaller and operate more efficiently with lower operating costs with a resulting decrease in the size and cost of ductwork and equipment. Open web floor trusses cost more than floor joists. However, it takes much less time, effort and money to install plumbing, electrical and ductwork within floor trusses than floor joists. A 2x6 costs more than a 2x4, but if you use far fewer 2x6's than 2x4's the building frame is put up faster and therefore less expensively.

Decisions relating to "keeping the building up" from the perspective of wind loading and seismic conditions should take a priority.

Resistance to shear loads due to wind must be provided (Figure 1.14). The choice of construction or framing approaches addressing shear loads should reflect the local conditions. For example, it is likely that homes built in low wind zones could be wood framed with non-structural sheathings and metal cross braces or wood "let-in" braces. Whereas a similar home built in a higher wind zone may be constructed from masonry, concrete or if constructed from wood framing would require shear panels (Figure 1.15) or plywood or OSB sheathing at exterior corners (Figure 1.16). Finally, if the home is built in a high wind costal wind zone, the exterior walls would likely need to be constructed from masonry, concrete or be completely sheathed with plywood, OSB or some other structural sheathing (Figure 1.17).

The recommendations on wind load resistance are extremely general and are not intended to substitute for professional design by a professional engineer or licensed architect and are not intended to conflict with specific code mandated requirements. The recommendations are intended to alert the reader that conditions vary significantly from one region to another and that due diligence should occur.

A similar discussion is also necessary for seismic design but is beyond the scope of this builder's guide. Local codes should be followed and the input of a licensed design professional is recommended.

Decisions relating to "keeping the rain water out" should also reflect the local conditions. Zones that see a great deal of rain (some coastal or mountain areas) will likely need a more "robust" approach to rain control than other zones. Specific wall assembly details for rain control can be found in Chapter 2, Rain, Drainage Planes and Flashings.

Once structural design decisions and rain water control decisions are made, more specific decisions relating to wall materials, roof design and foundation design can be made. These decisions also need to reflect local conditions.

Vapor Barriers

Vapor barriers or vapor retarders (intentional or unintentional) on the interior of wall assemblies and roof assemblies are not recommended in hot-humid climates. Interior relative humidity control should also vary based on local conditions to control interior condensation and mold growth. This interior relative humidity control can occur through the use of a dehumidification system, enhanced air conditioning or via controlled ventilation systems that provide dilution of interior moisture via induced air change. In hot-humid climates some supplemental dehumidification is typically required especially to control relative humidity during swing seasons.

Perimeter Slab Edge Insulation

Perimeter slab edge insulation is typically not necessary for moisture control or comfort reasons in hot-humid climates and can significantly complicate insect (termite) control strategies.

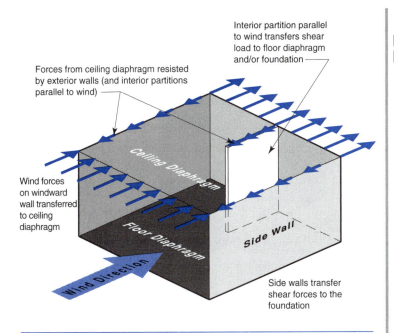

Interior partition parallel to wind transfers shear load to floor diaphragm and/or foundation

Forces from ceiling diaphragm resisted by exterior walls (and interior partitions parallel to wind)

Wind forces on windward wall transferred to ceiling diaphragm

Side walls transfer shear forces to the foundation

Figure 1.14
Shear Forces Due to Wind

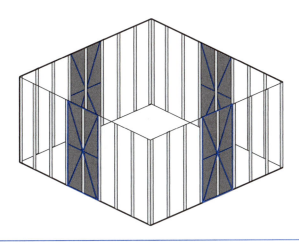

Figure 1.15
Shear Panels

- Proprietary prefabricated shear panels can be used for both wind and seismic loads

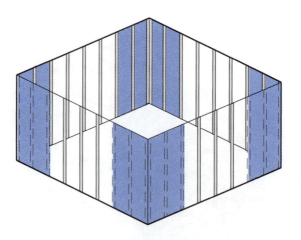

Figure 1.16
Shear Panels
- Plywood or OSB sheathing at exterior corners

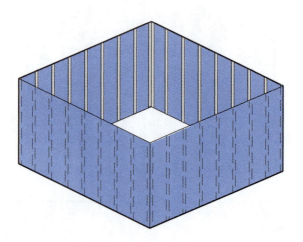

Figure 1.17
Structural Sheathing

Roofs

Vented and unvented roofs are both recommended in hot-humid climates.

The advantage of an unvented conditioned attic is that it provides a conditioned space to locate ductwork and air handlers.

Vented roofs are also recommended in hot-humid climates to reduce solar heat gain across the roof assembly. A vented roof reduces the total solar heat gain on the typical home approximately 3 to 5 percent as compared to an unvented roof. Clay tile or concrete tile roof shingles in place of asphalt shingles are also recommended for this reason. However, clay tile or concrete tile roof shingles provide a more significant reduction in total solar heat gain than roof venting. It is important to not install mechanical systems or ductwork in vented attic assemblies to fully realize the benefits of attic ventilation on solar heat gain.

Clay tile or concrete tile roof shingles in place of asphalt shingles can reduce the total solar heat gain on a typical home on the order of 10 to 15 percent and are recommended for this reason. The reason for this dramatic effect is the thermal mass and heat absorption characteristics of the clay and concrete tile roof shingles coupled with the air gaps between the tiles and the roof sheathing and as such are "back vented" unlike asphalt shingles.

The color of roofing materials does not have as much of an impact on solar gain as one would expect. The choice of roofing material tends to be more significant. The color of an asphalt shingle is less important than whether the shingle is asphalt or clay or concrete tile. Dark colored clay or concrete tiles significantly outperform the lightest colored asphalt shingles. However, all other things being equal, lighter colored roofs perform better than dark colored roofs.

Radiant barriers can also have a significant effect in reducing total solar heat gain on a typical home when a vented roof is constructed with asphalt shingles and where low levels of attic insulation (R-19 or less) are used. However, the benefit of radiant barriers is reduced when clay tile or concrete tile roof shingles are used with more than R-19 ceiling insulation. Additionally, radiant barriers provide less benefit in unvented roof assemblies. For these reasons radiant barriers are often considered a more attractive new construction strategy than a retrofit strategy. It is generally easier to retrofit with additional insulation than with a radiant barrier. As a final caution, it should also be noted that radiant barriers lose their effectiveness if the radiant surfaces become dusty or corroded.

Based on the foregoing it can be argued that the optimum roof assemblies in hot-humid climates are unvented roofs with clay tile or concrete tile roof shingles and R-19 or greater ceiling insulation or vented roofs where all ducts and mechanical equipment are located within the conditioned space, again, with R-19 or greater ceiling insulation.

It is typically desirable to build an unvented roof (Figure 1.18) to provide a conditioned attic for the location of ductwork and air handlers. A small energy penalty occurs from not venting a roof (about 3-to-5 percent of total cooling). As a result heat gain across the attic insulation is increased slightly if roof assembly is not vented. However, this is offset by a huge benefit from reducing duct leakage to the exterior and eliminating conductive gains across ductwork and, more significantly, reducing the moisture load.

The energy penalty also needs to be put into perspective. Locating ductwork in vented attics results in a 5-to-40 percent energy penalty due to the leakiness of ductwork, air handlers and conductive gains across ductwork and air handlers located in hot attic spaces. Constructing an unvented conditioned attic that now encloses ductwork (attic insulation is installed at the underside of the roof deck) results in significant energy savings over attics that are vented and leaky contain ductwork and leaky air handlers.

Constructing an unvented, unconditioned attic and locating ductwork and air handlers within the conditioned space is slightly more energy inefficient (3-to-5 percent) then constructing a vented attic where ductwork and air handlers are also located within the conditioned space.

Asphalt shingle life is not significantly affected by unvented roof construction since the single most important factor affecting asphalt shingle life is exposure to ultraviolet light and moisture from rainwater. Venting a roof does not affect UV or rain exposure.

The argument that asphalt shingles "cook" over unvented roofs is not supported by experiments and research. Having said that, shingles installed on unvented decks do operate at a slightly higher temperature — and therefore shingle life is affected. However, this needs to be put into perspective. Asphalt shingle operating temperature is principally determined by radiation heat transfer from the sun rather than conductive heat transfer from the bottom of the asphalt shingle to the roof sheathing. Selecting the asphalt shingle color, roof orientation and roof slope are much more significant in determining asphalt shingle operating temperatures than roof venting.

Where A/C ductwork or air handlers are located in conditioned spaces within the building thermal barrier, vented roofs in hot-humid climates with clay tile or concrete tile roof shingles and R-30 or greater ceiling insulation thermally outperform any other type of building enclosure approach. However, the moment A/C ductwork or air handlers are located in vented, unconditioned attics, the benefits of roof venting are completely lost. If A/C ductwork or air handlers must be located in attic spaces due to space limitations or other reasons, an unvented roof construction creating a conditioned attic space is recommended.

Additional benefits from constructing unvented conditioned attics are improved resistance to hurricane loads, particularly in reducing uplift forces ("roof/building pressurization"), and wind-driven rain entry through "off-ridge" vents and soffit vents. Finally, significant improvements in wildfire resistance are garnered with unvented conditioned attic construction.

Unvented Crawlspaces

Vented crawlspaces are not recommended in hot-dry or mixed-dry climates or in any climate for that matter. Crawlspaces should be sealed and enclosed like any other conditioned space such as a bedroom or a living room. They should be heated during the winter and cooled during the summer. They should be insulated around the perimeter and not between the crawlspace and the main floor. Good drainage is required. More discussion on crawlspaces can be found in Chapter 7.

Mechanical Systems

Mechanical equipment and ductwork should not be located outside of a home's thermal barrier and air pressure boundary. The building's thermal barrier and pressure boundary enclose the conditioned space. The pressure boundary is typically defined by the air barrier. Therefore, mechanical equipment and ductwork should not be located in exterior walls, vented attics, vented crawlspaces, garages or at any location exterior to a building's air barrier and thermal insulation. All air distribution systems should be located within the conditioned space. Additionally, ductwork should never be installed in or under floor slabs due to soil gas, radon, moisture condensation issues, and flooding of ductwork.

Figure 1.18 through Figure 1.24 illustrate six approaches that locate air handlers and ductwork completely within conditioned spaces, inside of the building thermal barrier and pressure boundary.

As discussed earlier, it is possible to construct unvented, conditioned attics that become ideal locations for air handlers and ductwork (Figure 1.19). The techniques used to construct unvented conditioned attics vary regionally based on climate. There are major differences constructing unvented attics in hot-humid climates as compared to constructing unvented attics in mixed-dry, mixed-humid and cold climates.

More discussion on vented attics and unvented conditioned attics can be found in Chapter 6.

The cold air portion of most air conditioning equipment is internally insulated with 1-inch of fiberglass insulation. This insulation thickness is not sufficient to prevent sweating of the exterior metal cabinet when it is located outside in vented attics, vented (unconditioned) crawlspaces or garages.

Similar comments can be made about insulated cold supply ducts. Although these ducts can be insulated with additional insulation, it is virtually impossible to install a 100% effective vapor barrier around these ducts. This means that water vapor finds its way past the vapor barrier and easily passes through the fiberglass insulation to the cold supply duct where it condenses. It is not recommended that ductwork be located outside in vented attics, vented (unconditioned) crawlspaces or garages.

The result is condensation, rust and mold. Units can rust-through in less than five years. Air handling units are not fit or suitable for installation in these locations in hot-humid climates.

Although it is not recommended that air handlers and ductwork be located in vented attics, vented crawlspaces or garages, this often occurs. The consequences of this often lead to major problems with thermal comfort, operating costs, durability, health and safety. If ductwork and air handlers located in vented attics are not installed in a leak free manner, significant negative air pressures can occur within the occupied spaces (Figures 1.25 and 1.26) that can result in the uncontrolled ingress of exterior air, moisture, radon, soil gas and pollutants as well as lead to backdrafting of combustion appliances such as fireplaces. Leaky air handlers and leaky ductwork located in vented attics typically result in a 20 percent to 40 percent energy penalty to the overall performance of the building and significant humidity control problems. Even if installed in a leak free manner, an approximate 5 percent energy penalty to the overall performance to the building results.

Techniques have been developed that allow ductwork and air handlers to be installed in a leak free manner (see Chapter 13). Leak-free ductwork and air handler installation is absolutely critical if ductwork and air handlers are installed outside of a building's thermal barrier and air pressure boundary such as in a vented attic. Recall that it is dumb to install them here — you've been warned. Leak-free ductwork and air handler installation is a good idea even if ductwork and air handlers are installed within building thermal barriers and air pressure boundaries. Leakage of ductwork within interior building interstitial cavities can result in carpet dust marking and pollutant transfer as well as affecting thermal performance and comfort.

Negative air pressures in common areas often occurs in conjunction with pressurization of bedrooms even with tight ductwork and leak-free air handlers due to insufficient return air paths (Figure 1.27). These air pressure differences can significantly increase uncontrolled air change and thereby increase energy costs, adversely affect comfort as well as result in carpet dust marking, spillage and backdrafting of combustion appliances. Clear return air paths must be provided via ducted returns, transfer grilles, jump ducts or a combination of all of the above.

Locating air handlers within closets and mechanical rooms also needs to be done in a leak-free manner, with adequate provision for return air in order to avoid depressurization (Figures 1.29 and 1.30).

Interior chases must be considered and provided during the schematic design phase to make it easy to install services within conditioned spaces. A designer must think about stair locations and openings while allowing ductwork and plumbing to get by them as it goes from one side of the house to the other.

Complaints on how plumbers and HVAC installers butcher framing during rough-in are often heard. However, if the designer does not leave them space, or make it easy for them to install equipment, piping and ductwork, it is often the only way to install services. The designer almost never talks to the plumber or HVAC contractor during the design phase about locating soil stacks, ductwork layouts and equipment locations. It's time we start the dialog, or stop complaining about cut floor framing, having to pad-out walls to hide plumbing and paying for tortured ductwork layouts that deliver only a fraction of the air that they should.

Controlled Ventilation and Equipment Selection

All buildings require controlled mechanical ventilation. Building intentionally leaky buildings and installing operable windows does not provide sufficient outside air in a consistent manner. Building enclosures must be "built tight and then ventilated right." Why? Because before you can control air you must enclose it. Once you eliminate big holes it becomes easy to control air exchange between the inside and the outside. Controlled mechanical ventilation can be provided in many forms (see Chapter 13), but what is important at this stage is that the designer recognize that a controlled mechanical ventilation system is necessary and that controlled mechanical ventilation works best in a tight building enclosure. You can't control anything in a leaky building.

Selecting a fuel source is usually based on availability, regional practices and customer preference. Electric heat pumps and natural gas are

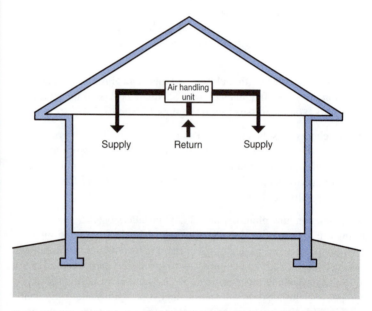

Note: Colored shading depicts the building's thermal barrier and pressure boundary. The thermal barrier and pressure boundary enclose the conditioned space.

Figure 1.18
Slab-on-Grade
Unvented, Conditioned Attic

- The air handling unit is located in an unvented, conditioned attic; the attic insulation is located at or above the roof deck.
- Low efficiency gas appliances that are prone to spillage or backdrafting are not recommended in this type of application; heat pumps, heat pump water heaters or sealed combustion furnaces and water heaters should be used
- A hot water-to-air coil in an air handling unit can be used to replace the gas furnace/heat exchanger. The coil can be connected to a sealed combustion (or power vented) water heater located within the conditioned space.
- Ductwork is not located in exterior walls.

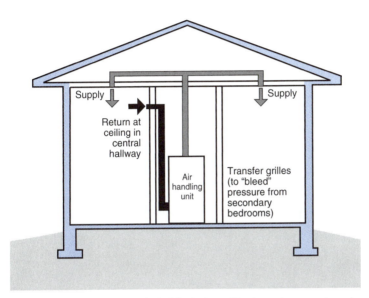

Note: Colored shading depicts the building's thermal barrier and pressure boundary. The thermal barrier and pressure boundary enclose the conditioned space.

Figure 1.19
Slab-on-Grade
Unvented, Conditioned Attic

- The air handling unit is located in an interior closet and the supply ductwork is located in the unvented, conditioned attic
- The attic insulation is located at or above the roof deck
- Transfer grilles "bleed" pressure from secondary bedrooms
- Ductwork does not have to extend to building perimeters when thermally efficient windows (low E, spectrally selective) and thermally efficient (well-insulated 2x6 frame walls with 1" of insulating sheathing) building enclosure construction is used; throw-type registers should be selected
- Low efficiency gas appliances that are prone to spillage or backdrafting are not recommended in this type of application; heat pumps, heat pump water heaters or sealed combustion furnaces and water heaters should be used
- A hot water-to-air coil in an air handling unit can be used to replace the gas furnace/heat exchanger. The coil can be connected to a sealed combustion (or power vented) water heater located within the conditioned space.
- Ductwork is not located in exterior walls

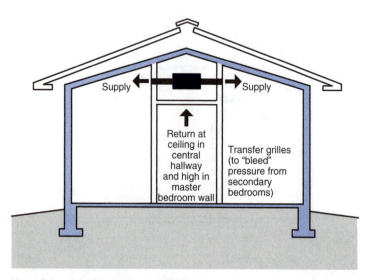

Note: Colored shading depicts the building's thermal barrier and pressure boundary. The thermal barrier and pressure boundary enclose the conditioned space.

Figure 1.20
Slab-on-Grade Dropped Ceiling
Vented , Unconditioned Attic

- The air handling unit is located in an interior closet and the supply and return ductwork are located in a dropped hallway
- Transfer grilles "bleed" pressure from secondary bedrooms
- Ductwork does not have to extend to building perimeters when thermally efficient windows (low E, spectrally selective) and thermally efficient (well-insulated 2x6 frame walls with 1" of insulating sheathing) building enclosure construction is used; throw-type registers should be selected
- Low efficiency gas appliances that are prone to spillage or backdrafting are not recommended in this type of application; heat pumps, heat pump water heaters or sealed combustion furnaces and water heaters should be used
- A hot water-to-air coil in an air handling unit can be used to replace the gas furnace/heat exchanger. The coil can be connected to a sealed combustion (or power vented) water heater located within the conditioned space.
- Ductwork is not located in exterior walls or in the vented attic.

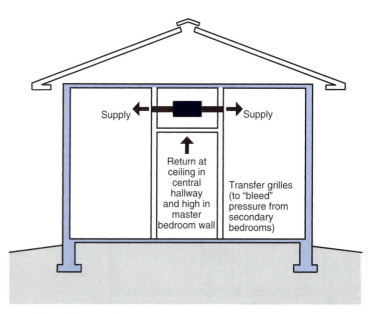

Note: Colored shading depicts the building's thermal barrier and pressure boundary. The thermal barrier and pressure boundary enclose the conditioned space.

Figure 1.21
Slab-on-Grade with Dropped Ceiling
Vented, Unconditioned Attic

- In this approach, exterior wall heights are typically increased to 9-feet or more leaving hallway ceiling heights at 8-feet
- The air handling unit is located in an interior closet and the supply and return ductwork are located in a dropped hallway
- Transfer grilles "bleed" pressure from secondary bedrooms
- Ductwork does not have to extend to building perimeters when thermally efficient windows (low E, spectrally selective) and thermally efficient (well-insulated 2x6 frame walls with 1" of insulating sheathing) building enclosure construction is used; throw-type registers should be selected
- Low efficiency gas appliances that are prone to spillage or backdrafting are not recommended in this type of application; heat pumps, heat pump water heaters or sealed combustion furnaces and water heaters should be used
- A hot water-to-air coil in an air handling unit can be used to replace the gas furnace/heat exchanger. The coil can be connected to a sealed combustion (or power vented) water heater located within the conditioned space.
- Ductwork is not located in exterior walls or in the vented attic.

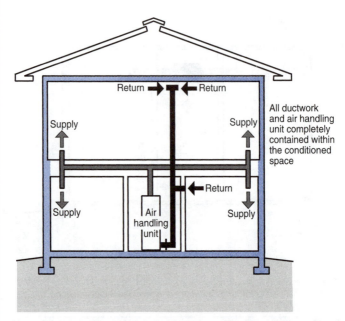

Note: Colored shading depicts the building's thermal barrier and pressure boundary. The thermal barrier and pressure boundary enclose the conditioned space.

Figure 1.22
Slab-on-Grade
Vented, Unconditioned Attic

- The air handling unit is located in an interior closet/utility room.
- Low efficiency gas appliances that are prone to spillage or backdrafting are not recommended in this type of application; heat pumps, heat pump water heaters or sealed combustion furnaces and water heaters should be used
- A hot water-to-air coil in an air handling unit can be used to replace the gas furnace/heat exchanger. The coil can be connected to a sealed combustion (or power vented) water heater located within the conditioned space.
- Ductwork is not located in exterior walls or in the vented attic.

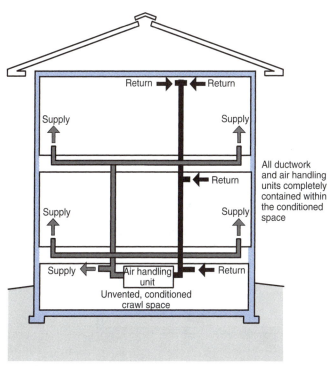

Note: Colored shading depicts the building's thermal barrier and pressure boundary. The thermal barrier and pressure boundary enclose the conditioned space.

Figure 1.23
Unvented, Conditioned Crawlspace
Vented, Unconditioned Attic

- The air handling unit is located in an unvented, conditioned crawlspace. The crawlspace has a supply duct and either a return or transfer grille; a transfer grille can be provided through the main floor to return air to the common area of the house and subsequently to the return grille on the main floor.
- Low efficiency gas appliances that are prone to spillage or backdrafting are not recommended in this type of application; heat pumps, heat pump water heaters or sealed combustion furnaces and water heaters should be used
- A hot water-to-air coil in an air handling unit can be used to replace the gas furnace/heat exchanger. The coil can be connected to a sealed combustion (or power vented) water heater located within the conditioned space.
- Although the air handling unit can be located in a conditioned crawlspace it is strongly recommended it be located within the occupied space for easy access

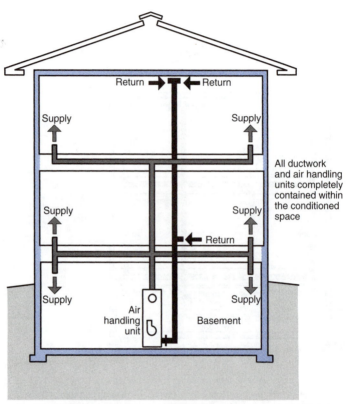

All ductwork
and air handling
units completely
contained within
the conditioned
space

Return → ← Return

Supply

Supply

Supply

Supply

← Return

Supply

Air
handling
unit

Supply

Basement

Note: Colored shading depicts the building's thermal barrier and pressure boundary.
The thermal barrier and pressure boundary enclose the conditioned space.

**Figure 1.24
Basement Foundation
Vented, Unconditioned Attic**
- The air handling unit is located in a conditioned basement.
- Low efficiency gas appliances that are prone to spillage or backdrafting are not recommended in this type of application; heat pumps, heat pump water heaters or sealed combustion furnaces and water heaters should be used
- A hot water-to-air coil in an air handling unit can be used to replace the gas furnace/heat exchanger. The coil can be connected to a sealed combustion (or power vented) water heater located within the conditioned space.
- Ductwork is not located in exterior walls or in the vented attic.

This Approach is Not Recommended

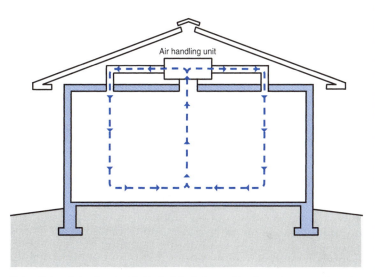

Air handling unit

Note: Colored shading depicts the building's thermal barrier and pressure boundary. The thermal barrier and pressure boundary enclose the conditioned space.

Figure 1.25
Ductwork and Air Handlers in Vented Attics
- No air pressure differences result in a house with an air handler and ductwork located in a vented attic if there are no leaks in the supply ducts, the return ducts or the air handler and if the amount of air delivered to each room equals the amount removed
- If no air leakage occurs in the ductwork and air handler, locating the ductwork and air handler in a vented attic typically results in an approximate 5% to 10% energy penalty to the overall performance of the building due to conductive gains and losses across the ductwork and air handler.
- This approach is prone to ice dams and should be avoided.
- If this approach is used, duct leakage testing should be used to assure leak-free installation

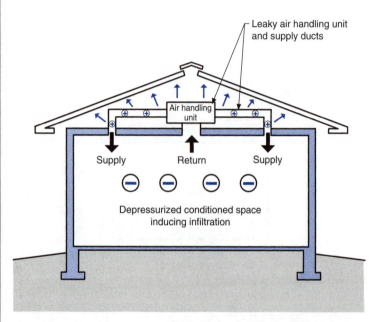

Note: Colored shading depicts the building's thermal barrier and pressure boundary.
The thermal barrier and pressure boundary enclose the conditioned space.

Figure 1.26
Ductwork and Air Handlers in Vented Attics

- Supply ductwork and air handler leakage is typically 20% or more of the flow through the system; in a 5 ton A/C system there is approximately 400 cfm of leakage, or almost the equivalent to 2 tons of capacity
- Leakage out of the supply system into the vented attic results in an equal quantity of infiltration; in other words, 400 cfm of leakage out of the supply system into the attic results in 400 cfm of infiltration (typically $^2/_3$ or more) through the attic ceiling. This guarantees ice dam creation.
- Air leakage in ductwork and air handlers located in vented attics typically results in an approximate 20% to 40% thermal penalty to the overall performance of the building

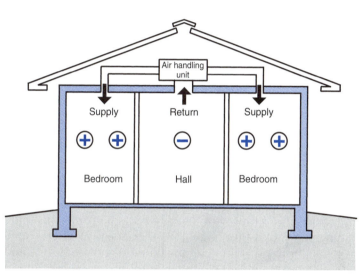

Note: Colored shading depicts the building's thermal barrier and pressure boundary.
        The thermal barrier and pressure boundary enclose the conditioned space.

Figure 1.27
Insufficient Return Air Paths
- Pressurization of bedrooms occurs if insufficient return air pathways are provided; undercutting bedroom doors is usually insufficient; transfer grilles (see Figure 13.34), jump ducts (Figure 13.35) or fully ducted returns may be necessary to prevent pressurization of bedrooms
- Master bedroom suites are often the most pressurized as they typically receive the most supply air; a fully ducted return in the master bedroom suite is generally recommended
- When bedrooms pressurize, common areas depressurize; this can have serious consequences when fireplaces are located in common areas and subsequently backdraft

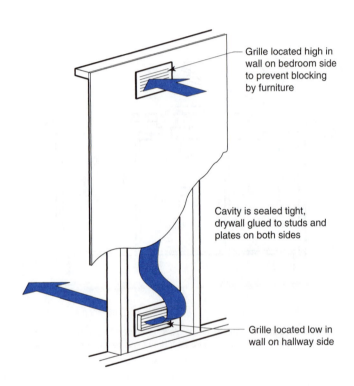

Grille located high in
wall on bedroom side
to prevent blocking
by furniture

Cavity is sealed tight,
drywall glued to studs and
plates on both sides

Grille located low in
wall on hallway side

Figure 1.28
New Construction Air Distribution Systems

- Use fully ducted returns or transfer grilles for return airflow paths. Do not use panned floor joists or other building cavities as returns.

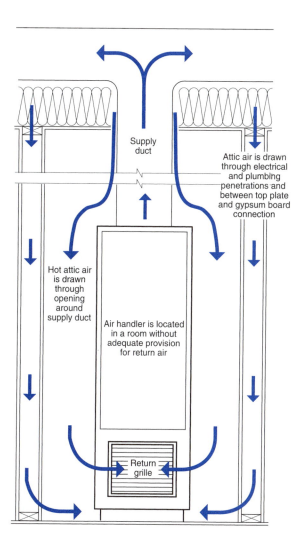

Supply duct

Attic air is drawn through electrical and plumbing penetrations and between top plate and gypsum board connection

Hot attic air is drawn through opening around supply duct

Air handler is located in a room without adequate provision for return air

Return grille

Figure 1.29
Air Handler Closet Depressurization
• Air handler typically located in closet with louvered door

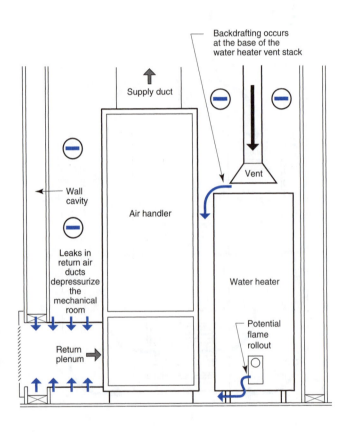

Backdrafting occurs at the base of the water heater vent stack

Supply duct

Vent

Wall cavity

Air handler

Leaks in return air ducts depressurize the mechanical room

Water heater

Potential flame rollout

Return plenum

Figure 1.30
Backdrafting in Mechanical Room/Utility Room
- Mechanical room depressurized by return system leakage of ductwork and air handling unit cabinet

the most common choice for space and domestic water heating. If gas heating or a gas water heater is selected, the appliances must be power vented, sealed combustion or installed external to the conditioned space (i.e. installed in a garage — except in cold or very cold climates). Gas appliances should not interact aerodynamically with the building (i.e. affected by interior air pressures or other air consuming devices). If a gas cook top or gas oven is installed, it must be installed in combination with a kitchen range hood directly ducted to the exterior (exhaust fan). Unvented gas fireplaces should never be installed.

Air change through mechanical ventilation should be used to control interior moisture levels year round. Dehumidification through the use of mechanical cooling (air conditioning) and supplemental dehumidification may also be necessary during cooling periods and part load conditions. Mechanical ventilation strategies are described in Chapter 13.

The need for mechanical cooling (air conditioning) should be reduced by displacing the cooling load as much as possible through the use of architectural design, spectrally selective glazing, heat absorbing cladding materials such as clay or concrete tile roof shingles, and the use of low heat producing appliances and lighting. Comfort levels can be enhanced at higher interior temperatures by inducing air movement through the use of ceiling fans thereby reducing the need for mechanical cooling.

Air conditioners, furnaces, air handlers and ductwork should be located within the conditioned space and provided with easy access to accommodate servicing, filter replacement, cooling coil and drain pan cleaning, future upgrading or replacement as technology improves. Hostile locations (extreme temperatures and moisture levels) such as vented attics and unconditioned (vented) crawlspaces or garages should be avoided.

Fireplaces

Fireplaces should not be located on exterior walls, they should be totally contained within the conditioned space. A warm chimney drafts much better than a cold one. In addition, masonry chimneys that are located within the conditioned space have thermal mass that stores the heat created by a fire and continues to warm the space even after the fire has gone out. Fireplaces should always be provided with ducted exterior combustion and make-up air.

Material Selection

There are only a few inherently bad materials, but there are many bad ways to use materials. The use of a material should be put into the con-

**Figure 1.31
Material Selection**

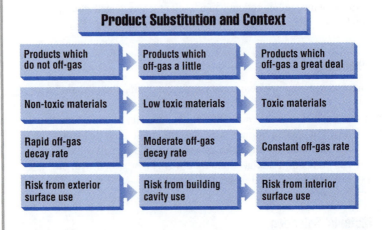

**Figure 1.32
Product Substitution and Context**

text of a system. In general, the system is more important than the material. Once the system is selected, you must determine if the material can perform its intended function as part of that system for the desired life of the system. What is the risk of using the material to the occupants, the building and to the local and global environment when used in that system (Figure 1.31)?

Extending the argument further, there are no truly benign materials, only degrees of impact. Nothing is completely risk free. However, risk can be managed. There may be no alternative to a particularly toxic material in a specific system, but the use of that material may pose little risk when used properly and provide significant benefits to that system. For example, bituminous dampproofing is a toxic material, but if it is installed on the exterior of a concrete foundation wall, there is little risk to the occupants, but there are substantial moisture control benefits to the foundation assembly.

The risk to occupants of a particular synthetic or natural agent in a building product, system or assembly is generally low where that agent is not inhaled or touched. In any case, building products and materials that do not off-gas are preferable to those that do. Less toxic alternatives should be used in place of more toxic materials (Figure 1.32). Remember that these material choices need to be placed in the context of the system or assembly of which it is a part. Is the toxic material being used in roofing? If so, it may pose little hazard to the occupants.

In addition to the specific concerns about material and product use on the interior environment of a building and the occupants, are the concerns relating to the local and global environment. Is it more appropriate to use a recycled, refurbished or re-manufactured product or material in place of a new material or product? Is the new material or product obtained or manufactured in a non-disruptive or the least-disruptive manner to the environment? Can the cost of a product justify its use? How far was the material transported? How much energy was used to make it?

Appliances

The designer is typically responsible for the selection of appliances, typically with home owner input. If home owners assume the responsibility of selecting some or all of the appliances, it is the responsibility of the designer to provide the necessary information to the home owner so that an informed decision be made within the context of the house system.

Appliances should be selected and installed in such a manner that they do not adversely affect the building enclosure or building sub-systems.

1

For example, gas cook tops and ovens should provide for their own exhaust of combustion products by installing vented range hoods or exhaust fans. They also need to be installed in a manner which prevents excessive depressurization (i.e. make-up air for indoor barbecues).

Fireplaces and wood stoves should be considered appliances. They should be provided with their own air supply independent of the other air requirements of the building enclosure. The location of combustion appliances should take into consideration air pressure differentials that may occur due to the stack effect or competition for air from other combustion appliances. Indoor barbecues should be considered similarly. Air supply, air pressure differentials and combustion product venting need to be addressed.

Energy consumption should be a prime consideration in the selection of refrigerators, freezers, light fixtures, washers and dryers. The yellow

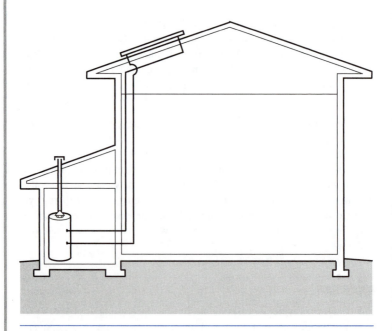

Figure 1.33
Solar Assisted Hot Water Heating
- Solar collector connected to hot water tank, solar heat supplements heating of hot water by gas or electricity (electric resistance or electric heat pump hot water heater)
- Solar collector can be a thermosiphon or use a water pump to circulate the hot water through the solar water heating system

D.O.E. Energy Guide label should be consulted. Dryers should be vented directly to the exterior and should not adversely affect the air pressure dynamics of the building enclosure when they are operating.

Solar Water Heating

Hot-humid climates can provide a cost effective opportunity for the application of solar water heating or solar assisted water heating. Two types of systems are common: active distribution or passive distribution. In active distribution, a pump is used to move water through the collector. In passive distribution, the buoyancy of heated water is used.

Solar systems for hot water can also be coupled with hydronic heating. A hydronic coil (water-to-air heat exchanger) can be combined with air conditioning and connected to either a gas or electric water heater that is also connected to a solar collector.

Various types of solar collectors are available including collectors with integral flashing systems that can be integrated into roofing assemblies in a manner similar to skylight installation.

Commissioning

The home designer must ensure that commissioning of the home occurs so that the home functions as intended by the design. Commissioning allows problems to be spotted through testing and then immediately remedied ("fix and tune"). The commissioning can be done by the designer, the contractor or some other competent person. At minimum, commissioning should include:

- testing of building enclosure leakage area

- testing of the leakage of duct systems

- testing of the air pressure relationships under all operating conditions

- testing for proper venting of all combustion appliances under all operating conditions

- testing of the carbon monoxide output of all combustion appliances (gas oven, range, water heater, gas fireplace)

- confirmation of air flow and refrigerant charge in HVAC systems

As part of the commissioning process, the home owner or occupants should be educated and informed as to the correct operation, maintenance and housekeeping requirements of the building and equipment.

What are appropriate temperature, relative humidity and ventilation ranges for the building? How do owners or occupants identify a system failure or the improper operation of a system? How do owners or occupants monitor the building conditions (temperature sensors, humidity sensors, ventilation sensors)? How often should filters be changed?

A system data sheet listing equipment, models, sizes and serial numbers should be placed in a permanent plastic documentation envelope along with equipment manuals, filter sizes and a copy of commissioning documentation.

There is information relating to air leakage testing of building enclosure leakage areas, testing of the leakage of duct systems, air pressure differentials and combustion safety in Appendix II.

Rain, Drainage Planes and Flashings

Rain is the single most important factor to control in order to construct a durable structure. Although controlling rain has preoccupied builders for thousands of years, significant insight into the physics of rain and its control was not developed until the middle of this century by the Norwegians and the Canadians. Both peoples are blessed by countries with miserable climates which no doubt made the issue pressing.

Experience from tradition-based practices combined with the physics of rain developed by the Norwegians and Canadians has provided us with effective strategies to control rain entry. The strategies are varied based on the frequency and severity of rain.

The amount of rain determines the amount of rain control needed. No rain, no rain control needed. Little rain, little rain control needed. Lots of rain, lots of rain control needed. Although this should be obvious, it is often overlooked by codes, designers, and builders. Strategies which work in Las Vegas do not necessarily work in Seattle. In simple terms, the amount of rainwater deposited on a surface determines the type of approach necessary to control rain.

The susceptibility of the building system to water damage also impacts the type of approach necessary to control rain. Buildings constructed from moisture-sensitive materials such as wood or steel studs, gypsum, OSB or plywood sheathings and fiberglass or cellulose cavity insulation require a different approach than masonry or concrete structures insulated with foam plastics. The more moisture sensitive the materials, the more rain control required.

Wind strength, wind direction, and rainfall intensity determine in a general way the amount of wind-driven rain deposited. These are factors governed by climate, not by design and construction. The actual distribution of rain on a building is determined by the pattern of wind flow around buildings. This, to a limited extent, can be influenced by design and construction.

2

Physics of Rain

Once rain is deposited on a building surface, its flow over the building surface will be determined by gravity, wind flow over the surface, and wall-surface features such as overhangs, flashings, sills, copings, and mullions. Gravity cannot be influenced by design and construction, and wind flow over building surfaces can only be influenced marginally. However, wall-surface features are completely within the control of the designer and builder. Tradition-based practice has a legacy of developing architectural detailing features that have been used to direct water along particular paths or to cause it to drip free of the wall. Overhangs were developed for a reason. Flashings with rigid drip edges protruding from building faces were specified for a reason. Extended window sills were installed for a reason.

Rain penetration into and through building surfaces is governed by capillarity, momentum, surface tension, gravity, and wind (air pressure) forces. Capillary forces draw rainwater into pores and tiny cracks, while the remaining forces direct rain water into larger openings.

In practice, capillarity can be controlled by capillary breaks, capillary resistant materials or by providing a receptor for capillary moisture. Momentum can be controlled by eliminating openings that go straight through the wall assembly. Rain entry by surface tension can be controlled by the use of drip edges and kerfs. Flashings and layering the wall assembly elements to drain water to the exterior (providing a "drainage plane") can be used to control rainwater from entering by gravity flow, along with simultaneously satisfying the requirements for control of momentum and surface tension forces. Sufficiently overlapping the wall assembly elements or layers comprising the drainage plane can also control entry of rainwater by air pressure differences. Finally, locating a ventilated or pressure moderated air space immediately behind the exterior cladding can be used to control entry of rainwater by air pressure differences by reducing those air pressure differences and providing moisture removal. This approach is called the "Rain Screen" approach.

The more susceptible or vulnerable to moisture damage the materials comprising an assembly are, the more "robust" the rain control strategy required. Additionally, assemblies have different drying characteristics based on material characteristics, insulation levels, interior and exterior climatic conditions and cross-assembly or inter-assembly air flows. In general, vulnerable assemblies with low drying characteristics require greater rain control than moisture insensitive assemblies with high drying potentials.

Coupling a ventilated or pressure moderated air space with a capillary resistant drainage plane ("Rain Screen") in an assembly constructed from moisture insensitive materials with a high drying potential represents the state-of-the-art for Norwegian and Canadian rain control practices. This approach addresses all of the driving forces responsible for rain penetration into and through building surfaces under the severest exposures.

This understanding of the physics of rain leads to the following general approach to rain control:

- reduce the amount of rainwater deposited and flowing on building surfaces

- control rainwater deposited and flowing on building surfaces

Water Management Approaches

The first part of the general approach to rain control involves locating buildings so that they are sheltered from prevailing winds, providing roof overhangs and massing features to shelter exterior walls and reduce wind flow over building surfaces, and finally, providing architectural detailing to shed rainwater from building faces.

The second part of the general approach to rain control involves dealing with capillarity, momentum, surface tension, gravity and air pressure forces acting on rainwater deposited on building surfaces.

The second part of the general approach to rain control employs two general design principles:

- Face Sealed/Barrier Approach
  Storage/Reservoir Systems/Durable Materials
    (all rain exposures)
  Non-Storage/Non-Reservoir Systems/Durable Materials
    (less than 20 inches average annual precipitation)

- Water Managed Approach
  Drain Screen Systems
    (less than 40 inches average annual precipitation)
  Rain Screen Systems
    (40 inches or greater average annual precipitation)

Rain is expected to enter through the cladding skin in the water managed systems: drain screen and rain screen systems. "Drain the rain" is the cornerstone of water managed systems. In the water managed systems, drainage of water is provided by a capillary resistant drainage plane or a capillary resistant drainage plane coupled with a ventilated air space behind the cladding.

2

In the face-sealed barrier approach, the exterior face is the only means to control rain entry. In storage/reservoir systems, some rain is also expected to enter and is stored in the mass of the wall assembly until drying occurs to either the exterior or interior. In non-storage/non-reservoir systems, no or negligible quantities of rain can be permitted to enter.

The performance of a specific system is determined by frequency of rain, severity of rain, system design, selection of materials, workmanship, and maintenance. In general, water managed systems outperform face-sealed/barrier systems due to their more forgiving nature. However, face-sealed/barrier systems constructed from water resistant materials that employ significant storage have a long historical track-record of exemplary performance even in the most severe rain exposures. These "massive" wall assemblies constructed out of masonry, limestone, granite and concrete, many of which are 18 inches or more thick, were typically used in public buildings such as courthouses, libraries, schools and hospitals.

The least forgiving and least water resistant assembly is a face-sealed/barrier wall constructed from water sensitive materials that does not have storage capacity. Most external insulation finish systems (EIFS) are of this type and are not generally recommended. They should be limited to climate zones which see little rain (less than 20 inches average annual precipitation).

The most forgiving and most water resistant assembly is a ventilated rain screen wall constructed from water resistant materials. These types of assemblies perform well in the most severe rain exposures (more than 60 inches average annual precipitation).

Water managed strategies are recommended in all climate regions, and are essential where average annual rainfall exceeds 20 inches. Drain-screen systems (drainage planes without ventilated air spaces) should be limited to regions where average annual rainfall is less than 40 inches and rain-screen systems (drainage planes with ventilated air spaces) should be used wherever average annual rainfall is greater than 40 inches.

Face-sealed/barrier strategies should be carefully considered. Non-storage/non-reservoir systems constructed out of water sensitive materials are not generally recommended and if used should be limited to regions where average annual rainfall is less than 20 inches. Storage/reservoir systems constructed with water resistant materials can be built anywhere. However, their performance is design, workmanship, and materials dependent. In general, these systems should be limited to regions or to designs with high drying potentials to the exterior, interior or, better still, to both.

Drainage Planes/Water Resistant Barriers

Drainage planes are water repellent materials (building paper, house-wrap, foam insulation, etc.), which are located behind the cladding and are designed and constructed to drain water that passes through the cladding. They are interconnected with flashings, window and door openings, and other penetrations of the building enclosure to provide drainage of water to the exterior of the building. The materials that form the drainage plane overlap each other shingle fashion or are sealed so that water drains down and out of the wall.

The most common drainage plane is "tar paper" or building paper. More recently, the term "housewrap" has been introduced to describe building papers that are not asphalt impregnated felts or coated papers. Drainage planes can also be created by sealing or layering water resistant sheathings such as a rigid insulation or a coated structural sheathing.

The drainage plane is also referred to as the "water resistant barrier" or WRB.

All exterior claddings pass some rainwater. Siding leaks, brick leaks, stucco leaks, stone leaks, etc. As such, some control of this penetrating rainwater is required. In most walls, this penetrating rainwater is controlled by the drainage plane that directs the penetrating rainwater downwards and outwards.

Since all exterior claddings "leak", all wood frame wall assemblies require a drainage plane coupled with a drainage space - where it rains. "Where it rains" is defined as locations in North America that receive more than 20 inches or rain annually. Traditionally, drainage planes consisted of tar paper installed shingle fashion behind exterior claddings coupled with a flashing at the base of each wall to direct rainwater that penetrated the cladding systems to the exterior. It was important that some form of air space or drainage space was also provided between the cladding system and the drainage plane to allow drainage.

With wood siding, the drainage space is typically intermittent and depends largely on the profile of the siding. Ideally, wood siding, should be installed over furring creating a drained (and vented) air space between the drainage plane and wood siding. With vinyl and aluminum siding, the drainage space is more pronounced and furring is not necessary.

With stucco claddings, the drainage space was traditionally provided by using two layers of asphalt impregnated felt paper. The water absorbed by the felt papers from the base coat of stucco caused the papers to swell and expand. When the assembly dried, the papers would shrink, wrinkle, and de-bond providing a tortuous, but reasonably effective drainage space. This drainage space was typically between $1/_{16}$-inch and $1/_8$-inch wide.

With brick veneers, the width of the drainage space has been based more on tradition rather than physics. A 1-inch airspace is more-or-less the width of a mason's fingers, hence, the typical requirement for a 1-inch airspace. However, from our experience with stucco and other cladding systems, spaces as small as $1/_{16}$-inch drain. However large the space it must be coupled with a functional drainage plane.

Rainwater Control

The fundamental principle of rainwater control is to shed water by layering materials in such a way that water is directed downwards and outwards from the building or away from the building. It applies to assemblies such as walls, roofs and foundations, as well as to the components that can be found in walls, roofs and foundations such as windows, doors and skylights. It also applies to assemblies that connect to walls, roofs and foundations such as balconies, decks, railings and dormers.

Layering materials to shed water applies to the building as a whole (see Figure 2.2). Overhangs can be used to keep water away from walls. Canopies can be used to keep water away from windows, and site grading can be used to keep water away from foundation perimeters.

When selecting building materials, take into account that building materials may be exposed to rain or other elements during construction. For example, walls without roofs on them will get wet. It is not a good idea to build these walls with exterior gypsum board that is paper-faced since they hold water. This is a major concern with party walls or fire walls in multifamily buildings. Glass-faced gypsum board or other water-resistant alternatives should be used.

Drainage is the key to rainwater control:
- Drain the site (see Figure 2.2)
- Drain the ground
- Drain the building (see Figure 2.3)
- Drain the assembly
- Drain the opening (see Figure 2.4)
- Drain the component
- Drain the material (see Figure 2.5)

Reservoirs on the outside of homes are a problem. What are reservoirs? Materials that store rainwater – sponges that get wet when it rains. Once the reservoirs get wet, the stored water can migrate elsewhere and cause problems. Common reservoirs are brick veneers, stuccos, wood siding, wood trim and fiber cement cladding.

How to handle reservoirs? Easy. Get rid of them or disconnect them from the building (see Figure 2.6). Back priming (painting all surfaces, back, front, edges and ends of wood siding, cement siding and all wood trim) gets rid of the moisture storage issue with these materials. No reservoir, no problem.

Back venting brick veneers and installing them over foam sheathings disconnects the brick veneer moisture reservoir from the home (see Figure 2.7). Installing stucco over two layers of building paper or over an appropriate capillary break, such as foam sheathing, similarly addresses stucco reservoirs.

Roofs should be designed to shed rainwater away from the building. Steep pitches are better than shallow pitches. Crickets should be used to divert water away from chimneys and architectural features.

Roofs should also be designed to protect walls. Large overhangs are better than small overhangs or no overhangs.

Ideally, roofs should have simple geometry. The more complex the roof, the more dormers, ridges and valleys, the more likely a roof will leak. Penetrations should also be minimized or avoided.

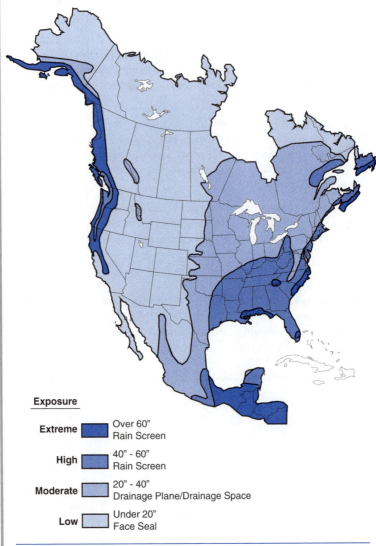

Exposure

Extreme		Over 60" Rain Screen
High		40" - 60" Rain Screen
Moderate		20" - 40" Drainage Plane/Drainage Space
Low		Under 20" Face Seal

Figure 2.1
Annual Rainfall Map

- Based on information from the U.S. Department of Agriculture and Environment Canada
- An example of a Drainage Plane is building paper (tar paper) installed shingle fashion
- An example of a Rain-Screen is building paper installed with a ventilated drainage space
- An example of a Face Seal is a non-drained EIFS

2

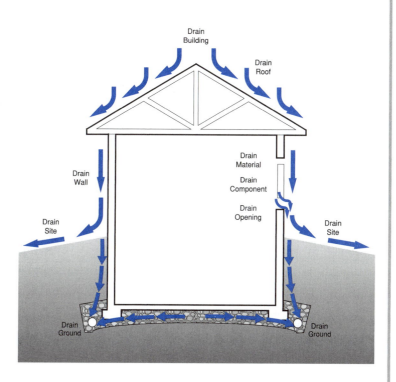

Figure 2.2
Layering Materials to Shed Water
  • Applies to whole building

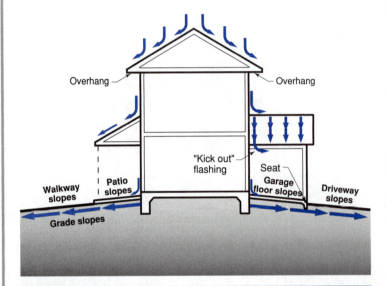

Figure 2.3
Drain the Building

- Patios and decks lower than floors and slope away from building
- Garage floor lower than main floor and slope away from building
- Driveway lower than garage floor and slope away from building
- Grade lower than main floor and slope away from building
- Stoops and walkways lower than main floor and slope away from building
- Kick-out flashings or diverters direct water away from walls at roof/wall intersections
- Overhangs protect walls

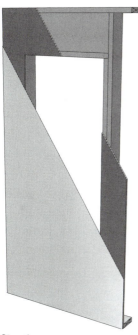

Step 1
Wood frame wall with OSB and housewrap

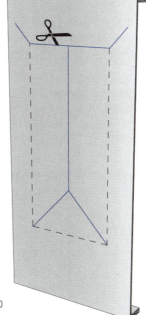

Step 2
Modified "I" cut in housewrap

Figure 2.4
Installing a Window with Housewrap on OSB Over a Wood Frame Wall in Nine Steps

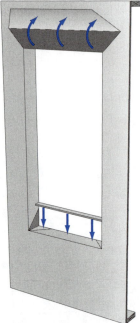

Step 3
Housewrap folded in at jambs and sill; Housewrap at head temporarily folded up or, alternatively, tucked under; Install backdam

Step 4
Install first piece of adhesive-backed sill flashing

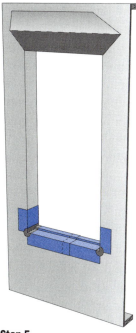

Step 5

Install second piece of adhesive-backed sill flashing and corner flashing patches at sill

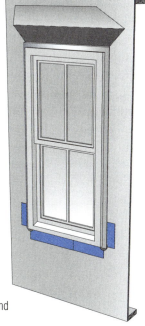

Step 6

Install window plumb, level and square per manufacturer's instructions

Step 7
Install jamb flashing first; Install a
drip cap (if applicable); Install
head flashing

Step 8
Fold down head housewrap

2

Step 9

Apply corner patches at head; Air seal window around entire perimeter on the interior with sealant or non-expanding foam

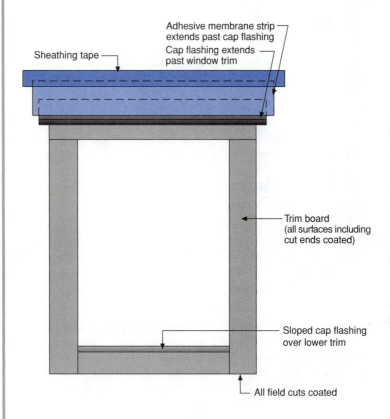

Figure 2.5
Flashing Over and Under Window Trim

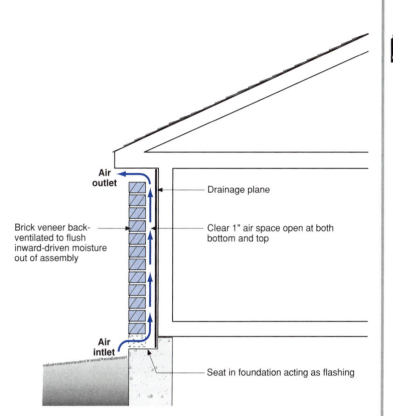

Air outlet

Brick veneer back-ventilated to flush inward-driven moisture out of assembly

Drainage plane

Clear 1" air space open at both bottom and top

Air intlet

Seat in foundation acting as flashing

Figure 2.6
Ventilated Cavity

- To effectively uncouple a brick veneer from a wall system by using back ventilation, a clear cavity must be provided along with both air inlets at the bottom and air outlets at the top

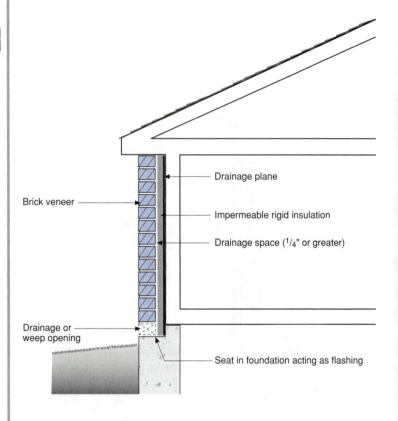

Figure 2.7
Drained Cavity with Condensing Surface

- To effectively uncouple a brick veneer from a wall system by using a condensing surface, the drainage plane must also be a vapor barrier or a vapor impermeable layer (i.e. rigid insulation) must be installed between the drainage plane and the brick veneer. Alternatively, the rigid insulation can be configured to act as both the drainage plane and vapor impermeable layer.
- When a condensing surface is used to uncouple a brick veneer from wall system, a ventilated air space is no longer necessary — i.e. the presence of mortar droppings is no longer an issue. Additionally, the width of the drainage space is almost irrelevant.

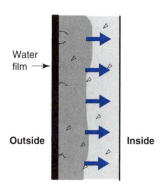

Water
film →

Outside   Inside

Capillary suction draws
water into porous material
and tiny cracks

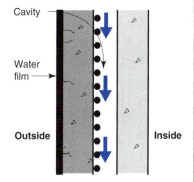

Cavity →

Water
film →

Outside   Inside

Cavity acts as capillary
break and receptor for
capillary water interrupting flow

Figure 2.8
Capillarity as a Driving Force for Rain Entry
- Capillary suction draws water into porous material and tiny cracks
- Cavity acts as capillary break and receptor for capillary water interrupting flow

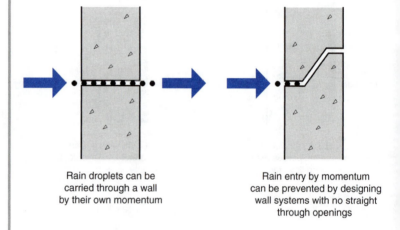

Rain droplets can be
carried through a wall
by their own momentum

Rain entry by momentum
can be prevented by designing
wall systems with no straight
through openings

Figure 2.9
Momentum as a Driving Force for Rain Entry
- Rain droplets can be carried through a wall by their own momentum
- Rain entry by momentum can be prevented by designing wall systems with no straight through openings
- Condition example: window sill

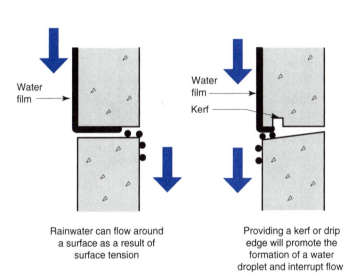

Water film →

Rainwater can flow around
a surface as a result of
surface tension

Water film →

Kerf →

Providing a kerf or drip
edge will promote the
formation of a water
droplet and interrupt flow

Figure 2.10
Surface Tension as a Driving Force for Rain Entry
- Rainwater can flow around a surface as a result of surface tension
- Providing a kerf or drip edge will promote the formation of a water droplet and interrupt flow
- Condition example: window head

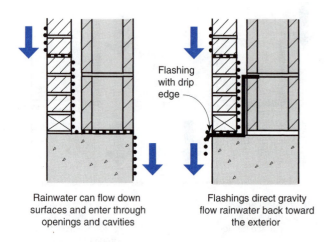

Rainwater can flow down
surfaces and enter through
openings and cavities

Flashing
with drip
edge

Flashings direct gravity
flow rainwater back toward
the exterior

Figure 2.11
Gravity as a Driving Force for Rain Entry
- Rainwater can flow down surfaces and enter through openings and cavities
- Flashings direct gravity flow rainwater back to the exterior

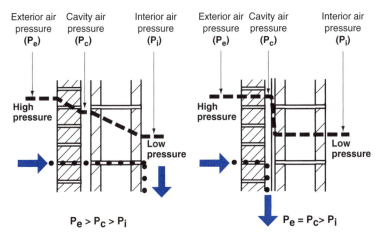

Exterior air pressure (P_e) Cavity air pressure (P_c) Interior air pressure (P_i)

High pressure

Low pressure

$$P_e > P_c > P_i$$

Driven by air pressure differences, rain droplets are drawn through wall openings from the exterior to the interior

Exterior air pressure (P_e) Cavity air pressure (P_c) Interior air pressure (P_i)

High pressure

Low pressure

$$P_e = P_c > P_i$$

By using pressure moderation between the exterior and cavity air, air pressure is diminished as a driving force for rain entry.

Figure 2.12
Air Pressure Difference as a Driving Force for Rain Entry
- Driven by air pressure differences, rain droplets are drawn through wall openings from the exterior to the interior
- By using pressure moderation between the exterior and cavity air, air pressure is diminished as a driving force for rain entry

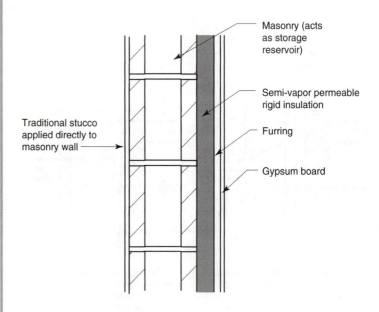

Masonry (acts
as storage
reservoir)

Semi-vapor permeable
rigid insulation

Furring

Gypsum board

Traditional stucco
applied directly to
masonry wall

Figure 2.13
Face-Sealed Barrier Wall
Storage Reservoir System
- Some rain entry past exterior face permitted
- Penetrating rain stored in mass of wall until drying occurs to interior or
  exterior
- Can be used in all rain exposure regions

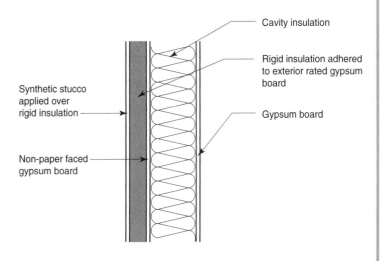

Cavity insulation

Rigid insulation adhered
to exterior rated gypsum
board

Gypsum board

Synthetic stucco
applied over
rigid insulation

Non-paper faced
gypsum board

**Figure 2.14
Face Sealed Barrier Wall
Non-Storage Non-Reservoir System**
- No rain entry past exterior face permitted
- Not generally recommended
- Should not be used in regions where the average annual precipitation exceeds 20 inches

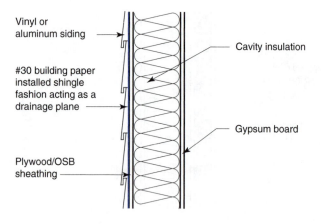

Vinyl or aluminum siding

#30 building paper installed shingle fashion acting as a drainage plane

Plywood/OSB sheathing

Cavity insulation

Gypsum board

Figure 2.15
Water Managed Wall
Drain Screen System (Drainage Plane)

- Can be used in regions where the average annual precipitation is less than 40 inches
- Should be limited to regions with a "moderate" or "low" rain exposure

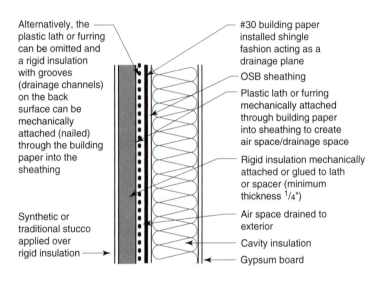

Alternatively, the plastic lath or furring can be omitted and a rigid insulation with grooves (drainage channels) on the back surface can be mechanically attached (nailed) through the building paper into the sheathing

Synthetic or traditional stucco applied over rigid insulation

#30 building paper installed shingle fashion acting as a drainage plane

OSB sheathing

Plastic lath or furring mechanically attached through building paper into sheathing to create air space/drainage space

Rigid insulation mechanically attached or glued to lath or spacer (minimum thickness $1/4$")

Air space drained to exterior

Cavity insulation

Gypsum board

Figure 2.16
Water Managed Wall

- Should be used in regions where the average annual precipitation exceeds 20 inches
- To convert this wall system to a "rain screen" system, the drainage space needs to be ventilated to the exterior rather than simply drained

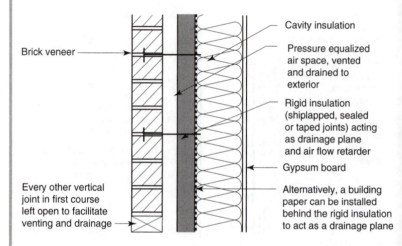

Cavity insulation

Pressure equalized air space, vented and drained to exterior

Rigid insulation (shiplapped, sealed or taped joints) acting as drainage plane and air flow retarder

Gypsum board

Alternatively, a building paper can be installed behind the rigid insulation to act as a drainage plane

Brick veneer

Every other vertical joint in first course left open to facilitate venting and drainage

Figure 2.17
Water Managed Wall
Rain Screen System (Drainage Plane with Ventilated Air Space)
- Should be used in regions where the average annual precipitation exceeds 40 inches
- Building papers used as drainage planes are typically more reliable than taped, sealed or shiplapped rigid insulation

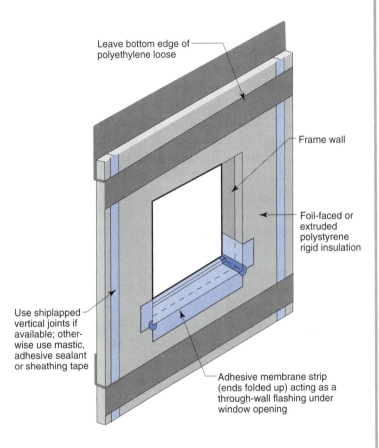

Leave bottom edge of polyethylene loose

Frame wall

Foil-faced or extruded polystyrene rigid insulation

Use shiplapped vertical joints if available; otherwise use mastic, adhesive sealant or sheathing tape

Adhesive membrane strip (ends folded up) acting as a through-wall flashing under window opening

Figure 2.18
Taped Rigid Insulation as Drainage Plane
- Flanged window inserted into opening, flanges back-caulked between rigid insulation and flange
- Tape or adhesive membrane strip installed over flanges to further seal flanges to rigid insulation
- Foil-faced insulations should have foil facing installed to exterior if only faced on one side

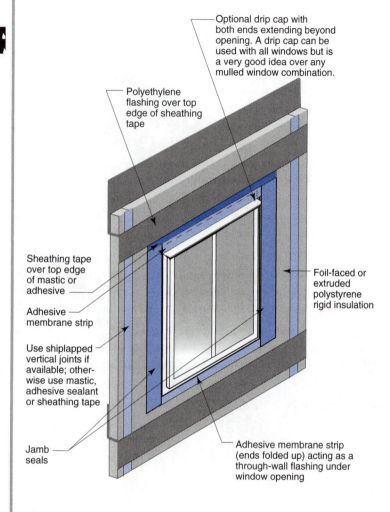

Optional drip cap with both ends extending beyond opening. A drip cap can be used with all windows but is a very good idea over any mulled window combination.

Polyethylene flashing over top edge of sheathing tape

Sheathing tape over top edge of mastic or adhesive

Adhesive membrane strip

Use shiplapped vertical joints if available; otherwise use mastic, adhesive sealant or sheathing tape

Jamb seals

Foil-faced or extruded polystyrene rigid insulation

Adhesive membrane strip (ends folded up) acting as a through-wall flashing under window opening

Figure 2.19
Window Installation with Foam Sheathing

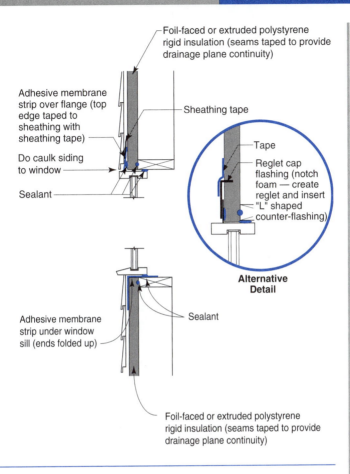

Foil-faced or extruded polystyrene
rigid insulation (seams taped to provide
drainage plane continuity)

Adhesive membrane
strip over flange (top
edge taped to
sheathing with
sheathing tape)

Sheathing tape

Do caulk siding
to window

Sealant

Tape

Reglet cap
flashing (notch
foam — create
reglet and insert
"L" shaped
counter-flashing)

**Alternative
Detail**

Adhesive membrane
strip under window
sill (ends folded up)

Sealant

Foil-faced or extruded polystyrene
rigid insulation (seams taped to provide
drainage plane continuity)

**Figure 2.20
Window Head and Window Sill**

• Foil-faced insulations should have foil facing installed to exterior if only
faced on one side

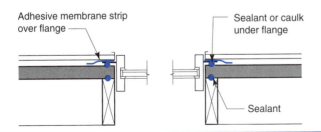

Adhesive membrane strip
over flange

Sealant or caulk
under flange

Sealant

**Figure 2.21
Window Jamb**

2

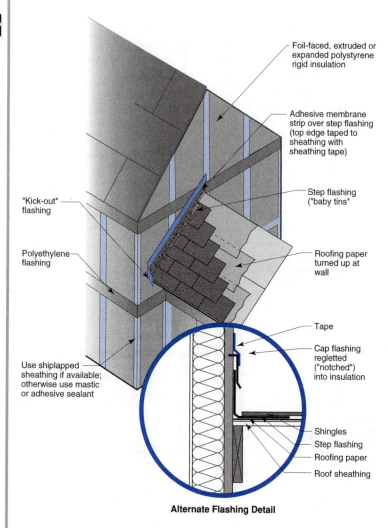

Foil-faced, extruded or expanded polystyrene rigid insulation

Adhesive membrane strip over step flashing (top edge taped to sheathing with sheathing tape)

"Kick-out" flashing

Step flashing ("baby tins")

Polyethylene flashing

Roofing paper turned up at wall

Use shiplapped sheathing if available; otherwise use mastic or adhesive sealant

Tape

Cap flashing regletted ("notched") into insulation

Shingles
Step flashing
Roofing paper
Roof sheathing

Alternate Flashing Detail

Figure 2.22
Step Flashing

2

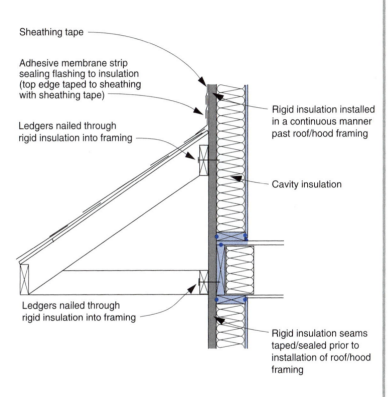

Sheathing tape

Adhesive membrane strip
sealing flashing to insulation
(top edge taped to sheathing
with sheathing tape)

Rigid insulation installed
in a continuous manner
past roof/hood framing

Ledgers nailed through
rigid insulation into framing

Cavity insulation

Ledgers nailed through
rigid insulation into framing

Rigid insulation seams
taped/sealed prior to
installation of roof/hood
framing

Figure 2.23
Flashing Above Shed Roof

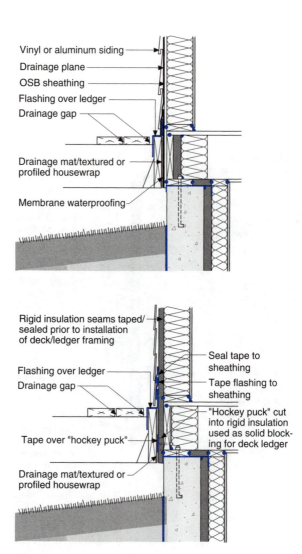

Vinyl or aluminum siding

Drainage plane

OSB sheathing

Flashing over ledger

Drainage gap

Drainage mat/textured or profiled housewrap

Membrane waterproofing

Rigid insulation seams taped/ sealed prior to installation of deck/ledger framing

Flashing over ledger

Drainage gap

Seal tape to sheathing

Tape flashing to sheathing

"Hockey puck" cut into rigid insulation used as solid blocking for deck ledger

Tape over "hockey puck"

Drainage mat/textured or profiled housewrap

Figure 2.24
Flashing Over Deck Ledger

2

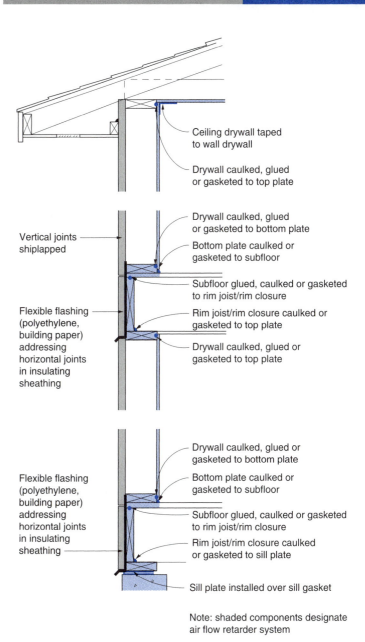

Ceiling drywall taped
to wall drywall

Drywall caulked, glued
or gasketed to top plate

Drywall caulked, glued
or gasketed to bottom plate

Vertical joints
shiplapped

Bottom plate caulked or
gasketed to subfloor

Subfloor glued, caulked or gasketed
to rim joist/rim closure

Flexible flashing
(polyethylene,
building paper)
addressing
horizontal joints
in insulating
sheathing

Rim joist/rim closure caulked or
gasketed to top plate

Drywall caulked, glued or
gasketed to top plate

Drywall caulked, glued or
gasketed to bottom plate

Flexible flashing
(polyethylene,
building paper)
addressing
horizontal joints
in insulating
sheathing

Bottom plate caulked or
gasketed to subfloor

Subfloor glued, caulked or gasketed
to rim joist/rim closure

Rim joist/rim closure caulked
or gasketed to sill plate

Sill plate installed over sill gasket

Note: shaded components designate
air flow retarder system

Figure 2.25
Flashing Horizontal Joints in Insulating Sheathing
- Eliminates tape at horizontal joints
- More durable than taped connection

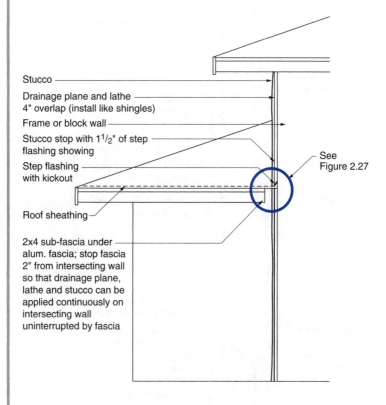

Stucco

Drainage plane and lathe
4" overlap (install like shingles)

Frame or block wall

Stucco stop with 1$1/_2$" of step
flashing showing

Step flashing
with kickout

See
Figure 2.27

Roof sheathing

2x4 sub-fascia under
alum. fascia; stop fascia
2" from intersecting wall
so that drainage plane,
lathe and stucco can be
applied continuously on
intersecting wall
uninterrupted by fascia

Figure 2.26
Roof Encroaching Wall Detail
• Prevents leaks into wall behind stucco

2

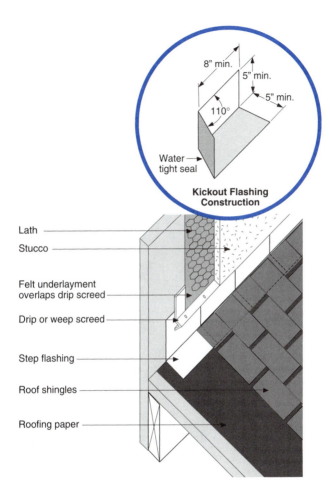

8" min.

5" min.

5" min.

110°

Water tight seal

Kickout Flashing Construction

Lath

Stucco

Felt underlayment overlaps drip screed

Drip or weep screed

Step flashing

Roof shingles

Roofing paper

Figure 2.27
Stucco at Step Flashing
- Kick-out flashing installed at bottom of intersecting roof

Air Barriers

Air barriers are systems of materials designed and constructed to control airflow between a conditioned space and an unconditioned space. The air barrier system is the primary air enclosure boundary that separates indoor (conditioned) air and outdoor (unconditioned) air. In multi-unit/townhouse/apartment construction the air barrier system also separates the conditioned air from any given unit and adjacent units. Air barrier systems also typically define the location of the pressure boundary of the building enclosure.

In multi-unit/townhouse/apartment construction the air barrier system is also the fire barrier and smoke barrier in inter-unit separations. In such assemblies the air barrier system must also meet the specific fire resistance rating requirement for the given separation.

The air barrier system also separates garages from conditioned spaces. In this regard the air barrier system is also the "gas barrier" and provides the gas-tight separation between a garage and the remainder of the house or building.

Air barriers are intended to resist the air pressure differences that act on them. Rigid materials such as gypsum board, exterior sheathing materials like plywood or OSB, and supported flexible barriers are typically effective air barrier systems if joints and seams are sealed. Spray foam systems can also act as effective air barrier systems either externally applied over structural elements or internally applied within cavity systems.

Air barrier systems keep outside air out of the building enclosure or inside air out of the building enclosure depending on climate or configuration. Sometimes, air barrier systems do both.

Air barrier systems can be located anywhere in the building enclosure — at the exterior surface, the interior surface, or at any location in between. In cold climates, interior air barrier systems control the exfiltration of interior, often moisture-laden air. Whereas exterior air barrier systems control the infiltration of exterior air and prevent wind-washing through cavity insulation systems.

Air barrier systems should be:

- impermeable to airflow;

- continuous over the entire building enclosure or continuous over the enclosure of any given unit;

- able to withstand the forces that may act on them during and after construction; and

- durable over the expected lifetime of the building.

Performance Requirements

Air barrier systems typically are assembled from **materials** incorporated in **assemblies** that are interconnected to create **enclosures**. Each of these three elements has measurable resistance to airflow. The recommended minimum resistances or air permeances for the three components are listed as follows:

- Material 0.02 l/(s-m^2)@75 Pa

- Assembly 0.20 l/(s-m^2)@75 Pa

- Enclosure $0.2.00$ l/(s-m^2)@75 Pa

Materials and assemblies that meet these performance requirements are said to be air barrier materials and air barrier assemblies. Air barrier materials incorporated in air barrier assemblies that in turn are interconnected to create enclosures are called air barrier systems.

Materials and assemblies that do not meet these performance requirements, but are nevertheless designed and constructed to control airflow are said to be air retarders.

The recommended minimum resistances are based on experience and current practice:

- gypsum board is a common air barrier material and it readily meets the material air permeance recommendation (0.0196 l/(s-m2)@75 Pa as tested by NRCC, 1997);

- The National Building Code of Canada specifies that the principle air barrier material must meet a minimum resistance or air permeance of 0.02 l/(s-m^2)@75 Pa;

- The U.S. Department of Energy Building America project sets an airtightness requirement of 1.65 l/(s-m^2)@75 Pa for residences (see Appendix III).

Design

Approaches

Four common approaches are used to provide air barrier systems in residential buildings:

- interior air barrier system using gypsum board and framing;
- interior air barrier system using polyethylene;
- exterior air barrier system using exterior sheathing; or
- exterior air barrier system using housewraps.

Spray applied foam insulations can be used as interstitial (cavity) air barrier systems. Damp spray applied cellulose does not meet the performance requirements of air barrier materials or assemblies — rather it is an air retarder.

An advantage of interior air barrier systems over exterior systems is that they control the entry of interior moisture-laden air into assembly cavities during heating periods. The significant disadvantage of interior air barrier systems is their inability to control wind-washing through cavity insulation and their inability to address the infiltration of exterior hot-humid air in hot-humid climates.

The significant advantage of exterior air barrier systems is the ease of installation and the lack of detailing issues related to intersecting partition walls and service penetrations. However, exterior air barrier systems must deal with transitions where roof assemblies intersect exterior walls. For example, an exterior housewrap should be sealed to the ceiling air barrier system across the top of the exterior perimeter walls.

An additional advantage of exterior air barrier systems is the control of wind-washing that an exterior air seal provides. The significant disadvantage of exterior air barrier systems is their inability to control the entry of air-transported moisture into cavities from the interior.

Installing both interior an exterior air barrier systems addresses the weakness of each.

Air barrier materials can also be provided with properties which also class them as vapor barriers. An example of this is polyethylene film which can be used as both an air barrier and a vapor barrier.

Keep in mind, however, polyethylene on the inside of building assemblies in cold, mixed-humid, marine, hot-dry and hot-humid climates is not generally a good idea; drying of building assemblies in these climates occurs to the inside as well as to the outside.

Interior drying is typically necessary in air conditioned enclosures. In other words, interior vapor barriers such as polyethylene should never be installed in an air conditioned building — even one located in a cold climate.

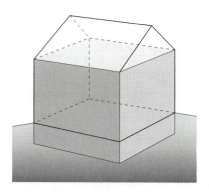

Figure 3.1
Air Barrier System — Single Family Detached
- Separates indoor (conditioned) and outdoor (unconditioned) air
- Separates garages from the conditioned spaces; acts as the "gas barrier"

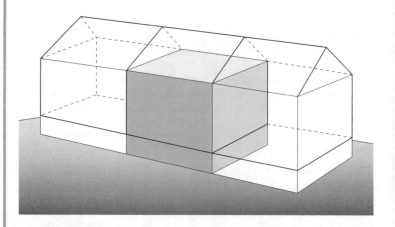

Figure 3.2
Air Barrier System — Multi-Family
- Separates indoor (conditioned) and outdoor (unconditioned) air
- Separates the conditioned air from any given unit and adjacent units

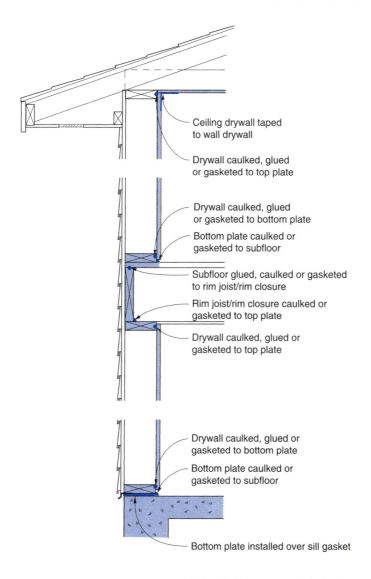

Ceiling drywall taped
to wall drywall

Drywall caulked, glued
or gasketed to top plate

Drywall caulked, glued
or gasketed to bottom plate

Bottom plate caulked or
gasketed to subfloor

Subfloor glued, caulked or gasketed
to rim joist/rim closure

Rim joist/rim closure caulked or
gasketed to top plate

Drywall caulked, glued or
gasketed to top plate

Drywall caulked, glued or
gasketed to bottom plate

Bottom plate caulked or
gasketed to subfloor

Bottom plate installed over sill gasket

Note: shaded components designate
air barrier system

Figure 3.3
Interior Air barrier Using Drywall and Framing
- Air Drywall Approach (ADA)

3

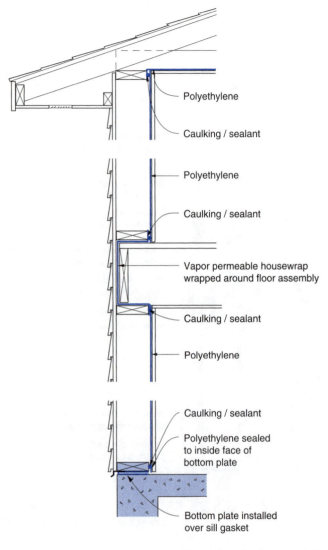

Polyethylene

Caulking / sealant

Polyethylene

Caulking / sealant

Vapor permeable housewrap wrapped around floor assembly

Caulking / sealant

Polyethylene

Caulking / sealant

Polyethylene sealed to inside face of bottom plate

Bottom plate installed over sill gasket

Note: shaded components designate air barrier system

Figure 3.4
Interior Air Barrier Using Polyethylene

- This approach is not recommended for cold, mixed-humid, marine, mixed-dry, hot-dry or hot-humid climates due to the impermeability of the polyethylene (see Chapter 4)
- Recommended for very cold climates only in non-air conditioned buildings

3

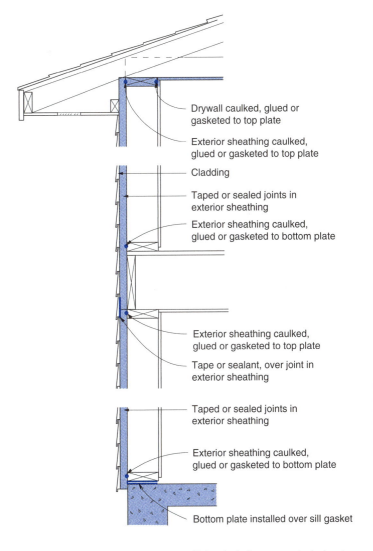

Drywall caulked, glued or gasketed to top plate

Exterior sheathing caulked, glued or gasketed to top plate

Cladding

Taped or sealed joints in exterior sheathing

Exterior sheathing caulked, glued or gasketed to bottom plate

Exterior sheathing caulked, glued or gasketed to top plate

Tape or sealant, over joint in exterior sheathing

Taped or sealed joints in exterior sheathing

Exterior sheathing caulked, glued or gasketed to bottom plate

Bottom plate installed over sill gasket

Note: shaded components designate air barrier system

Figure 3.5
Exterior Air Barrier Using Exterior Sheathing

3

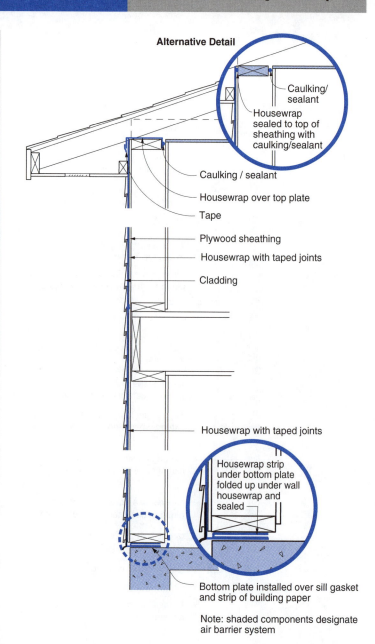

Alternative Detail

Caulking/
sealant

Housewrap
sealed to top of
sheathing with
caulking/sealant

Caulking / sealant

Housewrap over top plate

Tape

Plywood sheathing

Housewrap with taped joints

Cladding

Housewrap with taped joints

Housewrap strip
under bottom plate
folded up under wall
housewrap and
sealed

Bottom plate installed over sill gasket
and strip of building paper

Note: shaded components designate
air barrier system

Figure 3.6
Exterior Air Barrier Using Housewrap
- Polyethylene in ceiling is part of the air barrier system

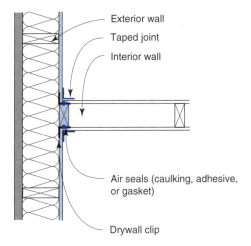

Exterior wall

Taped joint

Interior wall

Air seals (caulking, adhesive, or gasket)

Drywall clip

Figure 3.7
Intersection of Interior Partition Wall and Exterior Wall
- Air Drywall Approach (ADA) - interior air barrier
- Seal drywall to framing at partition wall

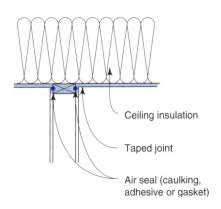

Ceiling insulation

Taped joint

Air seal (caulking, adhesive or gasket)

Figure 3.8
Intersection of Interior Partition Wall and Insulated Ceiling
- Air Drywall Approach (ADA) - interior air barrier
- Seal drywall to framing top plate

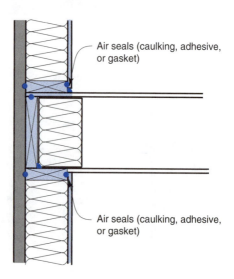

Air seals (caulking, adhesive, or gasket)

Air seals (caulking, adhesive, or gasket)

Figure 3.9
Intersection of Floor Joists and Exterior Wall
• Air Drywall Approach (ADA) - interior air barrier

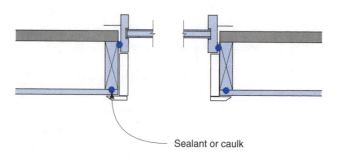

Sealant or caulk

Figure 3.10
Window Jamb with Wood Trim
• Air Drywall Approach (ADA) - interior air barrier

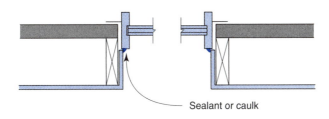

Sealant or caulk

Figure 3.11
Window Jamb with Drywall Returns
- Air Drywall Approach (ADA) - interior air barrier

Insulations, Sheathings and Vapor Retarders

Two seemingly innocuous requirements for building enclosure assemblies bedevil builders and designers almost endlessly:

- keep water vapor out
- let the water vapor out if it gets in

It gets complicated because, sometimes, the best strategies to keep water vapor out also trap water vapor in. This can be a real problem if the assemblies start out wet because of rain or the use of wet materials (wet framing, concrete, masonry or damp spray cellulose, fiberglass or rock wool cavity insulation).

It gets even more complicated because of climate. In general, water vapor moves from the warm side of building assemblies to the cold side of building assemblies. This means we need different strategies for different climates. We also have to take into account differences between summer and winter.

The good news is that water vapor moves only two ways - vapor diffusion and air transport. If we understand the two ways, and know where we are (climate zone) we can solve the problem.

The bad news is that techniques that are effective at controlling vapor diffusion can be ineffective at controlling air transported moisture, and vice versa.

Building assemblies, regardless of climate zone, need to control the migration of moisture as a result of both vapor diffusion and air transport. Techniques that are effective in controlling vapor diffusion can be very different from those that control air transported moisture.

Vapor Diffusion and Air Transport of Vapor

Vapor diffusion is the movement of moisture in the vapor state through a material as a result of a vapor pressure difference (concentration gradient) or a temperature difference (thermal gradient). It is often con-

4

fused with the movement of moisture in the vapor state into building assemblies as a result of air movement. Vapor diffusion moves moisture from an area of higher vapor pressure to an area of lower vapor pressure as well as from the warm side of an assembly to the cold side. Air transport of moisture will move moisture from an area of higher air pressure to an area of lower air pressure if moisture is contained in the moving air (Figure 4.1).

Vapor pressure is directly related to the concentration of moisture at a specific location. It also refers to the density of water molecules in air. For example, a cubic foot of air containing 2 trillion molecules of water in the vapor state has a higher vapor pressure (or higher water vapor density) than a cubic foot of air containing 1 trillion molecules of water in the vapor state. Moisture will migrate by diffusion from where there is more moisture to where there is less. Hence, moisture in the vapor state migrates by diffusion from areas of higher vapor pressure to areas of lower vapor pressure.

Moisture in the vapor state also moves from the warm side of an assembly to the cold side of an assembly. This type of moisture transport is called thermally driven diffusion.

The second law of thermodynamics governs the exchange of energy and can be used to explain the concept of both vapor pressure driven diffusion and thermally driven diffusion. The movement of moisture from an area of higher vapor pressure to an area of lower vapor pressure as well as from the warm side of an assembly to the cold side of an assembly is a minimization of available "system" energy (or an increase in entropy).

When temperature differences become large, water vapor can condense on cold surfaces. When condensation occurs, water vapor is removed from the air and converted to liquid moisture on the surface resulting in a reduction in water vapor density in the air near the cold surface (i.e. a lower vapor pressure). These cold surfaces now act as "dehumidifiers" pulling more moisture towards them.

Vapor diffusion and air transport of water vapor act independently of one another. Vapor diffusion will transport moisture through materials and assemblies in the absence of an air pressure difference if a vapor pressure or temperature difference exists. Furthermore, vapor diffusion will transport moisture in the opposite direction of small air pressure differences, if an opposing vapor pressure or temperature difference exists. For example, in a hot-humid climate, the exterior is typically at a high vapor pressure and high temperature during the summer. In addition, it is common for an interior air conditioned space to be maintained at a cool temperature and at a low vapor pressure through the dehumidification characteristics of the air conditioning system. This causes vapor diffusion to move water vapor from the exterior towards the interior. This will occur even if the interior conditioned space is maintained at a higher air pressure (a pressurized enclosure) relative to the exterior (Figure 4.2).

Vapor Retarders

The function of a vapor retarder is to control the entry of water vapor into building assemblies by the mechanism of vapor diffusion. The vapor retarder may be required to control the diffusion entry of water vapor into building assemblies from the interior of a building, from the exterior of a building or from both the interior and exterior.

Vapor retarders should not be confused with air barriers whose function is to control the movement of air through building assemblies. In some instances, air barrier systems may also have specific material properties which also allow them to perform as vapor retarders. For example, a rubber membrane on the exterior of a masonry wall installed in a continuous manner is a very effective air barrier. The physical properties of rubber also give it the characteristics of a vapor retarder; in fact, it can be considered a vapor "barrier." Similarly, a continuous, sealed polyethylene ground cover installed in an unvented, conditioned crawlspace acts as both an air barrier and a vapor retarder; and, in this case, it is also a vapor "barrier." The opposite situation is also common. For example, a building paper or a housewrap installed in a continuous manner can be a very effective air barrier. However, the physical properties of most building papers and housewraps (they are vapor permeable - they "breathe") do not allow them to act as effective vapor retarders.

Water Vapor Permeability

The key physical property which distinguishes vapor retarders from other materials, is permeability to water vapor. Materials which retard water vapor flow are said to be impermeable. Materials which allow water vapor to pass through them are said to be permeable. However, there are degrees of impermeability and permeability and the classification of materials typically is quite arbitrary. Furthermore, under changing conditions, some materials that initially are "impermeable," can become "permeable." Hygroscopic materials change their permeability characteristics as relative humidity increases. For example, plywood sheathing under typical conditions is relatively impermeable. However, once plywood becomes wet, it can become relatively permeable. As a result we tend to refer to plywood as a vapor semi-permeable material.

Non-hygroscopic materials such as polyethylene or plastic housewraps do not change their permeability as a function of relative humidity.

The unit of measurement typically used in characterizing permeability is a "perm." Many building codes define a vapor retarder as a material that has a permeability of one perm or less as tested under dry cup test method.

Materials are typically tested in two ways to determine permeability: dry cup testing and wet cup testing. Some confusion occurs when considering the difference between wet cup perm ratings and dry cup perm ratings. A wet cup test is conducted with 50 percent relative humidity maintained on one side of the test sample and 100 percent relative humidity maintained on the other side of the test sample. A dry cup test is conducted with 0 percent relative humidity maintained on one side of the test sample and 50 percent relative humidity maintained on the other side of the test sample.

Different values are typical between the two tests for materials that absorb and adsorb water — materials that are hygroscopic. As the quantity of adsorbed water on the surface of hygroscopic materials increases, the vapor permeability of the materials also increases. In other words, for hygroscopic materials, the vapor permeability goes up as the relative humidity goes up.

In general, for hygroscopic materials, the wet cup test provides perm ratings many times the dry cup test values. For non-hygroscopic materials, materials that are hydrophobic, there is typically no difference between wet cup and dry cup test results. For plywood, a hygroscopic material, a dry cup permeability of 0.5 perms is common. However, as the plywood gets wet, it "breathes" and wet cup permeabilities of 3 perms or higher are common.

Materials can be separated into four general classes based on their permeance:

- vapor impermeable   0.1 perm or less

- vapor semi-impermeable   1.0 perms or less and greater than 0.1 perm

- vapor semi-permeable   10 perms or less and greater than 1.0 perm

- vapor permeable   greater than 10 perms

Materials that are generally classed as impermeable to water vapor are:

- rubber membranes,
- polyethylene film,
- glass,
- aluminum foil,
- sheet metal,
- foil-faced insulating sheathings, and
- foil-faced non-insulating sheathings.

4

Materials that are generally classed as vapor semi-impermeable to water vapor are:

- oil-based paints,
- most vinyl wall coverings,
- unfaced extruded polystyrene greater than 1-inch thick, and
- traditional hard-coat stucco applied over building paper and OSB sheathing.

Materials that are generally classed as vapor semi-permeable to water vapor are:

- plywood,
- bitumen impregnated kraft paper,
- OSB,
- unfaced expanded polystyrene (EPS),
- unfaced extruded polystyrene (XPS) — 1-inch thick or less,
- fiber-faced isocyanurate,
- heavy asphalt impregnated building papers (#30 building paper), and
- most latex-based paints.

Depending on the specific assembly design, construction and climate, all of these materials may or may not be considered to act as vapor retarders. Typically, these materials are considered to be more vapor permeable than vapor impermeable. Again, however, the classifications tend to be quite arbitrary.

Materials that are generally classed as permeable to water vapor are:

- unpainted gypsum board and plaster,

- unfaced fiberglass insulation,

- cellulose insulation,

- synthetic stucco,

- some latex-based paints,

- lightweight asphalt impregnated building papers (#15 building paper),

- asphalt impregnated fiberboard sheathings, and

- "housewraps."

Part of the problem is that we struggle with names and terms. We use the terms vapor retarder and vapor barrier interchangeably. This can get us into serious trouble. Defining these terms is important.

A vapor retarder is the element that is designed and installed in an assembly to retard the movement of water by vapor diffusion. There are several classes of vapor retarders:

Class I vapor retarder   0.1 perm or less

Class II vapor retarder   1.0 perm or less and greater than 0.1 perm

Class III vapor retarder   10 perms or less and greater than 1.0 perm

(Test procedure for vapor retarders: ASTM E-96 Test Method A — the desiccant or dry cup method.)

Finally, a vapor barrier is defined as:

Vapor barrier   A Class I vapor retarder

The current International Building Code (and its derivative codes) defines a vapor retarder as 1.0 perms or less using the same test procedure. In other words, the current code definition of a vapor retarder is equivalent to the definition of a Class II vapor retarder used here.

Air Barriers

The key physical properties which distinguish air barriers from other materials are continuity and the ability to resist air pressure differences. Continuity refers to holes, openings and penetrations. Large quantities of moisture can be transported through relatively small openings by air transport if the moving air contains moisture and if an air pressure difference also exists. For this reason, air barriers must be installed in such a manner that even small holes, openings and penetrations are eliminated.

Air barriers must also resist the air pressure differences that act across them. These air pressure differences occur as a combination of wind, stack and mechanical system effects. Rigid materials such as interior gypsum board, exterior sheathing and rigid draftstopping materials are effective air barriers due to their ability to resist air pressure differences.

Magnitude of Vapor Diffusion and Air Transport of Vapor

The differences in the significance and magnitude vapor diffusion and air transported moisture are typically misunderstood. Air movement as a moisture transport mechanism is typically far more important than vapor diffusion in many (but not all) conditions. The movement of water vapor through a 1-inch square hole as a result of a 10 Pascal air pressure differential is 100 times greater than the movement of water vapor as a result of vapor diffusion through a 32-square-foot sheet of gypsum board under normal heating or cooling conditions (see Figure 4.4).

In most climates, if the movement of moisture-laden air into a wall or building assembly is eliminated, movement of moisture by vapor diffusion is not likely to be significant. The notable exceptions are hot-humid climates or rain wetted walls experiencing solar heating.

Furthermore, the amount of vapor which diffuses through a building component is a direct function of area. That is, if 90 percent of the building enclosure surface area is covered with a vapor retarder, then that vapor retarder is 90 percent effective. In other words, continuity of the vapor retarder is not as significant as the continuity of the air barrier. For instance, polyethylene film which may have tears and numerous punctures present will act as an effective vapor barrier, whereas at the same time it is a poor air barrier. Similarly, the kraft-facing on fiberglass batts installed in exterior walls acts as an effective vapor retarder, in spite of the numerous gaps and joints in the kraft-facing.

It is possible and often practical to use one material as the air barrier and a different material as the vapor retarder. However, the air barrier must be continuous and free from holes, whereas the vapor retarder need not be.

In practice, it is not possible to eliminate all holes and install a "perfect" air barrier. Most strategies to control air transported moisture depend on the combination of an air barrier, air pressure differential control and interior/exterior moisture condition control in order to be effective. Air barriers are often utilized to eliminate the major openings in building enclosures in order to allow the practical control of air pressure differentials. It is easier to pressurize or depressurize a building enclosure made tight through the installation of an air barrier than a leaky building enclosure. The interior moisture levels in a tight building enclosure are also much easier to control by ventilation and dehumidification than those in a leaky building enclosure.

Combining Approaches

In most building assemblies, various combinations of materials and approaches are often incorporated to provide for both vapor diffusion control and air transported moisture control. For example, controlling air transported moisture can be accomplished by controlling the air pressure acting across a building assembly. The air pressure control is facilitated by installing an air barrier such as glued (or gasketed) interior gypsum board in conjunction with draftstopping. For example, in cold climates during heating periods, maintaining a slight negative air pressure within the conditioned space will control the exfiltration of interior moisture-laden air. However, this control of air transported moisture will not control the migration of water vapor as a result of vapor diffusion. Accordingly, installing a vapor retarder towards the interior of the building assembly, such as the kraft paper backing on fiberglass batts is also typically necessary. Alternatives to the kraft paper backing are low permeability paint on the interior gypsum board surfaces.

In the above example, control of both vapor diffusion and air transported moisture in cold climates during heating periods can be enhanced by maintaining the interior conditioned space at relatively low moisture levels through the use of controlled ventilation and source control. Also, in the above example, control of air transported moisture during cooling periods (when moisture flow is typically from the exterior towards the interior) can be facilitated by maintaining a slight positive air pressure across the building enclosure thereby preventing the infiltration of exterior, hot, humid air.

Overall Strategy

Building assemblies need to be protected from wetting by air transport and vapor diffusion. The typical strategies used involve vapor retarders, air barriers, air pressure control, and control of interior moisture levels through ventilation and dehumidification via air conditioning. The location of air barriers and vapor retarders, pressurization versus depressurization, and ventilation versus dehumidification depend on climate location and season.

The overall strategy is to keep building assemblies from getting wet from the interior, from getting wet from the exterior, and allowing them to dry to either the interior, exterior or both should they get wet or start out wet as a result of the construction process or through the use of wet materials.

In general moisture moves from warm to cold. In cold climates, moisture from the interior conditioned spaces attempts to get to the exterior by passing through the building enclosure. In hot climates, moisture from the exterior attempts to get to the cooled interior by passing through the building enclosure.

Cold Climates

In cold climates and during heating periods, building assemblies need to be protected from getting wet from the interior. As such, vapor retarders and air barriers are installed towards the interior warm surfaces. Furthermore, conditioned spaces should be maintained at relatively low moisture levels through the use of controlled ventilation (dilution) and source control.

In cold climates the goal is to make it as difficult as possible for the building assemblies to get wet from the interior. The first line of defense is the control of moisture entry from the interior by installing interior vapor retarders, interior air barriers along with ventilation (dilution with exterior air) and source control to limit interior moisture levels. Since it is likely that building assemblies will get wet, a degree of forgiveness should also be designed into building assemblies allowing them to dry should they get wet. In cold climates and during heating periods, building assemblies dry towards the exterior. Therefore, permeable ("breathable") materials are often specified as exterior sheathings.

In general, in cold climates, air barriers and vapor retarders are installed on the interior of building assemblies, and building assemblies are allowed to dry to the exterior by installing permeable sheathings and building papers/housewraps towards the exterior. A "classic" cold climate wall assembly is presented in Figure 4.5.

Hot Climates

In hot climates and during cooling periods the opposite is true. Building assemblies need to be protected from getting wet from the exterior, and allowed to dry towards the interior. Accordingly, air barriers and vapor retarders are installed on the exterior of building assemblies, and building assemblies are allowed to dry towards the interior by using

4

permeable interior wall finishes, installing cavity insulations without vapor retarders (unfaced fiberglass batts) and avoiding interior "non-breathable" wall coverings such as vinyl wallpaper. Furthermore, conditioned spaces are maintained at a slight positive air pressure with conditioned (dehumidified) air in order to limit the infiltration of exterior, warm, potentially humid air (in hot, humid climates rather than hot, dry climates). A "classic" hot climate wall assembly is presented in Figure 4.6.

Mixed Climates

In mixed climates, the situation becomes more complicated. Building assemblies need to be protected from getting wet from both the interior and exterior, and be allowed to dry to either the exterior, interior or both. Three general strategies are typically employed:

- Selecting either a classic cold climate assembly or classic hot climate assembly, using air pressure control (typically only pressurization during cooling), using interior moisture control (ventilation/air change during heating, dehumidification/air conditioning during cooling) and relying on the forgiveness of the classic approaches to dry the accumulated moisture (from opposite season exposure) to either the interior or exterior. In other words the moisture accumulated in a cold climate wall assembly exposed to hot climate conditions is anticipated to dry towards the exterior when the cold climate assembly finally sees heating conditions, and vice versa for hot climate building assemblies;

- Adopting a "flow-through" approach by using permeable building materials on both the interior and exterior surfaces of building assemblies to allow water vapor by diffusion to "flow-through" the building assembly without accumulating. Flow would be from the interior to exterior during heating periods, and from the exterior towards the interior during cooling periods. In this approach air pressure control and using interior moisture control would also occur. The location of the air barrier can be towards the interior (sealed interior gypsum board), or towards the exterior (sealed exterior sheathing). A "classic" flow-through wall assembly is presented in Figure 4.8; or

- Installing the vapor retarder roughly in the middle of the assembly from a thermal perspective. This is typically accomplished by installing impermeable or semi-impermeable insulating sheathing on the exterior of a frame cavity wall (see Figure 4.9). For example, installing 1.5 inches of foil-faced insulating sheathing (approximately R-10) on the exterior of a 2x6 frame cavity wall insulated with unfaced fiberglass batt insulation (approximately R-19). The vapor retarder is the interior face of the exterior impermeable insulating sheathing (Figure 4.8). If the wall assembly total thermal resistance is R-29 (R-19 plus R-10), the location of the vapor retarder is 66 percent of the way (thermally) towards the exterior (19/29 = .66). In this approach air pressure control and utilizing interior moisture control would also occur. The location of the air barrier can be towards the interior or exterior.

The advantage of the wall assembly described in Figure 4.9 is that an interior vapor retarder is not necessary. In fact, locating an interior vapor retarder at this location would be detrimental, as it would not allow the wall assembly to dry towards the interior during cooling periods. The wall assembly is more forgiving without the interior vapor retarder than if one were installed. If an interior vapor retarder were installed, this would result in a vapor retarder on both sides of the assembly significantly impairing durability.

Note that this discussion relates to a wall located in a mixed climate with an exterior impermeable or semi-impermeable insulating sheathing. Could a similar argument be made for a heating climate wall assembly? Could we construct a wall in a heating climate without an interior vapor retarder? How about a wall in a heating climate with an exterior vapor retarder and no interior vapor retarder? The answer is yes to both questions, but with caveats.

Control of Condensing Surface Temperatures

The performance of a wall assembly in a cold climate without an interior vapor retarder (such as the wall described in Figure 4.8) can be more easily understood in terms of condensation potentials and the control of condensing surface temperatures.

Figure 4.10 illustrates the performance of a 2x6 wall with semi-permeable OSB sheathing (perm rating of about 1.0 perms, dry cup; 2.0 perms, wet cup) covered with building paper and vinyl siding located in Chicago, IL. The interior conditioned space is maintained at a relative humidity of 35 percent at 70 degrees Fahrenheit. For the purposes of this example, it is assumed that no interior vapor retarder is installed (unpainted drywall as an interior finish over unfaced fiberglass, yech!). This illustrates a case we would never want to construct in a cold climate, a wall with a vapor retarder on the exterior (semi-permeable OSB sheathing and no vapor retarder on the interior.

The mean daily ambient temperature over a one-year period is plotted (Figure 4.10). The temperature of the insulation/OSB sheathing interface (back side of the OSB sheathing) is approximately equivalent to the mean daily ambient temperature, since the thermal resistance values of the siding, building paper and the OSB sheathing are small compared to the thermal resistance of the insulation in the wall cavity. The dew point temperature of the interior air/water vapor mix is approximately 40 degrees Fahrenheit (this can be found from examining a psychrometric chart). In other words, whenever the back side of the OSB sheathing drops below 40 degrees Fahrenheit, the potential for condensation exists at that interface should moisture migrate from the interior conditioned space via vapor diffusion or air movement.

From the plot it is clear that the mean daily temperature of the back side of the OSB sheathing drops below the dew point temperature of the interior air at the beginning of November and does not go above the dewpoint temperature until early March. The shaded area under the dewpoint line is the potential for condensation, or wetting potential for this assembly should moisture from the interior reach the back side of the OSB sheathing. With no interior vapor retarder, moisture from the interior will reach the back side of the plywood sheathing.

Figure 4.11 illustrates the performance of the wall assembly described in Figure 4.9, a 2x6 wall insulated on the exterior with 1.5 inches of rigid foil-faced impermeable insulating sheathing (approximately R-10, perm rating of about 0.1 perms, wet cup and dry cup), located in Chicago, IL. The wall cavity is insulated with unfaced fiberglass batt insulation (approximately R-19). Unpainted drywall is again the interior finish (no interior vapor retarder). Now this wall assembly also has a vapor retarder — in fact, it has a vapor barrier — on the exterior, but with a huge difference. This exterior vapor retarder (vapor barrier) has a significant insulating value since it is a rigid insulation. The temperature of the first condensing surface within the wall assembly, namely the cavity insulation/rigid insulation interface (the back side of the rigid insulation), is raised above the interior dewpoint temperature because of the insulating value of the rigid insulation. This illustrates a case we could construct in a cold climate, a wall with a "warm" vapor retarder (vapor barrier) on the exterior and no vapor retarder on the interior.

The temperature of the condensing surface (back side of the rigid insulation) is calculated in the following manner. Divide the thermal resistance to the exterior of the condensing surface by the total thermal resistance of the wall. Then multiply this ratio by the temperature difference between the interior and exterior. Finally, add this to the outside base temperature.

$$T_{(interface)} = R_{(exterior)} / R_{(total)} \times (T_{in} - T_{out}) + T_{out}$$

where:

$T_{(interface)}$ = the temperature at the sheathing/insulation interface or the temperature of the first condensing surface

$R_{(exterior)}$ = the R-value of the exterior sheathing

$R_{(total)}$ = the total R-value of the entire wall assembly

T_{in} = the interior temperature

T_{out} = the exterior temperature

The R-10 insulating sheathing raises the dew point temperature at the first condensing surface so that no condensation will occur with interior conditions of 35 percent relative humidity at 70 degrees Fahrenheit. In other words, no interior vapor retarder of any kind is necessary with this wall assembly if the interior relative humidity is kept below 35 percent. This is a "caveat" for this wall assembly. Now remember, this wall is located in Chicago, IL. This is another "caveat" for this wall assembly.

What happens if we move this wall to Minneapolis? Big change. Minneapolis is a miserable place in the winter. The interior relative humidity would have to be kept below 25 percent to prevent condensation at the first condensing surface. What happens if we move the wall back to Chicago and install a modest interior vapor retarder, such as one coat of a standard interior latex paint (perm rating of about 5 perms) over the previously unpainted drywall (perm rating of 20)? If we control air leakage, interior relative humidities can be raised above 50 percent before condensation occurs.

What happens if we move this wall to Tupelo, MS, and reduce the thickness of the rigid insulation? Another big change. Tupelo has a moderate winter. Figure 4.12 illustrates the performance of a 2x6 wall insulated on the exterior with 1 inch of rigid extruded polystyrene insulating sheathing (approximately R-5, perm rating of about 1.0 perms, wet cup and dry cup), located in Tupelo.

In Tupelo, with no interior vapor retarder of any kind, condensation will not occur in this wall assembly until interior moisture levels are raised above 45 percent, 70 degrees Fahrenheit during the coldest part of the heating season. Since these interior conditions are not likely (or desirable), the potential for condensation in this wall assembly is small.

What happens if we move the wall assembly described in Figure 4.10 that experienced condensation in Chicago to Charleston, SC? No condensation results (see Figure 4.13) until interior moisture levels exceed 40 percent relative humidity at 70 degrees F. In Charleston, an interior vapor retarder is not necessary to control winter condensation where interior moisture levels are maintained below 40 percent relative humidity.

Sheathings and Cavity Insulations

Exterior sheathings can be permeable, semi-permeable, semi-impermeable, impermeable, insulating and non-insulating. Mixing and matching sheathings, building papers and cavity insulations should be based on climate location and therefore can be challenging. The following guidelines are offered:

- Impermeable and semi-impermeable non-insulating sheathings are not recommended in cold climates (inward drying reduced due to requirement for interior vapor retarder, condensing surface temperature not controlled due to use of non-insulating sheathing).

- Impermeable and semi-impermeable non-insulating sheathings are not recommended for use with damp spray cellulose cavity insulations in cold climates.

- Impermeable insulating sheathings should be of sufficient thermal resistance to control condensation at cavity insulation/sheathing interfaces.

- Permeable sheathings are not recommended for use with brick veneers and stuccos due to moisture flow reversal from solar radiation (sun heats wet brick driving moisture into wall assembly through permeable sheathing).

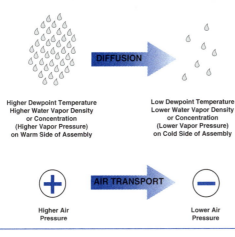

Figure 4.1
Water Vapor Movement
- Vapor diffusion is the movement of moisture in the vapor state as a result of a vapor pressure difference (concentration gradient) or a temperature difference (thermal gradient)
- Air transport is the movement of moisture in the vapor state as a result of an air pressure difference

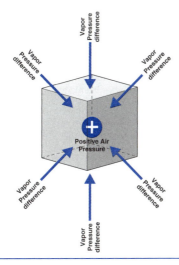

Figure 4.2
Opposing Air and Vapor Pressure Differences
- The atmosphere within the cube is under higher air pressure but lower vapor pressure relative to surroundings
- Vapor pressure acts inward in this example
- Air pressure acts outward in this example

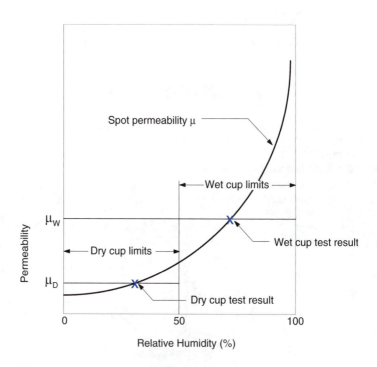

Figure 4.3
Permeability vs. Relative Humidity

- Typical relationship between dry- and wet-cup methods and spot permeability for many hygroscopic building materials such as asphalt impregnated felt building papers, plywood, OSB and kraft facings on insulation batts
- $\mu_w \cong 2$ to 5 times greater than μ_D
- Wet cup testing occurs with 50% RH on one side of test specimen and 100% RH on other side
- Dry cup testing occurs with 0% RH on one side of test specimen and 50% RH on other side

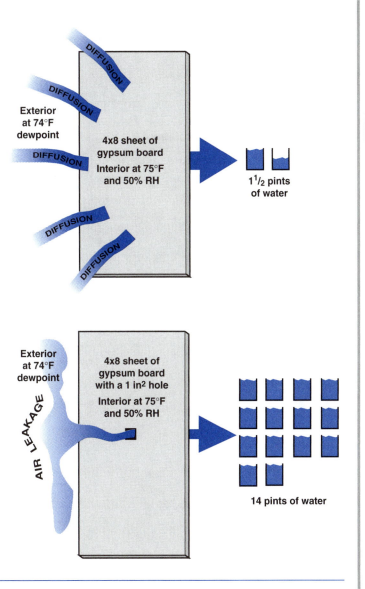

Figure 4.4
Diffusion vs. Air Leakage

- In a hot-humid climate over a one-week period, $1\frac{1}{2}$ pints of water can be collected by diffusion through painted gypsum board ($\cong$ 5 perms); 14 pints of water can be collected through air leakage through a 1-in^2 hole under a 5 Pascal air pressure differential

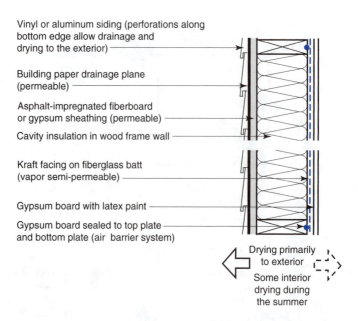

Vinyl or aluminum siding (perforations along bottom edge allow drainage and drying to the exterior)

Building paper drainage plane (permeable)

Asphalt-impregnated fiberboard or gypsum sheathing (permeable)

Cavity insulation in wood frame wall

Kraft facing on fiberglass batt (vapor semi-permeable)

Gypsum board with latex paint

Gypsum board sealed to top plate and bottom plate (air barrier system)

Drying primarily to exterior

Some interior drying during the summer

Figure 4.5
Classic Cold Climate Wall Assembly
- Vapor retarder to the interior (the kraft facing on the fiberglass batt)
- Air barrier to the interior
- Permeable exterior sheathing and permeable building paper drainage plane
- Ventilation provides air change (dilution) and also limits the interior moisture levels

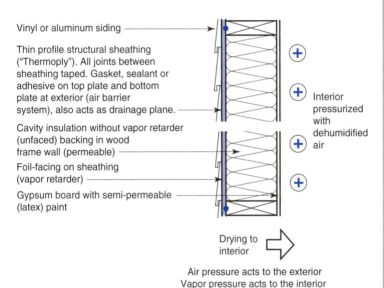

Vinyl or aluminum siding

Thin profile structural sheathing ("Thermoply"). All joints between sheathing taped. Gasket, sealant or adhesive on top plate and bottom plate at exterior (air barrier system), also acts as drainage plane.

Cavity insulation without vapor retarder (unfaced) backing in wood frame wall (permeable)

Foil-facing on sheathing (vapor retarder)

Gypsum board with semi-permeable (latex) paint

Interior pressurized with dehumidified air

Drying to interior

Air pressure acts to the exterior
Vapor pressure acts to the interior

Figure 4.6
Classic Hot-Humid Climate Wall Assembly — Frame Wall

- Vapor retarder to the exterior
- Air barrier to the exterior
- Pressurization of conditioned space
- Impermeable exterior sheathing also acts as drainage plane
- Permeable interior wall finish
- Interior conditioned space is maintained at a slight positive air pressure with respect to the exterior to limit the infiltration of exterior, hot, humid air
- Properly-sized air conditioning may not provide sufficient dehumidification (moisture removal) from interior; supplemental dehumidification may be required

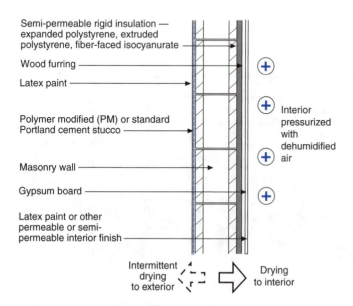

Figure 4.7
Classic Hot-Humid Climate Wall Assembly — Masonry Wall

- Vapor retarder is on the interior of the masonry and is the warm (exterior) surface of the interior semi-permeable rigid insulation
- Air barrier to the exterior (stucco)
- Pressurization of conditioned space
- Permeable interior wall finish
- Moisture in masonry released to interior in a controlled manner through rigid insulation
- The key to the performance of this wall assembly is reducing rain water absorption and penetration of the stucco rendering
- Interior conditioned space is maintained at a slight positive air pressure with respect to the exterior to limit the infiltration of exterior, hot, humid air
- Properly sized air conditioning may not provide sufficient dehumidification (moisture removal) from interior — supplementary dehumidification may be required

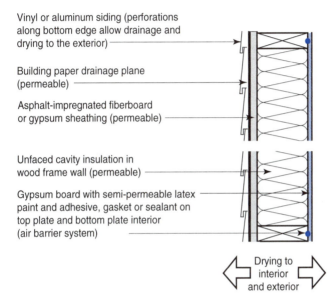

Vinyl or aluminum siding (perforations along bottom edge allow drainage and drying to the exterior)

Building paper drainage plane (permeable)

Asphalt-impregnated fiberboard or gypsum sheathing (permeable)

Unfaced cavity insulation in wood frame wall (permeable)

Gypsum board with semi-permeable latex paint and adhesive, gasket or sealant on top plate and bottom plate interior (air barrier system)

Drying to interior and exterior

Figure 4.8
Classic Flow-Through Wall Assembly
- Permeable interior surface and finish and permeable exterior sheathing and permeable building paper drainage plane
- Interior conditioned space is maintained at a slight positive air pressure with respect to the exterior to limit the infiltration of exterior moisture-laden air during cooling
- Ventilation provides air change (dilution) and also limits the interior moisture levels during heating
- Air conditioning/dehumidification limits the interior moisture levels during cooling

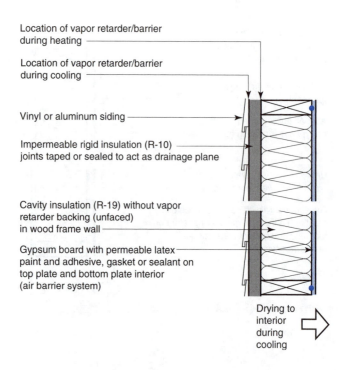

Location of vapor retarder/barrier during heating

Location of vapor retarder/barrier during cooling

Vinyl or aluminum siding

Impermeable rigid insulation (R-10) joints taped or sealed to act as drainage plane

Cavity insulation (R-19) without vapor retarder backing (unfaced) in wood frame wall

Gypsum board with permeable latex paint and adhesive, gasket or sealant on top plate and bottom plate interior (air barrier system)

Drying to interior during cooling

Figure 4.9
Vapor Retarder in the Middle of the Wall
- Air barrier to the interior
- Permeable interior wall finish
- Interior conditioned space is maintained at a slight positive air pressure with respect to the exterior to limit the infiltration of exterior moisture-laden air during cooling
- Ventilation provides air change (dilution) and also limits the interior moisture levels during heating
- Air conditioning/dehumidification limits the interior moisture levels during cooling
- Impermeable exterior sheathing also acts as drainage plane

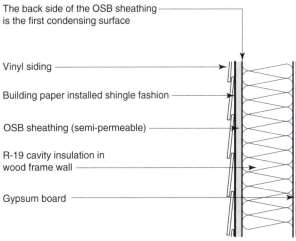

The back side of the OSB sheathing is the first condensing surface

Vinyl siding

Building paper installed shingle fashion

OSB sheathing (semi-permeable)

R-19 cavity insulation in wood frame wall

Gypsum board

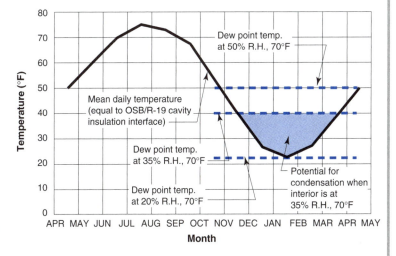

Figure 4.10
Potential for Condensation in a Wood Frame Wall Cavity in Chicago, Illinois

- By reducing interior moisture levels, the potential condensation is reduced or eliminated
- No condensation occurs if interior moisture levels are maintained below 20% RH at 70°F

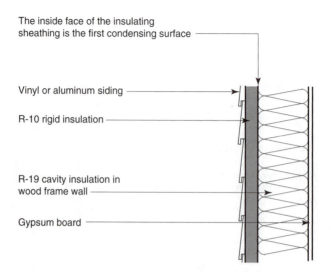

The inside face of the insulating sheathing is the first condensing surface

Vinyl or aluminum siding

R-10 rigid insulation

R-19 cavity insulation in wood frame wall

Gypsum board

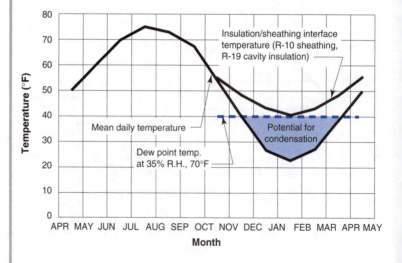

Figure 4.11
Potential for Condensation in a Wood Frame Wall Cavity Without an Interior Vapor Retarder in Chicago, Illinois

- The R-10 insulating sheathing raises the dew point temperature at the first condensing surface (cavity side of the foam sheathing) so that no condensation will occur when interior moisture levels are less than 35% relative humidity at 70°F

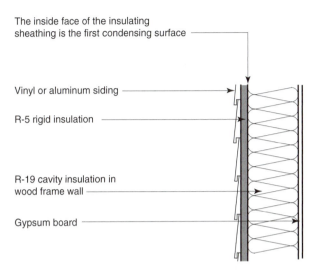

The inside face of the insulating sheathing is the first condensing surface

Vinyl or aluminum siding

R-5 rigid insulation

R-19 cavity insulation in wood frame wall

Gypsum board

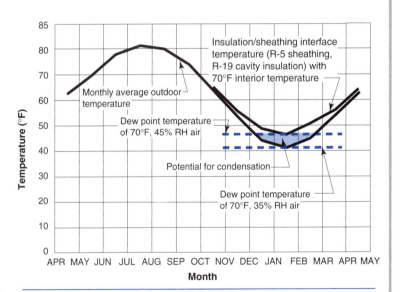

Insulation/sheathing interface temperature (R-5 sheathing, R-19 cavity insulation) with 70°F interior temperature

Monthly average outdoor temperature

Dew point temperature of 70°F, 45% RH air

Potential for condensation

Dew point temperature of 70°F, 35% RH air

Figure 4.12
Potential for Condensation in Tupelo, Mississippi
- 3,150 heating degree days
- Winter design temperature 19°F
- Summer design temperature 94°F dry bulb; 78°F wet bulb

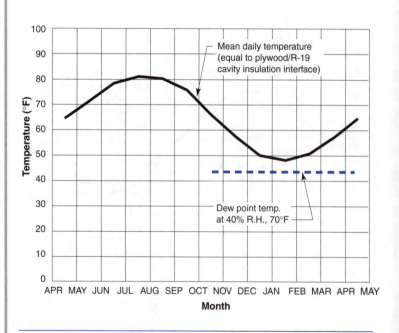

The back side of the plywood sheathing is the first condensing surface

Painted wood siding

Building paper installed shingle fashion

Plywood sheathing (semi-permeable)

R-19 cavity insulation in wood frame wall

Gypsum board

Mean daily temperature (equal to plywood/R-19 cavity insulation interface)

Dew point temp. at 40% R.H., 70°F

Figure 4.13
Potential for Condensation in Charleston, South Carolina
- There is no potential for condensation until interior moisture levels exceed 40% RH at 70°F during the coldest months of the year
- An interior vapor retarder is not necessary in building assemblies in Charleston where interior moisture levels are maintained below 40% RH at 70°F during the heating period

Wall Design

Ideally, building assemblies would always be built with dry materials under dry conditions, and would never get wet from imperfect design, poor workmanship or occupants. Unfortunately, these conditions do not exist.

It has been accepted by the building industry that many building assemblies become wet during service, and in many cases start out wet. Furthermore, the industry has recognized that in many circumstances it may be impractical to design and build building assemblies which never get wet. This has given rise to the concept of acceptable performance. Acceptable performance implies the design and construction of building assemblies which may periodically get wet, or start out wet yet are still durable and provide a long, useful service life. Repeated wetting followed by repeated drying can provide acceptable performance if during the wet period, materials do not stay wet long enough under adverse conditions to deteriorate.

Good design and practice involve controlling the wetting of building assemblies from both the exterior and interior. They may also involve the drying of building assemblies should they become wet during service or as a result of building with wet materials or under wet conditions.

Moisture Balance

Moisture accumulates when the rate of moisture entry into an assembly exceeds the rate of moisture removal. When moisture accumulation exceeds the ability of the assembly materials to store the moisture without significantly degrading performance or long term service life, moisture problems result.

Building assemblies can get wet from the building interior or exterior, or they can start out wet as a result of the construction process due to wet building materials or construction under wet conditions. Good design and practice address these wetting mechanisms.

Various strategies can be implemented to minimize the risk of moisture damage. The strategies fall into following three groups:

- control of moisture entry

- control of moisture accumulation

- removal of moisture

Strategies in the three groupings can be utilized in combination and have been proven to be most effective in that manner. Strategies effective in the control of moisture entry, however, are often not effective if building assemblies start out wet, and in fact can be detrimental. If a technique is effective at preventing moisture from entering an assembly, it is also likely to be effective at preventing moisture from leaving an assembly. Conversely, a technique effective at removing moisture may also allow moisture to enter. Balance between entry and removal is the key in many assemblies.

Historically successful approaches to moisture control have typically been based on the following strategy: prevent building assemblies and surfaces from getting wet from the exterior, prevent building assemblies and surfaces from getting wet from the interior, and should building assemblies or surfaces get wet, or start out wet, allow them to dry to either the exterior or the interior.

The wall assemblies presented in this chapter are designed to have an element of drying to either the interior or exterior or both.

One of the most important concepts presented in this section is the use of insulating sheathing (rigid insulation) to warm up the temperature of potential condensing surfaces.

Figure 5.1 is a plot of monthly average outdoor temperatures for Jacksonville, FL. The plot shows that there is no potential for condensation of interior moisture on the interior of the exterior sheathing, in this case plywood, until the interior relative humidity exceeds 55 percent. This is not a likely winter condition inside of a residence in Jacksonville. Accordingly, an interior vapor retarder is not necessary, nor recommended.

As climatic locations change, monthly average outdoor temperatures can become sufficiently low during winter months for condensation of interior moisture within building assemblies to occur. Such is the case for Atlanta, GA. However, these conditions can be controlled by elevating the temperature of condensing surfaces.

Figure 5.2 shows the use of insulating sheathing on the exterior of a wall assembly in Atlanta. The temperature of the condensing surface is plotted in Figure 5.3. The graph shows that condensation within a wall cavity in Atlanta will not occur if at least 1-inch of insulating sheathing is used.

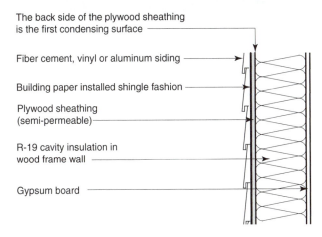

The back side of the plywood sheathing is the first condensing surface

Fiber cement, vinyl or aluminum siding

Building paper installed shingle fashion

Plywood sheathing (semi-permeable)

R-19 cavity insulation in wood frame wall

Gypsum board

5

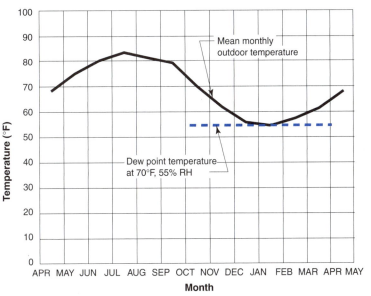

Mean monthly outdoor temperature

Dew point temperature at 70°F, 55% RH

Figure 5.1
Potential for Condensation in Jacksonville, Florida with Plywood Sheathing
- There is no potential for condensation until interior moisture levels exceed 55% RH at 70°F during the coldest month of the year
- Fiber cement should be backprimed
- If wood sing is used, it should be installed over furring strips and be backprimed; all cut edges sealed
- An interior vapor retarder is not necessary nor recommended in building assemblies in Jacksonville, Florida where interior moisture levels are maintained below 55% RH at 70°F during the heating period

5

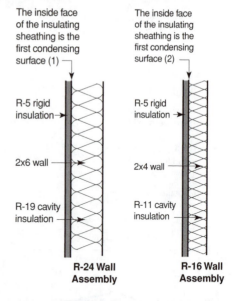

The inside face of the insulating sheathing is the first condensing surface (1)

R-5 rigid insulation

2x6 wall

R-19 cavity insulation

R-24 Wall Assembly

The inside face of the insulating sheathing is the first condensing surface (2)

R-5 rigid insulation

2x4 wall

R-11 cavity insulation

R-16 Wall Assembly

	Inside Temp.	Monthly Average Outdoor Temp.	ΔT	Temperature of Condensing Surface (1)	Temperature of Condensing Surface (2)
Oct	70	63	7	64.5	65
Nov	70	52	18	56	57.5
Dec	70	45	25	50	53
Jan	70	45	25	50	53
Feb	70	46	24	51	53.5
Mar	70	53	17	56.5	58
Apr	70	62	8	63.5	64.5

Ratio (1) $5 \div 24 = .208$
Temperature of Condensing Surface (1) = (ΔT x 0.208) + Outdoor Temperature

Ratio (2) $5 \div 16 = .3125$
Temperature of Condensing Surface (2) = (ΔT x 0.3125) + Outdoor Temperature

Figure 5.2
Condensing Surface Temperatures in Atlanta, Georgia
- The more insulation in the cavity, the lower the temperature of the condensing surface
- The thicker the insulating sheathing, the higher the temperature of the condensing surface

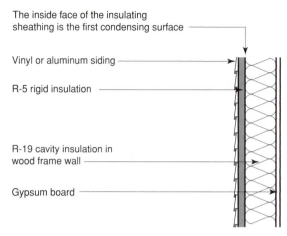

The inside face of the insulating sheathing is the first condensing surface

Vinyl or aluminum siding

R-5 rigid insulation

R-19 cavity insulation in wood frame wall

Gypsum board

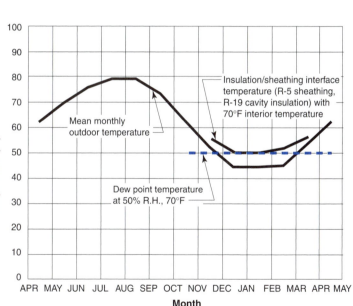

Insulation/sheathing interface temperature (R-5 sheathing, R-19 cavity insulation) with 70°F interior temperature

Mean monthly outdoor temperature

Dew point temperature at 50% R.H., 70°F

Month

Temperature (°F)

Figure 5.3
Potential for Condensation in Atlanta, Georgia with Insulating Sheathing

- There is no potential for condensation until interior moisture levels exceed 50% RH at 70°F during the coldest month of the year if R-5 insulating sheathing is used
- An interior vapor retarder is not necessary nor recommended in building assemblies in Atlanta, Georgia where interior moisture levels are maintained below 50% RH at 70°F during the heating period

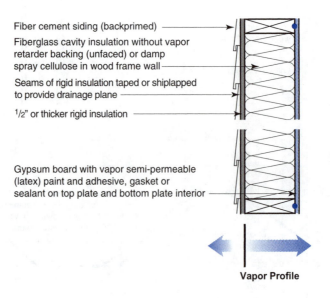

Fiber cement siding (backprimed)

Fiberglass cavity insulation without vapor retarder backing (unfaced) or damp spray cellulose in wood frame wall

Seams of rigid insulation taped or shiplapped to provide drainage plane

1/2" or thicker rigid insulation

Gypsum board with vapor semi-permeable (latex) paint and adhesive, gasket or sealant on top plate and bottom plate interior

Vapor Profile

Figure 5.4
Drying to Interior in Hot-Humid Climates
Frame Wall — Exterior Rigid Insulation — Siding
- The vapor semi-permeable latex paint permits drying to the interior
- If wood siding is used, it should be installed over furring strips and be backprimed, all cut edges sealed

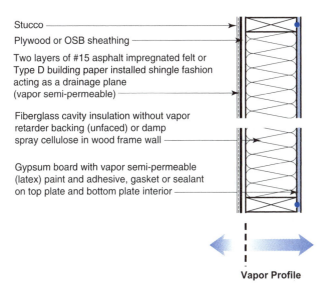

Stucco

Plywood or OSB sheathing

Two layers of #15 asphalt impregnated felt or
Type D building paper installed shingle fashion
acting as a drainage plane
(vapor semi-permeable)

Fiberglass cavity insulation without vapor
retarder backing (unfaced) or damp
spray cellulose in wood frame wall

Gypsum board with vapor semi-permeable
(latex) paint and adhesive, gasket or sealant
on top plate and bottom plate interior

Vapor Profile

5

Figure 5.5
Drying to Interior in Hot-Humid Climates
Frame Wall — Plywood/OSB Sheathing — Stucco
- The vapor semi-permeable latex paint permits drying to the interior
- If fiber cement siding is used, it back primed
- Drainage is provided between two layers of building paper

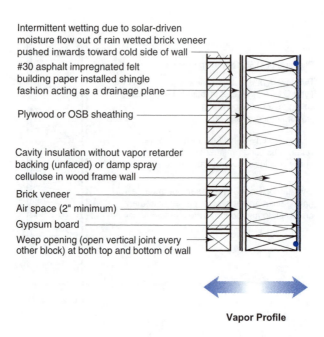

Intermittent wetting due to solar-driven moisture flow out of rain wetted brick veneer pushed inwards toward cold side of wall

#30 asphalt impregnated felt building paper installed shingle fashion acting as a drainage plane

Plywood or OSB sheathing

Cavity insulation without vapor retarder backing (unfaced) or damp spray cellulose in wood frame wall

Brick veneer

Air space (2" minimum)

Gypsum board

Weep opening (open vertical joint every other block) at both top and bottom of wall

Vapor Profile

Figure 5.6
Drying to Exterior and Interior in Hot-Humid Climates
Frame Wall — Plywood/OSB Sheathing — Brick Veneer
- A rigid, impermeable or vapor semi-permeable insulating sheathing can be used in place of the #30 asphalt impregnated felt to prevent the wall cavity from getting wet due to solar-driven moisture (seams taped or shiplapped to provide drainage plane)

5

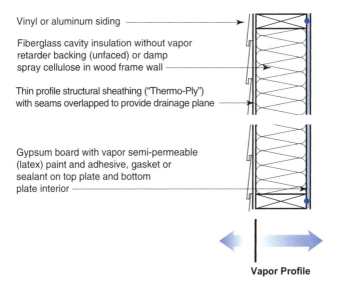

Vinyl or aluminum siding

Fiberglass cavity insulation without vapor
retarder backing (unfaced) or damp
spray cellulose in wood frame wall

Thin profile structural sheathing ("Thermo-Ply")
with seams overlapped to provide drainage plane

Gypsum board with vapor semi-permeable
(latex) paint and adhesive, gasket or
sealant on top plate and bottom
plate interior

Vapor Profile

Figure 5.7
Drying to Interior in Hot-Humid Climates
Frame Wall — Thin Profile Sheathing — Siding
- The vapor semi-permeable latex paint permits drying to the interior
- If wood siding is used, it should be installed over furring strips and be
  backprimed, all cut edges sealed
- If fiber cement siding is used, it should be backprimed

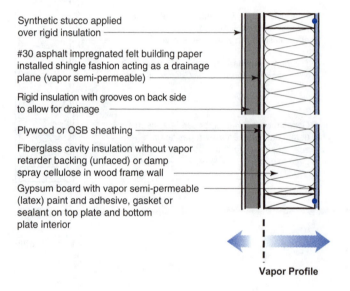

Synthetic stucco applied over rigid insulation

#30 asphalt impregnated felt building paper installed shingle fashion acting as a drainage plane (vapor semi-permeable)

Rigid insulation with grooves on back side to allow for drainage

Plywood or OSB sheathing

Fiberglass cavity insulation without vapor retarder backing (unfaced) or damp spray cellulose in wood frame wall

Gypsum board with vapor semi-permeable (latex) paint and adhesive, gasket or sealant on top plate and bottom plate interior

Vapor Profile

Figure 5.8
Drying to Interior in Hot-Humid Climates
Frame Wall — EIFS

• The vapor semi-permeable latex paint permits drying to the interior

5

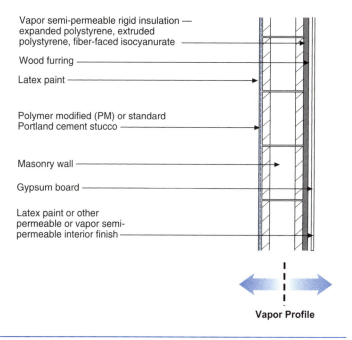

Vapor semi-permeable rigid insulation — expanded polystyrene, extruded polystyrene, fiber-faced isocyanurate

Wood furring

Latex paint

Polymer modified (PM) or standard Portland cement stucco

Masonry wall

Gypsum board

Latex paint or other permeable or vapor semi-permeable interior finish

Vapor Profile

Figure 5.9
Drying to Interior and Exterior in Hot-Humid Climates
Masonry Wall — Interior Rigid Insulation — Stucco

- The vapor semi-permeable rigid insulations and interior latex paint permit drying to the interior
- Vapor semi-permeable rigid insulations used on the interior should be unfaced or faced with permeable skins; foil facings, polypropylene skins should be avoided
- Avoid use of metal furring or "hat" channels due to thermal bridging and impermeability; use only wood furring
- Wood furring should be installed over rigid insulation; rigid insulation should not be installed between wood furring installed directly on interior of masonry

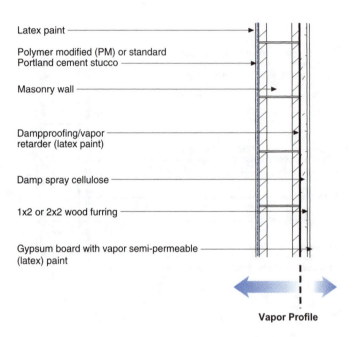

Latex paint

Polymer modified (PM) or standard
Portland cement stucco

Masonry wall

Dampproofing/vapor
retarder (latex paint)

Damp spray cellulose

1x2 or 2x2 wood furring

Gypsum board with vapor semi-permeable
(latex) paint

Vapor Profile

Figure 5.10
Drying to Interior and Exterior in Hot-Humid Climates
Masonry Wall — Interior Cellulose — Stucco

- A vapor barrier such as sheet polyethylene should not be used in place of the vapor semi-permeable dampproofing or latex paint vapor retarder
- Avoid use of metal furring or "hat" channels due to thermal bridging and impermeability; use only wood furring

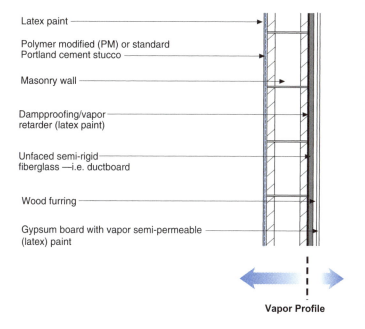

Latex paint

Polymer modified (PM) or standard Portland cement stucco

Masonry wall

Dampproofing/vapor retarder (latex paint)

Unfaced semi-rigid fiberglass —i.e. ductboard

Wood furring

Gypsum board with vapor semi-permeable (latex) paint

Vapor Profile

Figure 5.11
Drying to Interior and Exterior in Hot-Humid Climates
Masonry Wall — Interior Rigid Fiberglass — Stucco
- A vapor barrier such as sheet polyethylene should not be used in place of the vapor semi-permeable dampproofing or latex paint vapor retarder
- Avoid use of metal furring or "hat" channels due to thermal bridging and impermeability; use only wood furring
- Wood furring should be installed over rigid insulation; rigid insulation should not be installed between wood furring installed directly on interior of masonry

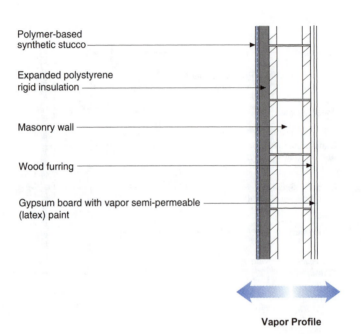

Polymer-based synthetic stucco

Expanded polystyrene rigid insulation

Masonry wall

Wood furring

Gypsum board with vapor semi-permeable (latex) paint

Vapor Profile

Figure 5.12
Drying to Interior and Exterior in Hot-Humid Climates
Masonry Wall — EIFS
• A vapor semi-permeable latex paint permits drying to the interior

Brick veneer

Fluid applied or sheet waterproofing

1-inch air space vented top and bottom

Vapor semi-permeable rigid insulation

Wood furring

Gypsum board with vapor semi-permeable (latex) paint

Vapor Profile

Figure 5.13
Drying to Interior in Hot-Humid Climates
Masonry Wall — Interior Rigid Insulation — Brick Veneer

- The vapor semi-permeable rigid insulations and interior latex paint permit drying to the interior
- Vapor semi-permeable rigid insulations used on the interior should be unfaced or faced with permeable skins; foil facings, polypropylene skins should be avoided
- Avoid use of metal furring or "hat" channels due to thermal bridging and impermeability; use only wood furring
- Wood furring should be installed over rigid insulation; rigid insulation should not be installed between wood furring installed directly on interior of masonry

Roof Design

Roofs can be designed and constructed to be either vented or unvented in any hygro-thermal zone.

In cold climates, the primary purpose of attic ventilation is to maintain a cold roof temperature to avoid ice dams created by melting snow, and to vent moisture that moves from the conditioned space to the attic. Melted snow, in this case, is caused by heat loss from the conditioned space.

In hot climates, the primary purpose of attic ventilation is to expel solar heated hot air from the attic to lessen the building cooling load.

The amount of roof cavity ventilation is specified by numerous ratios of free vent area to insulated ceiling area ranging from 1:150 to 1:600 depending on which building code is consulted, the 1:300 ratio being the most common.

Control of ice dams, moisture accumulation and heat gain can also be facilitated by unvented roof design.

Approach

In cold climates the main strategy that should be utilized when designing roofs to be free from moisture problems and ice dams along with control of heat gain or heat loss regardless of ventilation approach is the elimination of air movement, particularly exfiltrating air. This can be accomplished by the installation of an air barrier system or by the control of the air pressure difference across the roof system.

Air barrier systems are typically the most common approach, however, air pressure control approaches are becoming more common especially in cases involving remedial work on existing structures.

Vapor diffusion should be considered as a secondary moisture transport mechanism when designing and building roofs. Specific vapor retarders are often unnecessary if appropriate air movement control is provided or if control of condensing surface temperatures is provided.

Vented Roof Design

Vented roofs should not communicate with the conditioned space — they are coupled to the exterior. Therefore, an air barrier at the ceiling line should be present to isolate the attic space from the conditioned space. No services such as HVAC distribution ducts, air handlers, plumbing or fire sprinkler systems should be located external to the air barrier (Figure 6.1).

The recommended ventilation ratio to provide in vented roof assemblies when an air barrier is present, is the 1:300 ratio (as specified by most codes).

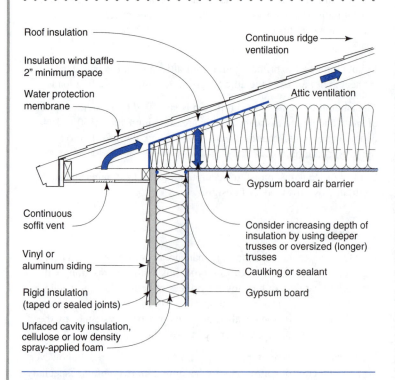

Figure 6.1
Vented Roof Assembly
- Roof insulation thermal resistance (depth) at truss heel (roof perimeter) should be equal or greater to thermal resistance of exterior wall
- 1:300 roof ventilation ratio recommended

In vented cathedral ceiling assemblies a minimum 2-inch clear airspace is recommended between the underside of the roof deck and the top of the cavity insulation. This is not a code requirement but ought to be (only 1-inch is typically specified).

In addition to an air barrier at the ceiling line, a vapor retarder should be installed in cold and very cold climate zones. No interior roof assembly-side vapor control is required or recommended in climate zones other than cold or very cold.

Unvented Roof Design

Unvented roof design falls into two categories: systems where condensing surface temperatures are controlled and systems where condensing surface temperatures are not controlled. The two categories essentially are the demarcation between regions where cold weather conditions occur with sufficient frequency and intensity that sufficient moisture accumulation from interior sources can occur on an uninsulated roof deck to risk mold, corrosion and decay problems.

The key is to keep the roof deck — the principle condensing surface in roof assemblies — sufficiently warm throughout the year. This can be accomplished either because of the local climate or as a result of design principally through the use of rigid insulation installed above the roof deck or air-impermeable spray foam insulation installed under the roof deck in direct contact with it.

Where rigid insulation is installed above the roof deck, or air-impermeable spray foam insulation is installed under the roof deck condensing surface temperatures are said to be controlled.

The demarcation is specified as a distinction between regions where the monthly average temperature remains above 45 degrees F throughout the year and where it drops below 45 degrees F during the year. An additional criteria is also necessary — that of keeping interior relative humidities below 45 percent.

These criterion were selected for two reasons. First, by keeping the roof deck above 45 degrees F, condensation can be minimized or eliminated. Condensation will not occur unless the dewpoint temperature of the interior air exceeds 45 degrees F and this air contacts the roof deck. This interior dewpoint temperature is approximately equal to an interior conditioned space temperature of 70 degrees F at an interior relative humidity of 45 percent. These are interior moisture conditions that can easily be avoided with air change/ventilation or the avoidance of over humidification during the coldest month of the year in the climate zones specified.

Second, a monthly average temperature was selected, rather than a design heating temperature, as it is more representative of building enclosure performance. Short term, intermittent "spikes" in parameters/environmental loads are of interest to structural engineers and in the sizing of equipment, but not typically relevant to moisture induced deterioration. Wood-based roof sheathing typical to residential construction has sufficient hygric buffer capacity to absorb, redistribute and re-release significant quantities of condensed moisture should intermittent condensation occur during cold nights when the sheathing temperature occasionally dips below 45 degrees F. The average monthly conditions more accurately reflect moisture content in wood-based assemblies.

Asphalt roofing shingles require special attention when installed on unvented roof assemblies in hot-humid, mixed-humid and marine climates due to inward vapor drive from incident solar radiation. A 1 perm or lower vapor retarder (Class II) as tested by the wet-cup procedure should be installed under the asphalt roofing shingles to control this inward drive.

Wood shingles or shakes, require a minimum $1/_4$-inch vented airspace that separates the shingles/shakes and the roofing felt placed over the roof sheathing for similar reasons.

The demarcation between regions that require the control of condensing surface temperatures and regions that do not can be obtained by using the hygro-thermal zones definitions in this builder's guide. Both hot-humid and hot-dry climate zones meet the 45°F roof deck criteria. However, the high interior humidities found in buildings located in hot-humid zones during the winter months do not always meet the 45% interior relative humidity criteria. Therefore, the only zone that meets both of these requirements is the hot-dry hygro-thermal region. Only hot-cry climates do not require the control of condensing surface temperatures. All other regions require some form of control.

Control of condensing surface temperatures typically involves the installation of insulating sheathing above the roof deck. In residential wood frame construction this involves installing rigid insulation between the roof shingles and the roof plywood or OSB (Figure 6.2). The installation of the rigid insulation elevates the temperature of the roof deck to minimize condensation.

Figure 6.3 and Figure 6.4 illustrate the differences between the two fundamental systems. Figure 6.3 shows the potential for condensation of an unvented roof assembly in Phoenix, AZ. Phoenix, AZ is located in a hot-dry climate zone. This roof assembly has no insulating sheathing installed above the roof deck.

Figure 6.4 shows the potential for condensation of an unvented roof assembly in Dallas, TX. Dallas, TX is located in a mixed-humid climate zone. Note that this roof assembly has rigid insulation installed above the roof deck in order to control the condensation potential. The thermal resistance of the rigid insulation (thickness) necessary to control condensation depends on the severity of the climate. The colder the climate, the greater the resistance of the rigid insulation required. Note that the thermal resistance of the rigid insulation is based on the ratio of the thermal resistance of the insulation above the roof deck as compared to the thermal resistance of the insulation below the roof deck. The key is to elevate the temperature of the condensing surface to 45 degrees F or higher during the coldest months of the year.

Figure 6.5 shows the use of rigid insulation in a cathedral ceiling assembly in Washington, DC. A calculation procedure is presented that determines the temperature of the condensing surface. This calculation procedure is similar to the one used in Chapter 4 to determine the sheathing temperature in wall assemblies.

Figure 6.6 plots the temperature of the condensing surface. The graph shows that condensation within the roof assembly will not occur if in-

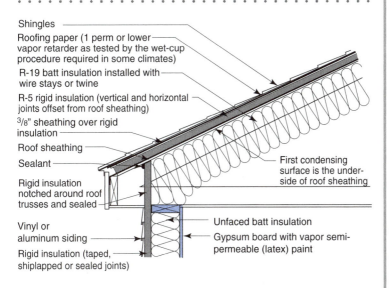

Shingles

Roofing paper (1 perm or lower vapor retarder as tested by the wet-cup procedure required in some climates)

R-19 batt insulation installed with wire stays or twine

R-5 rigid insulation (vertical and horizontal joints offset from roof sheathing)

$^3/_8$" sheathing over rigid insulation

Roof sheathing

Sealant

Rigid insulation notched around roof trusses and sealed

Vinyl or aluminum siding

Rigid insulation (taped, shiplapped or sealed joints)

First condensing surface is the underside of roof sheathing

Unfaced batt insulation

Gypsum board with vapor semi-permeable (latex) paint

Figure 6.2
Rigid Insulation Used to Control Condensing Surface Temperatures
- Rigid insulation installed above roof deck
- Ratio of R-value between rigid insulation and batt insulation is climate-dependent

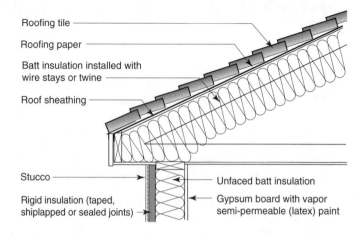

Roofing tile

Roofing paper

Batt insulation installed with wire stays or twine

Roof sheathing

Stucco

Rigid insulation (taped, shiplapped or sealed joints)

Unfaced batt insulation

Gypsum board with vapor semi-permeable (latex) paint

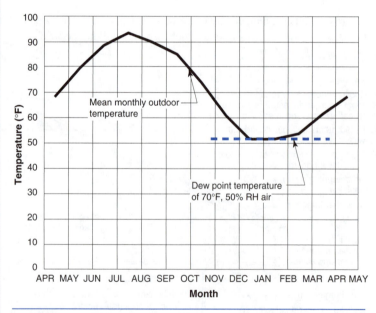

Mean monthly outdoor temperature

Dew point temperature of 70°F, 50% RH air

Figure 6.3
Potential for Condensation in Phoenix, Arizona with Unvented Roof

- 1,750 heating degree days
- Winter design temperature 34°F
- Summer design temperature 107°F dry bulb; 71°F wet bulb
- There is no potential for condensation on the underside of the roof sheathing until interior moisture levels exceed 50% RH at 70°F

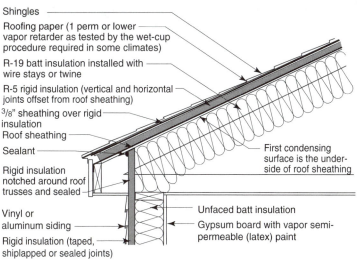

Shingles

Roofing paper (1 perm or lower vapor retarder as tested by the wet-cup procedure required in some climates)

R-19 batt insulation installed with wire stays or twine

R-5 rigid insulation (vertical and horizontal joints offset from roof sheathing)

$^3/_8$" sheathing over rigid insulation

Roof sheathing

Sealant

Rigid insulation notched around roof trusses and sealed

Vinyl or aluminum siding

Rigid insulation (taped, shiplapped or sealed joints)

First condensing surface is the underside of roof sheathing

 6

Unfaced batt insulation

Gypsum board with vapor semi-permeable (latex) paint

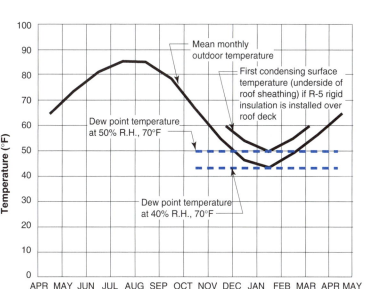

Mean monthly outdoor temperature

First condensing surface temperature (underside of roof sheathing) if R-5 rigid insulation is installed over roof deck

Dew point temperature at 50% R.H., 70°F

Dew point temperature at 40% R.H., 70°F

Temperature (°F)

Month

Figure 6.4
Potential for Condensation in Dallas, Texas with Unvented Roof and Insulating Sheathing
- There is no potential for condensation on the underside of the roof sheathing until moisture levels exceed 40% RH at 70°F
- Rigid insulation is recommended in this roof assembly to raise the condensation potential above 50% RH at 70°F
- In hot-humid, mixed-humid and marine climates a 1 perm or lower vapor retarder as tested by the wet-cup procedure should be installed under the asphalt roofing shingles

terior conditions are maintained at 45 percent relative humidity or less at 70 degrees F during the coldest month of the year.

Figure 6.7 shows a roof design that is not as dependent on controlling interior moisture levels as the other roof designs presented. The absence of cavity insulation yields the highest condensing surface temperature of any of the designs presented. In this particular design, the condensing surface is the air barrier membrane installed over the wood decking. With this design interior relative humidities should be kept below 60 percent in order to control surface mold. In cold and very cold climate zones where there is likely snow accumulation on roof surfaces, there is also the likelihood of ice-damming. In order to control ice-damming, heat flow form the interior to the roof cladding must be minimized. In cold climate zones the minimum total R-value for the entire unvented roof assembly should be R-40. In very cold climate zones this minimum R-value should be increased to R-50.

In extreme snow regions it is typical to add a vented air space between the roof cladding (shingles) and the rigid insulation in Figure 6.7 to flush heat away trapped due to the insulating value of the snow (the snow becomes an insulating "blanket"). In essence creating a vented-unvented hybrid roof assembly.

Note that in these types of unvented roof assemblies (except Figure 6.7), interior vapor barriers are not recommended as these assemblies are expected to be able to "dry" towards the interior.

Instead of installing rigid insulation above the roof deck to control condensing surface temperature, air-impermeable spray foam insulation can be installed in direct contact to the underside of the roof deck to accomplish the same thing.

Figure 6.8 shows a roof design where air-impermeable spray foam insulation is installed in direct contact to the underside of the structural roof deck.

In Figures 6.9 and 6.10, gypsum board is used to provide a thermal barrier to meet code requirements when an attic space is occupiable.

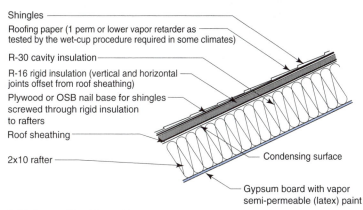

Shingles

Roofing paper (1 perm or lower vapor retarder as tested by the wet-cup procedure required in some climates)

R-30 cavity insulation

R-16 rigid insulation (vertical and horizontal joints offset from roof sheathing)

Plywood or OSB nail base for shingles screwed through rigid insulation to rafters

Roof sheathing

2x10 rafter

Condensing surface

Gypsum board with vapor semi-permeable (latex) paint

R-46 Unvented Roof Assembly

- Two layers of R-6.5/inch rigid insulation yielding a total combined thickness of 2.5 inches
- Layers of rigid insulation have staggered joints to facilitate air tightness; two layers are preferable to one layer due to increase in air flow resistance

	Inside Temp.	Monthly Average Outdoor Temp.	ΔT	Temperature of Condensing Surface
Oct	70	55	15	60
Nov	70	45	25	54
Dec	70	36	34	48
Jan	70	31	39	45
Feb	70	34	36	47
Mar	70	43	27	52
Apr	70	49	21	56

Ratio of thermal resistance above condensing surface to total thermal resistance:

$$16 \div 46 = 0.348$$

Temperature of Condensing Surface = (ΔT x 0.348) + Outdoor Temperature

Figure 6.5
Condensing Surface Temperatures in Washington, D.C.

- The greater the thermal resistance in the cavity, the lower the temperature of the condensing surface
- The greater the thermal resistance of the rigid insulation, the higher the temperature of the condensing surface
- In designs with no cavity insulation, only rigid insulation, yield the highest (warmest) condensing surface temperatures
- In hot-humid, mixed-humid and marine climates a 1 perm or lower vapor retarder as tested by the wet-cup procedure should be installed under the asphalt roofing shingles

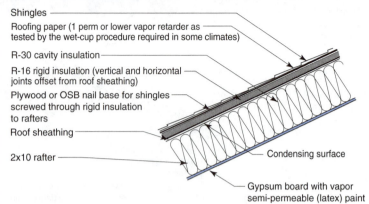

Shingles

Roofing paper (1 perm or lower vapor retarder as tested by the wet-cup procedure required in some climates)

R-30 cavity insulation

R-16 rigid insulation (vertical and horizontal joints offset from roof sheathing)

Plywood or OSB nail base for shingles screwed through rigid insulation to rafters

Roof sheathing

2x10 rafter

Condensing surface

Gypsum board with vapor semi-permeable (latex) paint

R-46 Unvented Roof Assembly

- Two layers of R-6.5/inch rigid insulation yielding a total combined thickness of 2.5 inches
- Layers of rigid insulation have staggered joints to facilitate air tightness; two layers are preferable to one layer due to increase in air flow resistance

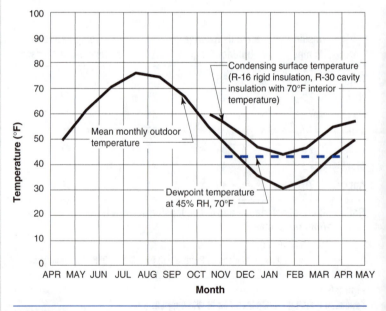

Condensing surface temperature (R-16 rigid insulation, R-30 cavity insulation with 70°F interior temperature)

Mean monthly outdoor temperature

Dewpoint temperature at 45% RH, 70°F

Figure 6.6
Potential for Condensation in Washington, D.C. with Rigid Insulation in Unvented Roof Assembly

- There is no potential for condensation until interior moisture levels exceed 45% RH at 70°F during the coldest month of the year if R-16 rigid insulation is installed over a rafter cavity insulation to R-30
- In hot-humid, mixed-humid and marine climates a 1 perm or lower vapor retarder as tested by the wet-cup procedure should be installed under the asphalt roofing shingles

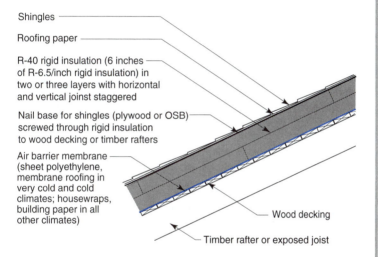

Shingles

Roofing paper

R-40 rigid insulation (6 inches of R-6.5/inch rigid insulation) in two or three layers with horizontal and vertical joinst staggered

Nail base for shingles (plywood or OSB) screwed through rigid insulation to wood decking or timber rafters

Air barrier membrane (sheet polyethylene, membrane roofing in very cold and cold climates; housewraps, building paper in all other climates)

Wood decking

Timber rafter or exposed joist

Figure 6.7
Compact Unvented Roof Assembly
- R-value increased to R-50 in very cold climate zones to control ice-damming
- Optimum roof assembly design to enclose pool areas and spas

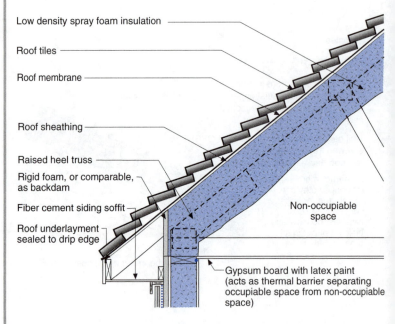

Low density spray foam insulation

Roof tiles

Roof membrane

Roof sheathing

Raised heel truss

Rigid foam, or comparable, as backdam

Fiber cement siding soffit

Roof underlayment sealed to drip edge

Non-occupiable space

Gypsum board with latex paint (acts as thermal barrier separating occupiable space from non-occupiable space)

Figure 6.8
Air Impermeable Spray Foam Insulation

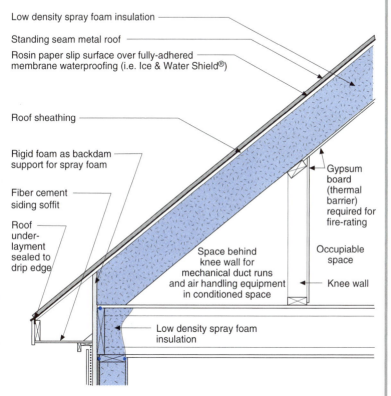

Low density spray foam insulation

Standing seam metal roof

Rosin paper slip surface over fully-adhered
membrane waterproofing (i.e. Ice & Water Shield®)

Roof sheathing

Rigid foam as backdam
support for spray foam

Fiber cement
siding soffit

Roof
under-
layment
sealed to
drip edge

Space behind
knee wall for
mechanical duct runs
and air handling equipment
in conditioned space

Low density spray foam
insulation

Gypsum
board
(thermal
barrier)
required for
fire-rating

Occupiable
space

Knee wall

Figure 6.9
Air Impermeable Spray Foam Insulation
• Spray foam protected with thermal barrier in occupied attic space

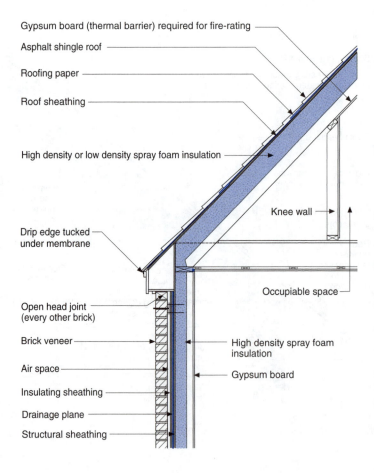

Gypsum board (thermal barrier) required for fire-rating

Asphalt shingle roof

Roofing paper

Roof sheathing

High density or low density spray foam insulation

Knee wall

Drip edge tucked under membrane

Occupiable space

Open head joint (every other brick)

Brick veneer

Air space

Insulating sheathing

Drainage plane

Structural sheathing

High density spray foam insulation

Gypsum board

Figure 6.10
Air Impermeable Spray Foam Insulation
- Spray foam protected with thermal barrier in occupied attic space
- High density or low density spray foam insulation partially fills roof rafter cavity and wall cavity

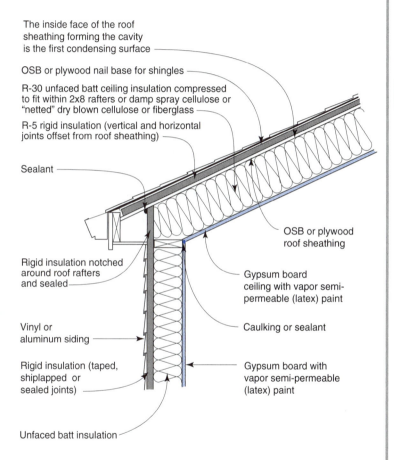

The inside face of the roof sheathing forming the cavity is the first condensing surface

OSB or plywood nail base for shingles

R-30 unfaced batt ceiling insulation compressed to fit within 2x8 rafters or damp spray cellulose or "netted" dry blown cellulose or fiberglass

R-5 rigid insulation (vertical and horizontal joints offset from roof sheathing)

Sealant

OSB or plywood roof sheathing

Rigid insulation notched around roof rafters and sealed

Gypsum board ceiling with vapor semi-permeable (latex) paint

Caulking or sealant

Vinyl or aluminum siding

Gypsum board with vapor semi-permeable (latex) paint

Rigid insulation (taped, shiplapped or sealed joints)

Unfaced batt insulation

Figure 6.11
Unvented Cathedral Ceiling for Hot-Humid Climates
Cellulose or Fiberglass Cavity Insulation
- The thickness of the insulating sheathing on the roof is 1" (R-7 polyisocyanurate) minimum

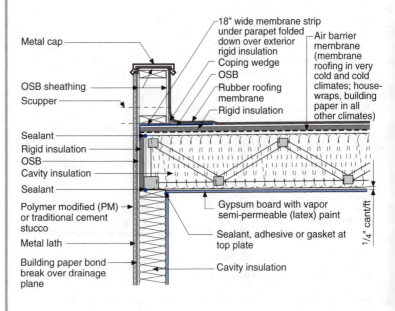

Metal cap

OSB sheathing

Scupper

Sealant

Rigid insulation

OSB

Cavity insulation

Sealant

Polymer modified (PM) or traditional cement stucco

Metal lath

Building paper bond break over drainage plane

18" wide membrane strip under parapet folded down over exterior rigid insulation

Coping wedge

OSB

Rubber roofing membrane

Rigid insulation

Air barrier membrane (membrane roofing in very cold and cold climates; house-wraps, building paper in all other climates)

Gypsum board with vapor semi-permeable (latex) paint

Sealant, adhesive or gasket at top plate

Cavity insulation

1/4" cant/ft

Figure 6.12
Unvented Flat Roof Assembly — Condensing Surface Temperature Controlled

- The thickness of the rigid insulation at both the roof deck and roof perimeter is 1" (R-7 polyisocyanurate) minimum

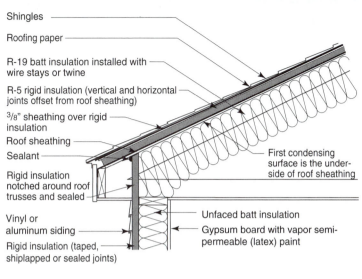

Shingles

Roofing paper

R-19 batt insulation installed with wire stays or twine

R-5 rigid insulation (vertical and horizontal joints offset from roof sheathing)

³/₈" sheathing over rigid insulation

Roof sheathing

Sealant

Rigid insulation notched around roof trusses and sealed

Vinyl or aluminum siding

Rigid insulation (taped, shiplapped or sealed joints)

First condensing surface is the underside of roof sheathing

Unfaced batt insulation

Gypsum board with vapor semipermeable (latex) paint

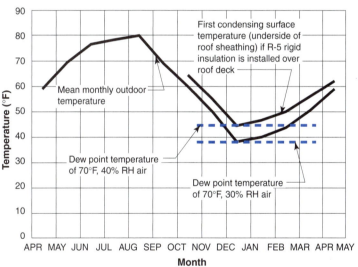

First condensing surface temperature (underside of roof sheathing) if R-5 rigid insulation is installed over roof deck

Mean monthly outdoor temperature

Dew point temperature of 70°F, 40% RH air

Dew point temperature of 70°F, 30% RH air

Figure 6.13
Potential for Condensation in Roswell, New Mexico with Unvented Roof and Insulating Sheathing

- 3,800 heating degree days
- Winter design temperature 18°F
- Summer design temperature 98°F dry bulb; 66°F wet bulb
- There is no potential for condensation on the underside of the roof sheathing when interior moisture levels exceed 30% RH unless rigid insulating sheathing is used to raise the temperature of the first condensing surface
- Rigid insulation is recommended in this roof assembly to raise the condensation potential above 40% RH

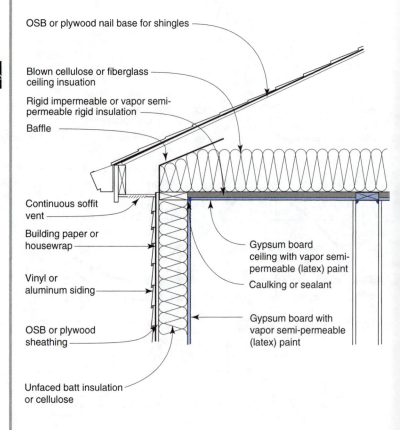

OSB or plywood nail base for shingles

Blown cellulose or fiberglass ceiling insuation

Rigid impermeable or vapor semi-permeable rigid insulation

Baffle

Continuous soffit vent

Building paper or housewrap

Vinyl or aluminum siding

OSB or plywood sheathing

Unfaced batt insulation or cellulose

Gypsum board ceiling with vapor semi-permeable (latex) paint

Caulking or sealant

Gypsum board with vapor semi-permeable (latex) paint

Figure 6.14
Vented Roof Assembly with Rigid Insulation

- Rigid insulation acts as "warm" interior vapor retarder under cellulose or fiberglass ceiling insulation
- Minimum R-5 rigid insulation under R-19 ceiling insulation or R-7.5 rigid insulation under R-28 ceiling insulation

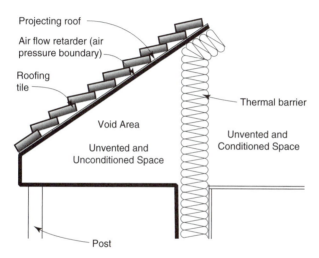

Projecting roof

Air flow retarder (air pressure boundary)

Roofing tile

Void Area

Thermal barrier

Unvented and Unconditioned Space

Unvented and Conditioned Space

Post

Figure 6.15
Unvented Roof

- Thermal barriers do not have to be adjacent to air barriers at projecting roofs, however, both air barrier and thermal barrier continuity is required

Foundations and Wall Assemblies

The three foundation approaches common to residential construction are crawlspaces, slabs and basements. Each can be built with concrete, masonry or wood. The two wall assembly approaches common to residential construction are masonry and wood frame. No matter which foundation approach or wall assembly is used they need to be integrated with each other. They all have to:

- hold the building up
- keep the groundwater out
- keep the rainwater out
- keep the soil gas out
- keep the wind out
- keep the water vapor out
- let the water vapor out if it gets inside
- keep the heat in during the winter
- keep the heat out during the summer

Water Managed Foundations

Water managed foundation systems rely on two fundamental principles (see Figure 7.1, Figure 7.2 and Figure 7.3):

- keep rain water away from the foundation wall perimeter
- drain groundwater with sub-grade perimeter footing drains before it gets to the foundation wall

Water managed foundation systems are different from waterproofing systems. Waterproofing relies on creating a watertight barrier without holes. It can't be done. Even boats need pumps. Water managed foundation systems prevent the buildup of water against foundation walls, thereby eliminating hydrostatic pressure. No pressure, no force to push water through a hole. Remember, we know the foundation wall will have holes.

Mixing control joints with water management is a fundamental requirement for functional foundation systems that provide an extended useful
service life.

Dampproofing should not be confused with waterproofing. Dampproofing protects foundation materials from absorbing ground moisture
by capillarity. Dampproofing is not intended to resist groundwater
forces (hydrostatic pressure). If water management is used, waterproofing is not necessary. However, control of capillary water is still required (dampproofing). Dampproofing is typically provided by coating
the exterior of a concrete foundation wall with a tar or bituminous
paint or coating.

Draining groundwater away from foundation wall perimeters is typically done with free-draining backfill such as sand, gravel or other wa-
ter-permeable material, or drainage boards or exterior foundation insulations with drainage properties.

Soil Gas

Keeping soil gas (radon, water vapor, herbicides, termiticides, meth-
ane, etc.) out of foundations cannot be done by building hole-free
foundations because hole-free foundations cannot be built. Soil gas
moves through holes due to a pressure difference. Since we cannot
eliminate the holes, the only thing we can do is control the pressure.

The granular drainage pad located under concrete slabs can be integrated into a sub-slab ventilation system to control soil gas migration
by creating a zone of negative pressure under the slab. A vent pipe connects the sub-slab gravel layer to the exterior through the roof (Figures
7.4, 7.5 and 7.6). An exhaust fan can be added later, if necessary.

Moisture

Controlling water vapor in foundations relies first on keeping it out,
and second, on letting it out when it gets in. Make no mistake, it will
get in. The issue is complicated by the use of concrete and masonry because there are thousands of pounds of water stored in freshly cast concrete and freshly laid masonry to begin with. This moisture of construction has to dry to somewhere, and it usually (but not always) dries
to the inside.

For example, we put coarse gravel (no fines) and a polyethylene vapor
retarder under a concrete slab to keep the water vapor and water in the
ground from getting into the slab from underneath. The gravel and
polyethylene do nothing for the water already in the slab. This water
can only dry into the building. Installing flooring, carpets or tile over
this concrete before it has dried sufficiently and in a manner that does

not permit drying, is a common mistake that leads to mold, buckled flooring and lifted tile, even in hot-dry and mixed-dry climates.

Similarly, we install dampproofing on the exterior of concrete foundation walls and provide a water managed foundation system to keep water vapor and water in the ground from getting into the foundation from the exterior. Again, this does nothing for the water already in the foundation wall. When we then install interior insulation and finishes on the interior of a foundation wall in a manner that does not permit drying to the interior, mold will grow.

Foundation wall and slab assemblies must be constructed so that they resist water vapor and water from getting in them, but they also must be constructed so that it is easy for water vapor to get out when it gets in or if the assembly was built wet to begin with (as they typically are).

Slab Construction

Capillary control is necessary for slab-on-grade construction and crawlspaces (see Figure 7.8). Monolithic slabs need plastic ground covers that extend under the perimeter grade beam and upwards to grade. Additionally, the exposed portion of the slab edge that is exposed to the outside must be painted with latex paint to reduce water absorption and a capillary break must be installed under perimeter wall framing.

Impermeable interior finishes should be avoided, such as vinyl floor coverings on slab-on-grade construction unless a low water-to-cement ratio concrete is used (less than 0.45) installed directly over a polyethylene vapor barrier — and only where slab edges are protected from capillary water (see Figure 7.8).

Slab Moisture Problems

One of the most common sources of interior mold in hot-humid climates is the irrigated plantings surrounding the house. The ground adjacent to the slab perimeter can become saturated and lead to the wetting of the slab and subsequently deteriorate the wall framing and encourage mold growth (see Figure 7.7). Slab perimeters need to be protected from ground dampness with capillary breaks.

In renovations, the conditions under a slab may be difficult to determine, or once they are determined, it is found that a stone layer or polyethylene is not present. It may be necessary to provide "top side" control of water and vapor. This can be done several ways. If salts are not present in the ground, epoxy coatings or chemical sealers may be used. Salts lead to osmosis and osmotic pressures are typically greater than the bond strength of most coatings and sometimes exceed the cohesive strength of concrete (i.e. the coating is pushed off the slab or the concrete spalls/flakes apart). If salts are present, spacer systems that provide vapor control and drainage can be used over the top of existing slabs (see Figure 7.9).

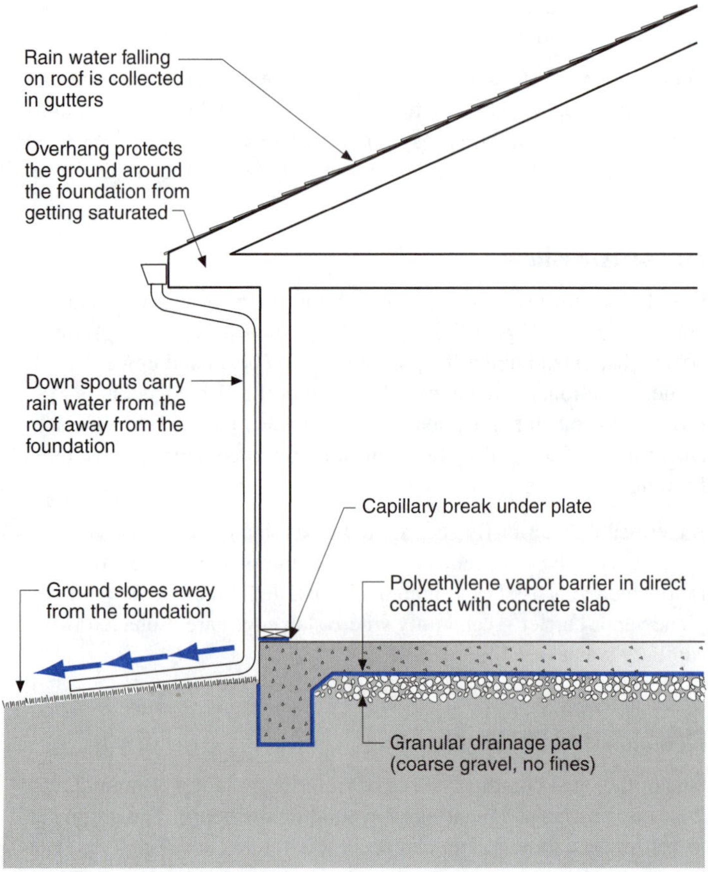

Rain water falling on roof is collected in gutters

Overhang protects the ground around the foundation from getting saturated

Down spouts carry rain water from the roof away from the foundation

Capillary break under plate

Ground slopes away from the foundation

Polyethylene vapor barrier in direct contact with concrete slab

Granular drainage pad (coarse gravel, no fines)

Figure 7.1
Groundwater Control with Slabs
- Keep rain water away from the foundation perimeter
- Do not place sand layer over polyethylene vapor barrier under concrete slab

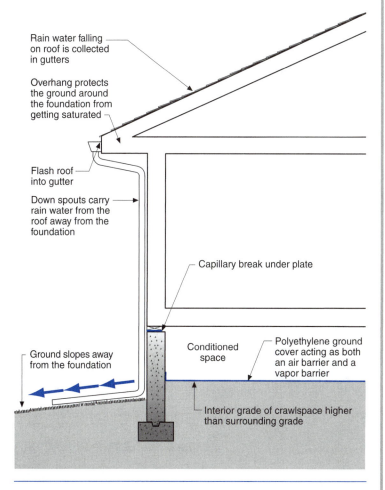

Rain water falling on roof is collected in gutters

Overhang protects the ground around the foundation from getting saturated

Flash roof into gutter

Down spouts carry rain water from the roof away from the foundation

Capillary break under plate

Ground slopes away from the foundation

Conditioned space

Polyethylene ground cover acting as both an air barrier and a vapor barrier

Interior grade of crawlspace higher than surrounding grade

Figure 7.2
Groundwater Control with Crawlspaces
- Keep rain water away from the foundation perimeter
- If the interior crawlspace is lower than the exterior grade, a sub-grade perimeter footing drain is necessary as in a basement foundation
- The crawlspace in this configuration is conditioned space; it is part of the "interior" of the building and should be heated, cooled and ventilated as part of the building's heating, cooling and ventilating strategy

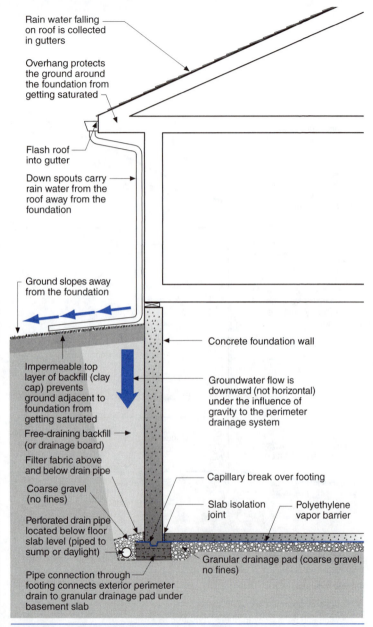

Rain water falling on roof is collected in gutters

Overhang protects the ground around the foundation from getting saturated

Flash roof into gutter

Down spouts carry rain water from the roof away from the foundation

Ground slopes away from the foundation

Impermeable top layer of backfill (clay cap) prevents ground adjacent to foundation from getting saturated

Free-draining backfill (or drainage board)

Filter fabric above and below drain pipe

Coarse gravel (no fines)

Perforated drain pipe located below floor slab level (piped to sump or daylight)

Pipe connection through footing connects exterior perimeter drain to granular drainage pad under basement slab

Concrete foundation wall

Groundwater flow is downward (not horizontal) under the influence of gravity to the perimeter drainage system

Capillary break over footing

Slab isolation joint

Polyethylene vapor barrier

Granular drainage pad (coarse gravel, no fines)

Figure 7.3
Groundwater Control with Basements

- Keep rain water away from the foundation perimeter
- Drain groundwater away in sub-grade perimeter footing drains before it gets to the foundation wall

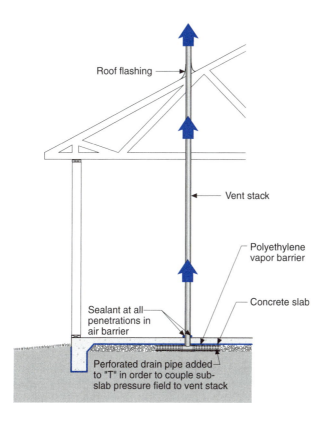

**Figure 7.4
Soil Gas Ventilation System — Slab Construction**

- Granular drainage pad depressurized by active fan located in attic or by passive stack action of warm vent stack located inside conditioned space or garage
- Communication to all sub-slab areas is required. Where slabs are divided by a thickened section to support a bearing wall, pipe connections through the thickened section will be necessary. Multiple connection points or interconnection piping may be required between multi-level slabs similar to those found in split level homes.
- Avoid offsets or elbows in vent stack to maximize airflow

7

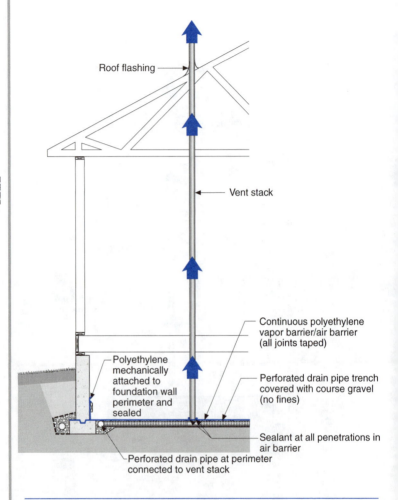

Figure 7.5
Soil Gas Ventilation System — Crawlspace Construction
- Perforated drain pipe in trenches covered with coarse gravel create depressurized zones under air barrier due to active fan located in attic or by passive stack action of warm vent stack located inside heated space
- Crawlspace is conditioned (heated during the winter, cooled during the summer) by a supply HVAC system duct
- Perforated drain pipe may not be necessary with tightly sealed polyethylene and coarse gravel
- Perimeter trench connected to centrally located vent stack
- Avoid offsets or elbows in vent stack to maximize airflow

7

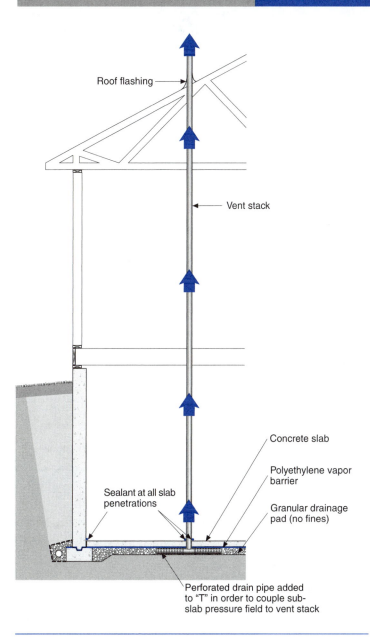

Roof flashing

Vent stack

Concrete slab

Polyethylene vapor barrier

Sealant at all slab penetrations

Granular drainage pad (no fines)

Perforated drain pipe added to "T" in order to couple sub-slab pressure field to vent stack

**Figure 7.6
Soil Gas Ventilation System — Basement Construction**

• Granular drainage pad depressurized by active fan located in attic or by passive stack action of warm vent stack located inside heated space
• Avoid offsets or elbows in vent stack to maximize air flow

A "floating floor" (see Figure 7.10) can also be used where moisture flow upwards is small – or where a finished wood floor (or carpet) is to be installed over a slab. Rigid insulation and plywood are installed on the top of the slab. In this assembly extruded polystyrene should be limited to $^{3}/_{4}$ - inch or less so that the slab can dry upwards (floors are different than walls with respect to permeability limits). Carpets should never be installed directly on below grade slabs unless slabs are insulated (below or on the top surface). Carpets on uninsulated slabs are cold resulting in sufficiently elevated relative humidities within the carpet to support dust mite and mold growth.

Carpets

Installing carpets on cold, damp concrete floor slabs can lead to serious allergic reactions and other health-related consequences. It is not recommended that carpets be installed on basement concrete slabs unless the carpets can be kept dry and warm. In practice, this is not possible unless basement floor slab assemblies are insulated and basement areas

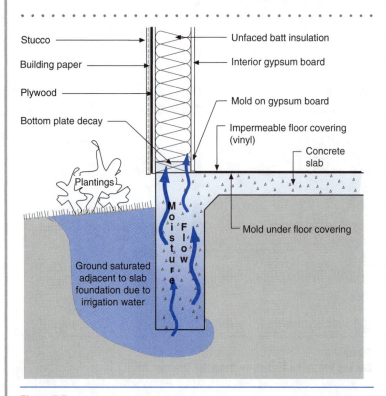

Figure 7.7
Slab Perimeter Moisture Problems

are conditioned. Installing carpets on concrete slab foundations located at grade typically does not pose a risk if the carpet and associated carpet pad are very vapor permeable. Slabs on grade are typically warmer and much dryer than basement slabs.

Crawlspaces

Crawlspaces should be designed and constructed to be dry and pest-free. A dry crawlspace is good for the inhabitants and good for building durability. A dry crawlspace is less likely to have pests and termites. You must control rainwater, groundwater and provide drainage for potential plumbing leaks or flooding incidents.

Crawlspaces should not be used for storage unless they are designed as storage areas with a concrete slab, a conditioning system and ready access. Otherwise, builders and contractors should use designs that discourage the use of crawlspaces for storage, and provide clear guidance to owners and occupants to avoid using this area for storage.

Crawlspaces should ideally be designed and constructed as mini-basements, part of the house – within the pressure boundary and thermal barrier. They should not be vented to the exterior. They should be insulated on their perimeters and should have a continuous sealed ground cover such as taped polyethylene. They should have perimeter drainage just like a basement (when the crawlspace ground level is below

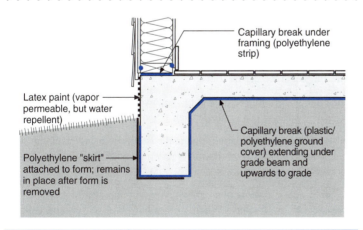

Capillary break under framing (polyethylene strip)

Latex paint (vapor permeable, but water repellent)

Polyethylene "skirt" attached to form; remains in place after form is removed

Capillary break (plastic/ polyethylene ground cover) extending under grade beam and upwards to grade

Figure 7.8
Capillary Control for Monolithic Slab
• Never install a sand layer between a polyethylene ground cover and a slab; the sand layer becomes wet and holds water indefinitely; the sand can only dry upwards, not downwards, due to the polyethylene

the ground level of the surrounding grade). Make sure there is good drainage away from crawlspaces.

While crawlspace venting has been viewed as good building practice and is still required by some codes, an unvented, conditioned crawlspace with insulation on the perimeter performs better in terms of moisture, durability and pest control.

Perimeter insulation rather than floor insulation performs better in all climates. The crawlspace temperatures and relative humidity track that of the house. Crawlspaces insulated on the perimeter are warmer and drier than crawlspaces insulated between the crawlspace and the house. Cold surfaces that can condense water are minimized.

Wintertime ventilation makes crawlspaces colder and increases the heat loss from the home – venting crawlspaces wastes energy, and can lead to freezing pipes and uncomfortable floors.

Note: The International Building Code (ICC) allows the construction of closed (unvented) —conditioned — crawlspaces. Contact code officials in the design phase to determine their requirements.

If it is not possible to treat the crawlspace as a part of the house such as in flood zones in costal areas – or where it is not necessary such as in dry climates, it is important to construct the house such that the crawlspace is isolated from the house – outside of the building boundary. This can be accomplished by air sealing the boundary between the crawlspace and the house and by installing a vapor barrier on the underside of the floor assembly (see Figure 7.25). This vapor barrier needs to have sufficient thermal resistance to control condensation (in both summer and winter)— as such insulating sheathing is recommended in this location. A similar approach is recommended for homes on piers (see Figure 7.26).

No heating and cooling equipment or ductwork should be in the crawlspace if it is treated as an outside (vented) space.

In parts of the country where radon and pesticides in soil gases can be found, sub-slab passive ventilation is recommended (see Figure 7.5). This also helps keep a crawlspace drier.

If possible, seal the vents in an existing crawlspace. Build new crawlspaces without vents. Where homes have both a crawlspace and a basement they should be connected together and treated together as a conditioned space.

It is always necessary to have a drying mechanism. One option is to passively connect the crawlspace to the house via floor registers or transfer grilles. The incidental air change that happens between the

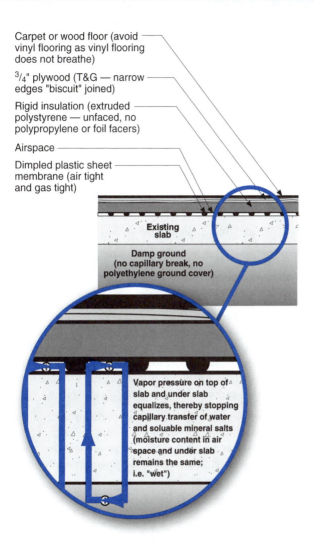

Carpet or wood floor (avoid vinyl flooring as vinyl flooring does not breathe)

$^3/_4$" plywood (T&G — narrow edges "biscuit" joined)

Rigid insulation (extruded polystyrene — unfaced, no polypropylene or foil facers)

Airspace

Dimpled plastic sheet membrane (air tight and gas tight)

Existing slab

Damp ground
(no capillary break, no
polyethylene ground cover)

Vapor pressure on top of slab and under slab equalizes, thereby stopping capillary transfer of water and soluable mineral salts (moisture content in air space and under slab remains the same; i.e. "wet")

Figure 7.9
Slab Top-Side Vapor Control — Airspace Approach
- Works in both new construction and rehabilitation
- Plywood glued (T&G edges) to itself not mechanically fastened (no screws or nails) through foam and dimpled plastic sheet membrane so that gas barrier/air barrier is not comprised
- Groundwater leakage can also be handled with this approach by draining the airspace to a sump or floor drain
- It is important to seal the sheet membrane around the foundation perimeter thereby isolating the airspace from the interior

crawlspace and the house in this manner typically provides sufficient drying. A second option is to heat and cool them as if they are included as part of the home. Air must be supplied to the crawlspace from the home. This air can be returned back to the home or it can be exhausted (see Figure 7.29A through Figure 7.29E). A third option is a dehumidifier plumbed to a sump pump or drain.

Some existing crawlspaces are sources of pollutants that cannot be satisfactorily removed or controlled. The most practical solution is to install a durable fan to exhaust air continuously from the crawlspace to the outside. The fan should be rated for continuous duty and sized according to either ASHRAE Standard 62.2 (so that it also provides ventilation for the house if desired) or at a minimum rate of 20 cfm/1000 square feet otherwise. This reverses the flow air, pulling air from the house into the crawlspace and then out of the building (Figure 7.29B).

7

To keep them dry all crawlspaces should have:

- Continuous, durable ground cover or liner

- Rainwater and groundwater control similar to a basement if the crawlspace is below the ground level of the surrounding grade

- Pest control measures as appropriate for the location

- Inside sloped to one or more low places for when a flooding incident occurs from a plumbing leak or rain entry – the low places should be either drained to daylight or a sump pump.

There are several ways to provide a durable ground cover or liner. The option used depends on the resources available, the frequency of people entering the crawlspace to either store possessions (not a good idea) or to maintain equipment and the severity of the pest problem.

Insects and Termites

There is no good way of dealing with termites. Borate-treated wood framing, cavity insulation (cellulose) and rigid foams are a promising approach, but long-term performance has yet to be demonstrated. Using a protective membrane or a stainless steel mesh with a polymer cement slurry as a termite barrier coupled with soil treatment seems to work on the few projects that have used the approach. However, there is no universal consensus on this matter as no long term performance information is available.

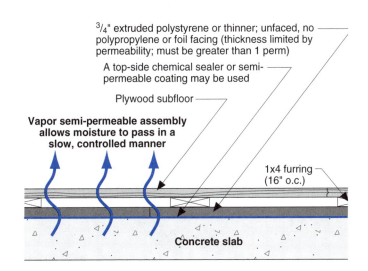

3/4" extruded polystyrene or thinner; unfaced, no
polypropylene or foil facing (thickness limited by
permeability; must be greater than 1 perm)

A top-side chemical sealer or semi-
permeable coating may be used

Plywood subfloor

**Vapor semi-permeable assembly
allows moisture to pass in a
slow, controlled manner**

1x4 furring
(16" o.c.)

Concrete slab

Figure 7.10
Slab Top-Side Vapor Control — Semi-Permeable Floating Floor
- Extruded polystyrene should be used due to its compressive strength (expanded
  polystyrene can be used if furring spacing is reduced to 12" o.c. or if plywood is
  supported directly on foam (i.e. no furring)
- Not applicable with visibly wet slabs and where efflorescence (salts) is visible
- Avoid vinyl flooring with this assembly as vinyl flooring does not breathe

7

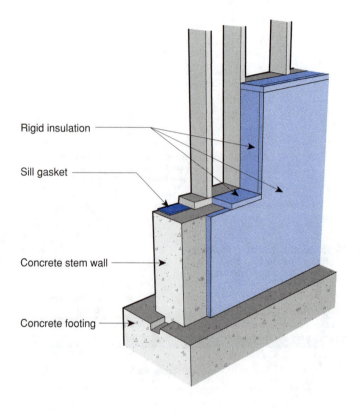

Rigid insulation

Sill gasket

Concrete stem wall

Concrete footing

Figure 7.11
Step Down Foundation Wall

- In basement assemblies and conditioned crawlspace assemblies that are internally insulated, it is important that rigid insulation "wraps" around exposed concrete/block assemblies at "step downs" in order to control condensation, particularly during summer months

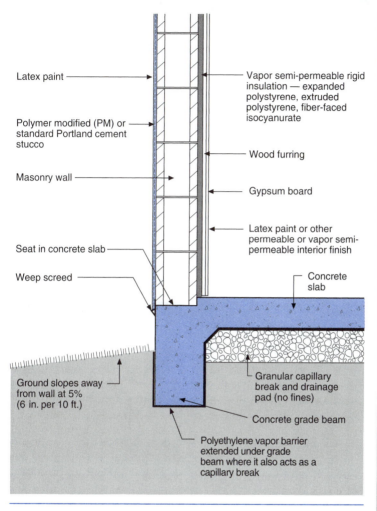

Figure 7.12
Slab Construction — Concrete Seat — Masonry Wall — Interior Rigid Insulation — Stucco

- "Seat" in concrete slab perimeter facilities rain control
- Vapor semi-permeable rigid insulations used on the interior of wall assemblies should be unfaced or faced with permeable skins; foil facings, polypropylene skins should be avoided
- Avoid use of metal furring or "hat" channels due to thermal bridging and impermeability; use wood furring
- Wood furring should be installed over rigid insulation; rigid insulation should not be installed between wood furring installed directly on interior of masonry

7

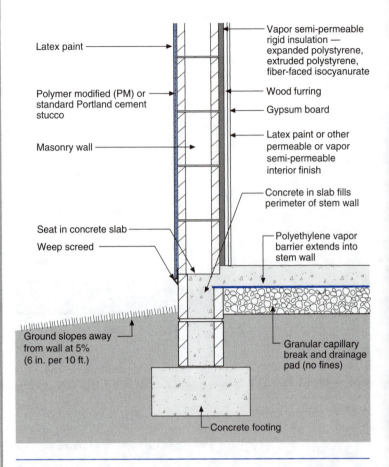

Figure 7.13

Stem Wall — Concrete Seat — Masonry Wall — Interior Rigid Insulation — Stucco

- "Seat" in concrete slab perimeter facilities rain control
- The vapor semi-permeable rigid insulation on the interior of this wall assembly protects the interior gypsum board from capillary water migrating up through the stem wall because the sub-slab polyethylene cannot be extended across the stem wall
- Vapor semi-permeable rigid insulations used on the interior of wall assemblies should be unfaced or faced with permeable skins; foil facings, polypropylene skins should be avoided
- Avoid use of metal furring or "hat" channels due to thermal bridging and impermeability; use wood furring
- Wood furring should be installed over rigid insulation; rigid insulation should not be installed between wood furring installed directly on interior of masonry

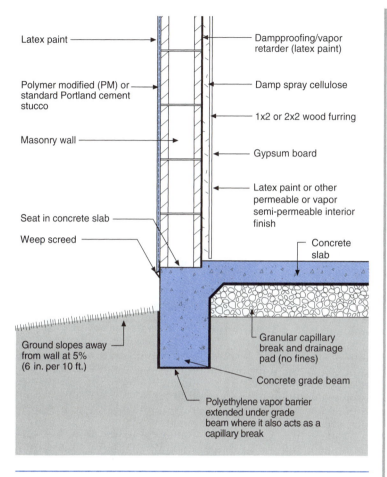

Latex paint

Polymer modified (PM) or standard Portland cement stucco

Masonry wall

Seat in concrete slab

Weep screed

Ground slopes away from wall at 5% (6 in. per 10 ft.)

Dampproofing/vapor retarder (latex paint)

Damp spray cellulose

1x2 or 2x2 wood furring

Gypsum board

Latex paint or other permeable or vapor semi-permeable interior finish

Concrete slab

Granular capillary break and drainage pad (no fines)

Concrete grade beam

Polyethylene vapor barrier extended under grade beam where it also acts as a capillary break

7

Figure 7.14
Slab Construction — Masonry Wall — Interior Cellulose Insulation — Stucco
- "Seat" in concrete slab perimeter facilities rain control
- A vapor barrier such as sheet polyethylene should not be used in place of the vapor semi-permeable dampproofing or latex paint vapor retarder
- Avoid use of metal furring or "hat" channels due to thermal bridging and impermeability; use wood furring

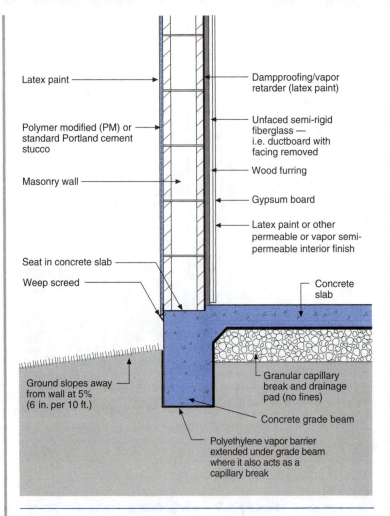

Latex paint

Dampproofing/vapor retarder (latex paint)

Polymer modified (PM) or standard Portland cement stucco

Unfaced semi-rigid fiberglass — i.e. ductboard with facing removed

Wood furring

Masonry wall

Gypsum board

Latex paint or other permeable or vapor semi-permeable interior finish

Seat in concrete slab

Weep screed

Concrete slab

Ground slopes away from wall at 5% (6 in. per 10 ft.)

Granular capillary break and drainage pad (no fines)

Concrete grade beam

Polyethylene vapor barrier extended under grade beam where it also acts as a capillary break

Figure 7.15
Slab Construction — Masonry Wall — Interior Rigid Fiberglass Insulation — Stucco

- "Seat" in concrete slab perimeter facilities rain control
- A vapor barrier such as sheet polyethylene should not be used in place of the vapor semi-permeable dampproofing or latex paint vapor retarder
- Avoid use of metal furring or "hat" channels due to thermal bridging and impermeability; use wood furring
- Wood furring should be installed over rigid insulation; rigid insulation should not be installed between wood furring installed directly on interior of masonry

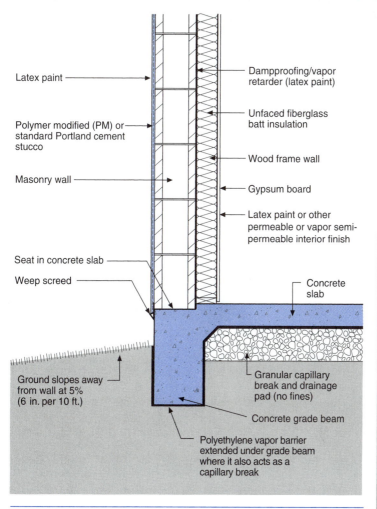

Latex paint

Polymer modified (PM) or standard Portland cement stucco

Masonry wall

Seat in concrete slab

Weep screed

Ground slopes away from wall at 5% (6 in. per 10 ft.)

Dampproofing/vapor retarder (latex paint)

Unfaced fiberglass batt insulation

Wood frame wall

Gypsum board

Latex paint or other permeable or vapor semi-permeable interior finish

Concrete slab

Granular capillary break and drainage pad (no fines)

Concrete grade beam

Polyethylene vapor barrier extended under grade beam where it also acts as a capillary break

7

Figure 7.16
Slab Construction — Masonry Wall — Interior Fiberglass Insulation — Stucco
- "Seat" in concrete slab perimeter facilities rain control
- A vapor barrier such as sheet polyethylene should not be used in place of the vapor semi-permeable dampproofing or latex paint vapor retarder
- Avoid use of metal frame wall to interior

7

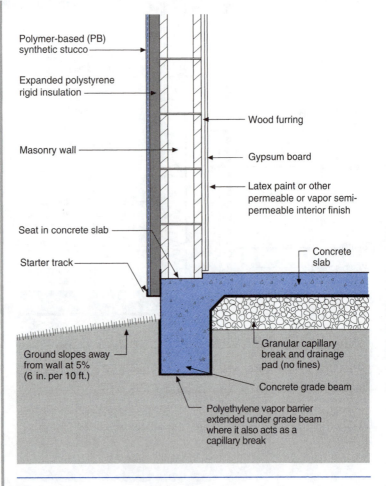

Polymer-based (PB) synthetic stucco

Expanded polystyrene rigid insulation

Masonry wall

Seat in concrete slab

Starter track

Ground slopes away from wall at 5% (6 in. per 10 ft.)

Wood furring

Gypsum board

Latex paint or other permeable or vapor semi-permeable interior finish

Concrete slab

Granular capillary break and drainage pad (no fines)

Concrete grade beam

Polyethylene vapor barrier extended under grade beam where it also acts as a capillary break

Figure 7.17
Slab Construction — Masonry Wall —EIFS
- "Seat" in concrete slab perimeter facilities rain control
- The vapor semi-permeable latex paint permits drying to the interior

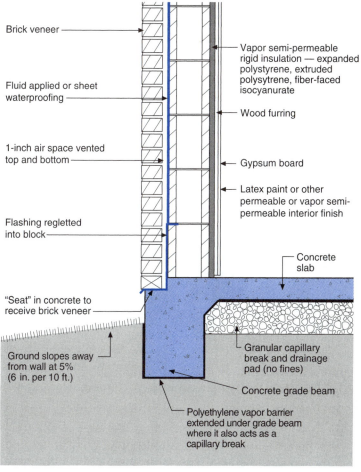

Brick veneer

Fluid applied or sheet waterproofing

1-inch air space vented top and bottom

Flashing regletted into block

"Seat" in concrete to receive brick veneer

Ground slopes away from wall at 5% (6 in. per 10 ft.)

Vapor semi-permeable rigid insulation — expanded polystyrene, extruded polysytrene, fiber-faced isocyanurate

Wood furring

Gypsum board

Latex paint or other permeable or vapor semi-permeable interior finish

Concrete slab

Granular capillary break and drainage pad (no fines)

Concrete grade beam

Polyethylene vapor barrier extended under grade beam where it also acts as a capillary break

Figure 7.18
Slab Construction — Masonry Wall — Interior Rigid Insulation — Brick Veneer

- Where vinyl flooring is installed over slabs, a low w/c ratio ($\cong 0.45$ or less is recommended) to reduce water content in the concrete; alternatively, the slab should be allowed to dry (less than 0.3 grams/24 hrs/ft^2) prior to flooring installation
- Vapor semi-permeable rigid insulations used on the interior of wall assemblies should be unfaced or faced with permeable skins; foil facings, polypropylene skins should be avoided
- Avoid use of metal furring or "hat" channels due to thermal bridging and impermeability; use wood furring
- Wood furring should be installed over rigid insulation; rigid insulation should not be installed between wood furring installed directly on interior of masonry

197

7

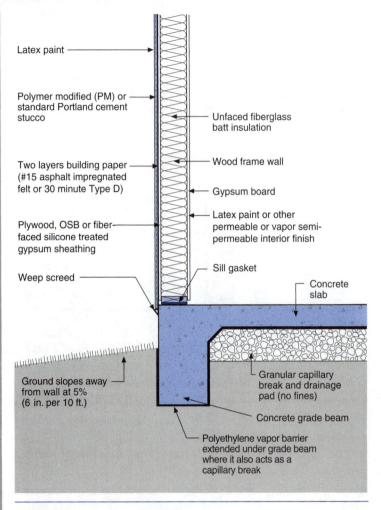

Latex paint

Polymer modified (PM) or standard Portland cement stucco

Unfaced fiberglass batt insulation

Two layers building paper (#15 asphalt impregnated felt or 30 minute Type D)

Wood frame wall

Gypsum board

Plywood, OSB or fiber-faced silicone treated gypsum sheathing

Latex paint or other permeable or vapor semi-permeable interior finish

Weep screed

Sill gasket

Concrete slab

Ground slopes away from wall at 5% (6 in. per 10 ft.)

Granular capillary break and drainage pad (no fines)

Concrete grade beam

Polyethylene vapor barrier extended under grade beam where it also acts as a capillary break

Figure 7.19
Slab Construction — Wood Frame Wall —Stucco
- The vapor semi-permeable latex permits drying to the interior
- Where vinyl flooring is installed over slabs, a low w/c ratio ($\cong 0.45$ or less is recommended) to reduce water content in the concrete; alternatively, the slab should be allowed to dry (less than 0.3 grams/24 hrs/ft^2) prior to flooring installation

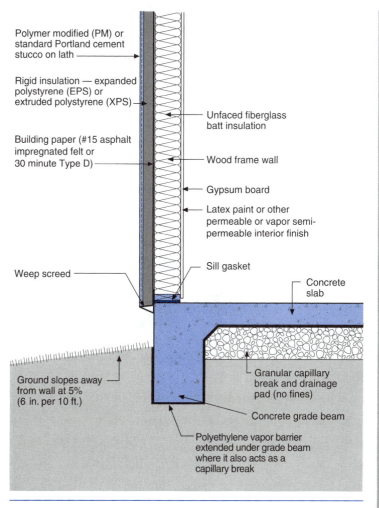

Polymer modified (PM) or standard Portland cement stucco on lath

Rigid insulation — expanded polystyrene (EPS) or extruded polystyrene (XPS)

Building paper (#15 asphalt impregnated felt or 30 minute Type D)

Unfaced fiberglass batt insulation

Wood frame wall

Gypsum board

Latex paint or other permeable or vapor semi-permeable interior finish

7

Weep screed

Sill gasket

Concrete slab

Ground slopes away from wall at 5% (6 in. per 10 ft.)

Granular capillary break and drainage pad (no fines)

Concrete grade beam

Polyethylene vapor barrier extended under grade beam where it also acts as a capillary break

Figure 7.20
Slab Construction — Wood Frame Wall — EIFS
- The vapor semi-permeable latex permits drying to the interior
- Where vinyl flooring is installed over slabs, a low w/c ratio ($\cong 0.45$ or less is recommended) to reduce water content in the concrete; alternatively, the slab should be allowed to dry (less than 0.3 grams/24 hrs/ft^2) prior to flooring installation

7

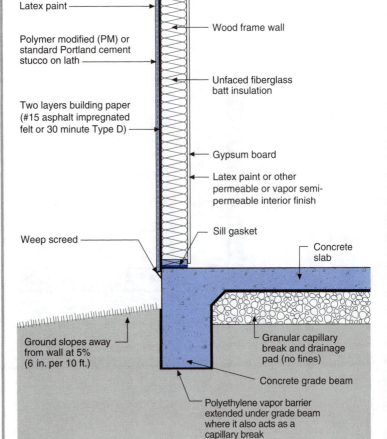

Latex paint

Polymer modified (PM) or standard Portland cement stucco on lath

Two layers building paper (#15 asphalt impregnated felt or 30 minute Type D)

Wood frame wall

Unfaced fiberglass batt insulation

Gypsum board

Latex paint or other permeable or vapor semi-permeable interior finish

Weep screed

Sill gasket

Concrete slab

Ground slopes away from wall at 5% (6 in. per 10 ft.)

Granular capillary break and drainage pad (no fines)

Concrete grade beam

Polyethylene vapor barrier extended under grade beam where it also acts as a capillary break

Figure 7.21
Slab Construction — Wood Frame Wall —Stucco

- The vapor semi-permeable latex permits drying to the interior
- Where vinyl flooring is installed over slabs, a low w/c ratio ($\cong 0.45$ or less is recommended) to reduce water content in the concrete; alternatively, the slab should be allowed to dry (less than 0.3 grams/24 hrs/ft^2) prior to flooring installation

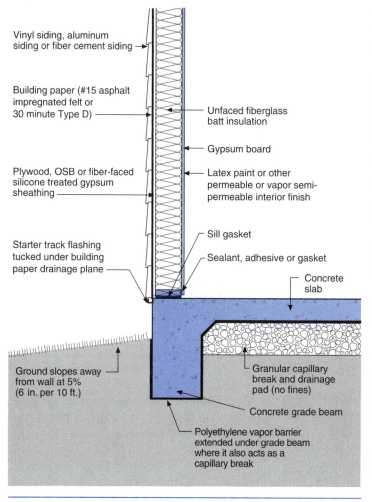

Vinyl siding, aluminum siding or fiber cement siding

Building paper (#15 asphalt impregnated felt or 30 minute Type D)

Unfaced fiberglass batt insulation

Gypsum board

Latex paint or other permeable or vapor semi-permeable interior finish

Plywood, OSB or fiber-faced silicone treated gypsum sheathing

Sill gasket

Sealant, adhesive or gasket

Concrete slab

Starter track flashing tucked under building paper drainage plane

Ground slopes away from wall at 5% (6 in. per 10 ft.)

Granular capillary break and drainage pad (no fines)

Concrete grade beam

Polyethylene vapor barrier extended under grade beam where it also acts as a capillary break

7

Figure 7.22
Slab Construction — Wood Frame Wall — Siding
- Fiber cement siding should be backprimed (coated on all six sides)
- Wood siding should only be installed on furring strips and backprimed (coated on all six sides)

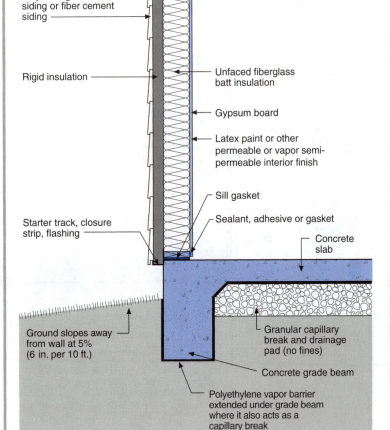

Vinyl siding, aluminum siding or fiber cement siding

Rigid insulation

Unfaced fiberglass batt insulation

Gypsum board

Latex paint or other permeable or vapor semi-permeable interior finish

Sill gasket

Starter track, closure strip, flashing

Sealant, adhesive or gasket

Concrete slab

Ground slopes away from wall at 5% (6 in. per 10 ft.)

Granular capillary break and drainage pad (no fines)

Concrete grade beam

Polyethylene vapor barrier extended under grade beam where it also acts as a capillary break

Figure 7.23
Slab Construction — Wood Frame Wall —Siding

- Fiber cement siding should be backprimed (coated on all six sides)
- Wood siding should only be installed on furring strips and backprimed (coated on all six sides)

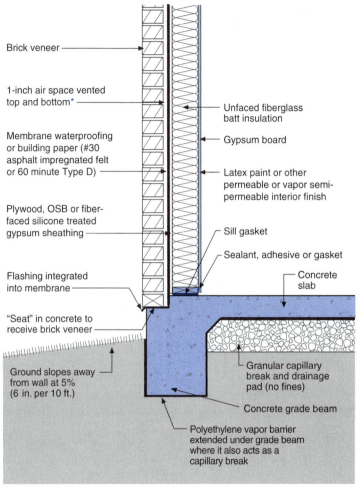

Brick veneer

1-inch air space vented top and bottom*

Membrane waterproofing or building paper (#30 asphalt impregnated felt or 60 minute Type D)

Plywood, OSB or fiber-faced silicone treated gypsum sheathing

Flashing integrated into membrane

"Seat" in concrete to receive brick veneer

Ground slopes away from wall at 5% (6 in. per 10 ft.)

Unfaced fiberglass batt insulation

Gypsum board

Latex paint or other permeable or vapor semi-permeable interior finish

Sill gasket

Sealant, adhesive or gasket

Concrete slab

Granular capillary break and drainage pad (no fines)

Concrete grade beam

Polyethylene vapor barrier extended under grade beam where it also acts as a capillary break

* Air space increased to 2-inches if building paper is used in place of membrane waterproofing

Figure 7.24
Slab Construction — Wood Frame Wall — Brick Veneer
- The vapor semi-permeable latex permits drying to the interior
- Where vinyl flooring is installed over slabs, a low w/c ratio ($\cong 0.45$ or less is recommended) to reduce water content in the concrete; alternatively, the slab should be allowed to dry (less than 0.3 grams/24 hrs/ft^2) prior to flooring installation

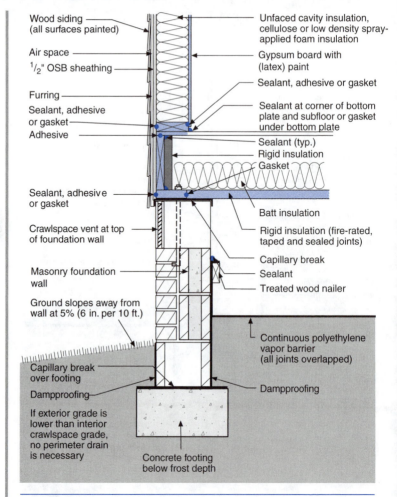

Wood siding (all surfaces painted)

Air space

1/2" OSB sheathing

Furring

Sealant, adhesive or gasket

Adhesive

Sealant, adhesive or gasket

Crawlspace vent at top of foundation wall

Masonry foundation wall

Ground slopes away from wall at 5% (6 in. per 10 ft.)

Capillary break over footing

Dampproofing

If exterior grade is lower than interior crawlspace grade, no perimeter drain is necessary

Unfaced cavity insulation, cellulose or low density spray-applied foam insulation

Gypsum board with (latex) paint

Sealant, adhesive or gasket

Sealant at corner of bottom plate and subfloor or gasket under bottom plate

Sealant (typ.)

Rigid insulation

Gasket

Batt insulation

Rigid insulation (fire-rated, taped and sealed joints)

Capillary break

Sealant

Treated wood nailer

Continuous polyethylene vapor barrier (all joints overlapped)

Dampproofing

Concrete footing below frost depth

Figure 7.25
Vented Crawlspace

- Rigid impermeable insulating sheathing protects underside of floor assembly from wetting during summer. Structural wood beams must also be similarly protected.
- Rigid insulation must be fire-rated if it is left exposed under the floor framing in the crawlspace
- Band joist assembly must be tight or entry of outside air will compromise the effectiveness of the floor cavity insulation
- Penetrations in bottom plates of interior and exterior partition walls should also be sealed to provide a degree of redundancy to the primary air barrier (the rigid insulation and band joist assembly)

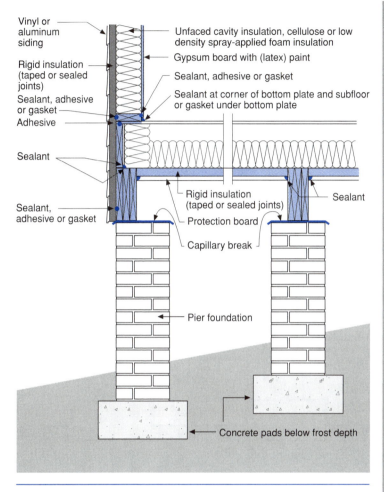

Vinyl or aluminum siding

Unfaced cavity insulation, cellulose or low density spray-applied foam insulation

Gypsum board with (latex) paint

Rigid insulation (taped or sealed joints)

Sealant, adhesive or gasket

Sealant, adhesive or gasket

Sealant at corner of bottom plate and subfloor or gasket under bottom plate

Adhesive

Sealant

Rigid insulation (taped or sealed joints)

Sealant

Sealant, adhesive or gasket

Protection board

Capillary break

Pier foundation

Concrete pads below frost depth

7

Figure 7.26
Pier Foundation
- Rigid impermeable insulating sheathing protects underside of floor assembly from wetting during summer
- Floor cavity insulation held down in contact with rigid insulation – air space above insulation provides warm floor during winter
- Band joist assembly must be tight or entry of outside air will compromise the effectiveness of the floor cavity insulation
- Penetrations in bottom plates of interior and exterior partition walls should also be sealed to provide a degree of redundancy to the primary air barrier (the rigid insulation and band joist assembly)

7

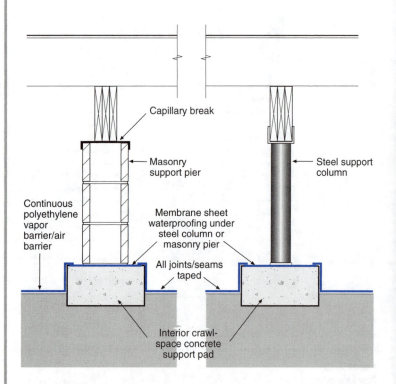

Capillary break

Masonry
support pier

Steel support
column

Continuous
polyethylene
vapor
barrier/air
barrier

Membrane sheet
waterproofing under
steel column or
masonry pier

All joints/seams
taped

Interior crawl-
space concrete
support pad

**Figure 7.27
Air Barrier Continuity at Piers**

- All joints and seams in polyethylene are taped
- Polyethylene ground cover taped to membrane sheet waterproofing at columns
  and piers

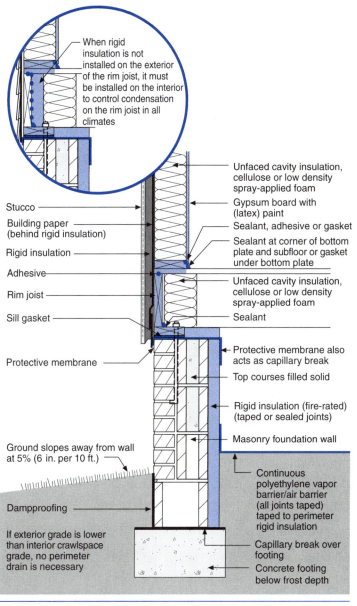

When rigid insulation is not installed on the exterior of the rim joist, it must be installed on the interior to control condensation on the rim joist in all climates

Unfaced cavity insulation, cellulose or low density spray-applied foam

Gypsum board with (latex) paint

Sealant, adhesive or gasket

Sealant at corner of bottom plate and subfloor or gasket under bottom plate

Unfaced cavity insulation, cellulose or low density spray-applied foam

Sealant

Stucco

Building paper (behind rigid insulation)

Rigid insulation

Adhesive

Rim joist

Sill gasket

Protective membrane

Protective membrane also acts as capillary break

Top courses filled solid

Rigid insulation (fire-rated) (taped or sealed joints)

Masonry foundation wall

Ground slopes away from wall at 5% (6 in. per 10 ft.)

Continuous polyethylene vapor barrier/air barrier (all joints taped) taped to perimeter rigid insulation

Dampproofing

If exterior grade is lower than interior crawlspace grade, no perimeter drain is necessary

Capillary break over footing

Concrete footing below frost depth

Figure 7.28
Internally Insulated Concrete Crawlspace with Stucco Wall Above
- Masonry wall cold; can dry to exterior; low likelihood of mold
- Protective membrane acts as termite barrier
- Rigid insulation must be fire-rated if it is left exposed on the interior
- Building paper installed shingle fashion acts as drainage plane located behind rigid insulation

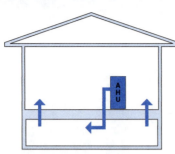

A: Supply air to crawlspace
- Minimum 2-4"x8" transfer grilles **to** house
- 50 cfm of flow per 1,000ft² of crawlspace
- Air handler cycled at 5 minutes per hour

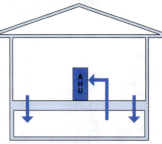

C: Return air from crawlspace
- Minimum 2-4"x8" transfer grilles **from** house
- 50 cfm of flow per 1,000ft² of crawlspace
- Air handler cycled at 5 minutes per hour

7

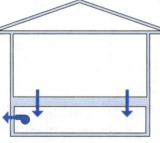

B: Exhaust fan in crawlspace
- Transfer air **from** house
- Fan sized at ASHRAE 62.2 whole house flow rates:
  7.5 cfm/person + 0.01 cfm/ft² of conditioned area
- For a 2,000ft² 3 bedroom house with 4 occupants:
  4 x 7.5 cfm = 30 cfm
  2,000ft² x 0.01 cfm = 20 cfm
  30 cfm + 20 cfm = 50 cfm (i.e. 50 cfm exhaust fan)
- Fan runs continuously

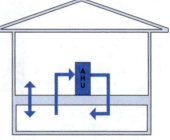

D: Supply and return to crawlspace
- Minimum 2-4"x8" transfer grilles **from** house through floor to equalize air pressure
- 50 cfm of flow per 1,000ft² of crawlspace
- Air handler cycled at 5 minutes per hour

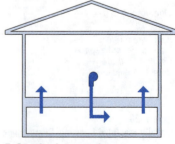

E: Supply air to crawlspace
- Minimum 2-4"x8" transfer grilles **to** house
- 50 cfm of flow per 1,000ft² of crawlspace

Figure 7.29
Conditioning Crawlspaces

7

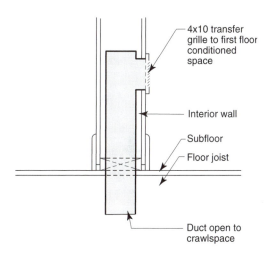

4x10 transfer grille to first floor conditioned space

Interior wall

Subfloor

Floor joist

Duct open to crawlspace

Figure 7.30
Transfer Grille to Crawlspace

7

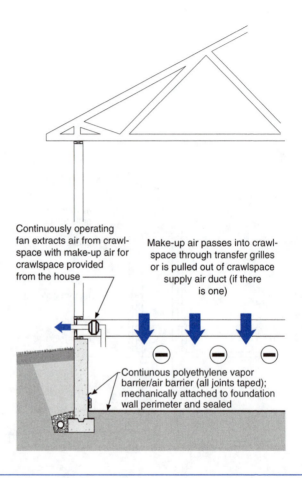

Continuously operating fan extracts air from crawl-space with make-up air for crawlspace provided from the house

Make-up air passes into crawl-space through transfer grilles or is pulled out of crawlspace supply air duct (if there is one)

Contiunous polyethylene vapor barrier/air barrier (all joints taped); mechanically attached to foundation wall perimeter and sealed

Figure 7.31
Controlled Mechanical Ventilation, Soil Gas and Crawlspace Ventilation System
- Crawlspace is conditioned (heated during the winter, cooled during the summer) either by make-up air pulled from the house or by a supply HVAC system duct
- Depressurization of crawlspace is facilitated by continuous exhaust from the crawlspace with make-up air for the crawlspace provided from the house common area
- Crawlspace ground cover is tighter than subfloor
- Crawlspace ventilation and house ventilation is provided by a single fan

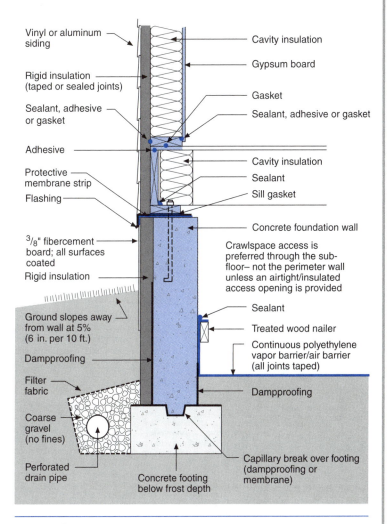

7

Figure 7.32
Externally Insulated Concrete Crawlspace with Vinyl or Aluminum Siding Above
- Concrete wall warm, can dry to the interior; extremely low likelihood of mold
- Protective membrane strip acts as termite barrier
- Perimeter drain is necessary since interior grade is lower than exterior grade
- Protective membrane strip can be adhesive-backed roll roofing. Below grade sheet waterproofing or ice-dam protection membranes can also be used.

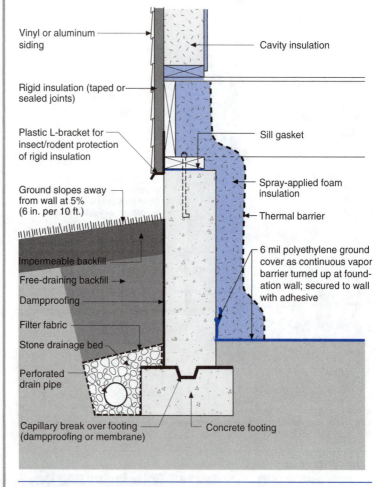

Vinyl or aluminum siding

Cavity insulation

Rigid insulation (taped or sealed joints)

Plastic L-bracket for insect/rodent protection of rigid insulation

Sill gasket

Ground slopes away from wall at 5% (6 in. per 10 ft.)

Spray-applied foam insulation

Thermal barrier

Impermeable backfill

Free-draining backfill

Dampproofing

6 mil polyethylene ground cover as continuous vapor barrier turned up at foundation wall; secured to wall with adhesive

Filter fabric

Stone drainage bed

Perforated drain pipe

Capillary break over footing (dampproofing or membrane)

Concrete footing

Figure 7.33
Internally Insulated Concrete Crawlspace with Spray Foam

• A thermal barrier must be installed over spray foam insulations in crawlspaces; cement stucco or intumescent paint can be used

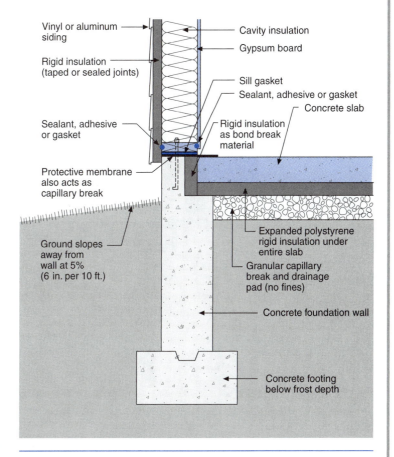

Vinyl or aluminum siding

Cavity insulation

Gypsum board

Rigid insulation (taped or sealed joints)

Sill gasket

Sealant, adhesive or gasket

Concrete slab

Sealant, adhesive or gasket

Rigid insulation as bond break material

Protective membrane also acts as capillary break

Ground slopes away from wall at 5% (6 in. per 10 ft.)

Expanded polystyrene rigid insulation under entire slab

Granular capillary break and drainage pad (no fines)

Concrete foundation wall

Concrete footing below frost depth

Figure 7.34
Slab with Concrete Perimeter and Vinyl or Aluminum Siding Above

- Protective membrane acts as termite barrier; sealed to slab
- Rigid insulation on frame wall extends downward below top of concrete foundation wall to shelter horizontal joint
- Floor slab is warm due to sub-slab rigid insulation under entire floor; can dry to the ground (since there is no under slab vapor barrier, retarder insulation selected is semi-permeable) as well as to the interior, lowest likelihood of mold

7

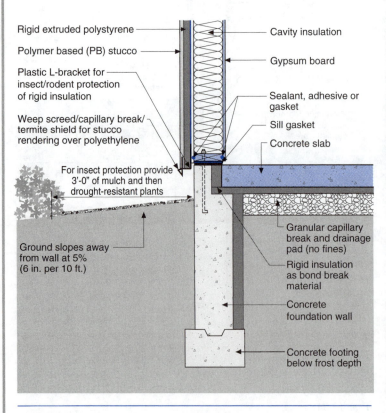

Rigid extruded polystyrene

Polymer based (PB) stucco

Plastic L-bracket for insect/rodent protection of rigid insulation

Weep screed/capillary break/ termite shield for stucco rendering over polyethylene

For insect protection provide 3'-0" of mulch and then drought-resistant plants

Ground slopes away from wall at 5% (6 in. per 10 ft.)

Cavity insulation

Gypsum board

Sealant, adhesive or gasket

Sill gasket

Concrete slab

Granular capillary break and drainage pad (no fines)

Rigid insulation as bond break material

Concrete foundation wall

Concrete footing below frost depth

Figure 7.35
Slab with Concrete Perimeter and Stucco Wall Above

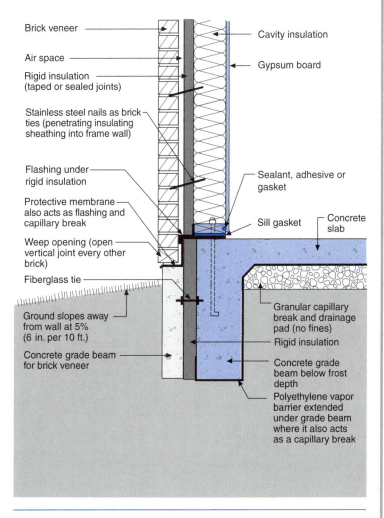

Brick veneer

Air space

Rigid insulation
(taped or sealed joints)

Stainless steel nails as brick
ties (penetrating insulating
sheathing into frame wall)

Flashing under
rigid insulation

Protective membrane
also acts as flashing and
capillary break

Weep opening (open
vertical joint every other
brick)

Fiberglass tie

Ground slopes away
from wall at 5%
(6 in. per 10 ft.)

Concrete grade beam
for brick veneer

Cavity insulation

Gypsum board

Sealant, adhesive or
gasket

Sill gasket

Concrete
slab

Granular capillary
break and drainage
pad (no fines)

Rigid insulation

Concrete grade
beam below frost
depth

Polyethylene vapor
barrier extended
under grade beam
where it also acts
as a capillary break

Figure 7.36
Monolithic Slab with Brick Veneer Above
- Protective membrane acts as termite barrier and acts as flashing at base of brick veneer
- Grade beam for brick veneer cast simultaneously with monolithic slab
- Exterior horizontal rigid insulation provides frost protection
- Airspace behind brick veneer can be as small as $3/8$"; 1" is typical

7

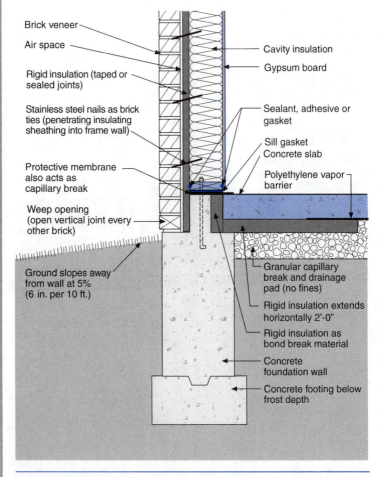

Brick veneer

Air space

Rigid insulation (taped or sealed joints)

Stainless steel nails as brick ties (penetrating insulating sheathing into frame wall)

Protective membrane also acts as capillary break

Weep opening (open vertical joint every other brick)

Ground slopes away from wall at 5% (6 in. per 10 ft.)

Cavity insulation

Gypsum board

Sealant, adhesive or gasket

Sill gasket
Concrete slab

Polyethylene vapor barrier

Granular capillary break and drainage pad (no fines)

Rigid insulation extends horizontally 2'-0"

Rigid insulation as bond break material

Concrete foundation wall

Concrete footing below frost depth

Figure 7.37
Slab with Concrete Perimeter and Brick Veneer Above

- Protective membrane acts as termite barrier; sealed to slab
- Airspace behind brick veneer can be as small as $^3/_8$"; 1" is typical

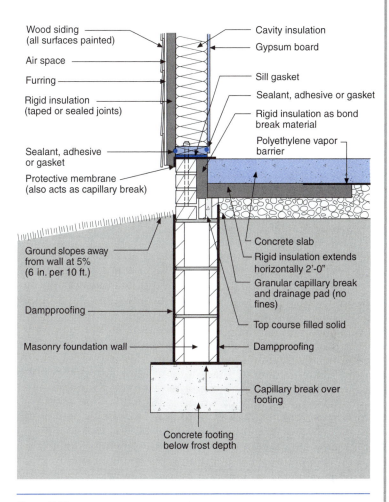

Wood siding (all surfaces painted)

Air space

Furring

Rigid insulation (taped or sealed joints)

Sealant, adhesive or gasket

Protective membrane (also acts as capillary break)

Cavity insulation

Gypsum board

Sill gasket

Sealant, adhesive or gasket

Rigid insulation as bond break material

Polyethylene vapor barrier

Ground slopes away from wall at 5% (6 in. per 10 ft.)

Dampproofing

Masonry foundation wall

Concrete slab

Rigid insulation extends horizontally 2'-0"

Granular capillary break and drainage pad (no fines)

Top course filled solid

Dampproofing

Capillary break over footing

Concrete footing below frost depth

Figure 7.38
Slab with Masonry Perimeter and Wood Siding Above
• Protective membrane acts as termite barrier; sealed to slab

7

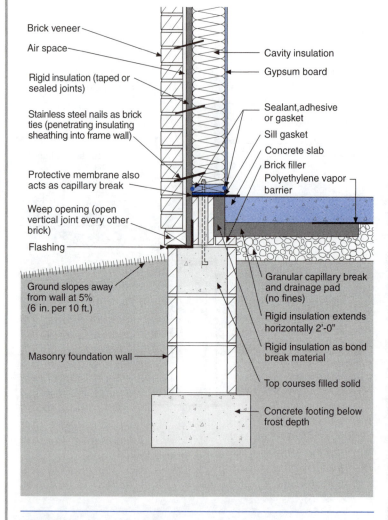

Figure 7.39
Slab with Masonry Perimeter and Brick Veneer Above
- Protective membrane acts as termite barrier; sealed to slab
- Brick veneer corbelled outwards at base to provide drainage space
- Airspace behind brick veneer can be as small as $3/8$"; 1" is typical

7

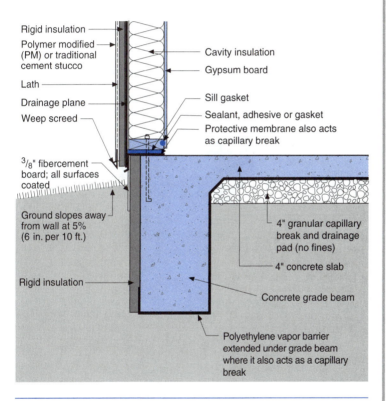

Rigid insulation

Polymer modified (PM) or traditional cement stucco

Cavity insulation

Gypsum board

Lath

Drainage plane

Weep screed

Sill gasket

Sealant, adhesive or gasket

Protective membrane also acts as capillary break

$3/8$" fibercement board; all surfaces coated

Ground slopes away from wall at 5% (6 in. per 10 ft.)

4" granular capillary break and drainage pad (no fines)

4" concrete slab

Rigid insulation

Concrete grade beam

Polyethylene vapor barrier extended under grade beam where it also acts as a capillary break

Figure 7.40
Externally Insulated Monolithic Slab and Stucco Wall Above

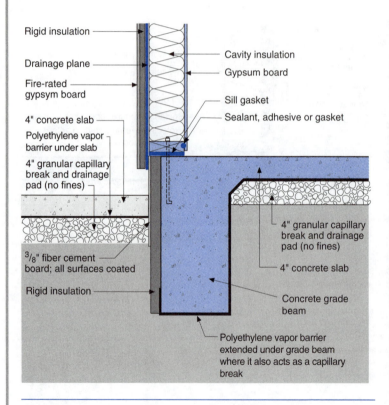

Rigid insulation

Drainage plane

Fire-rated
gypsym board

4" concrete slab

Polyethylene vapor
barrier under slab

4" granular capillary
break and drainage
pad (no fines)

Cavity insulation

Gypsum board

Sill gasket

Sealant, adhesive or gasket

$^3/_8$" fiber cement
board; all surfaces coated

Rigid insulation

4" granular capillary
break and drainage
pad (no fines)

4" concrete slab

Concrete grade
beam

Polyethylene vapor barrier
extended under grade beam
where it also acts as a capillary
break

Figure 7.41
Garage Slab

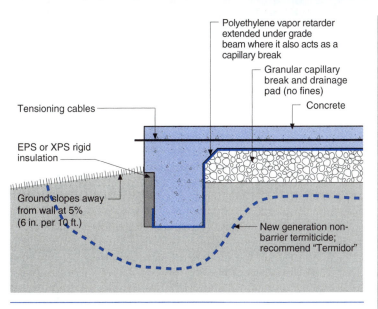

Polyethylene vapor retarder extended under grade beam where it also acts as a capillary break

Granular capillary break and drainage pad (no fines)

Concrete

Tensioning cables

EPS or XPS rigid insulation

Ground slopes away from wall at 5% (6 in. per 10 ft.)

New generation non-barrier termiticide; recommend "Termidor"

Figure 7.42
Post-tensioned Slab — Single Concrete Pour

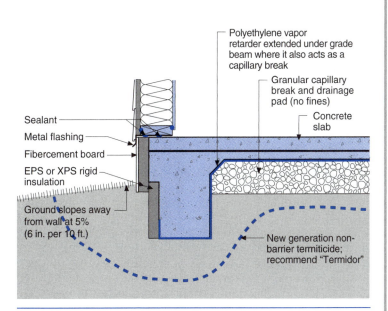

Polyethylene vapor retarder extended under grade beam where it also acts as a capillary break

Granular capillary break and drainage pad (no fines)

Concrete slab

Sealant

Metal flashing

Fibercement board

EPS or XPS rigid insulation

Ground slopes away from wall at 5% (6 in. per 10 ft.)

New generation non-barrier termiticide; recommend "Termidor"

Figure 7.43
Post-tensioned Slab

7

221

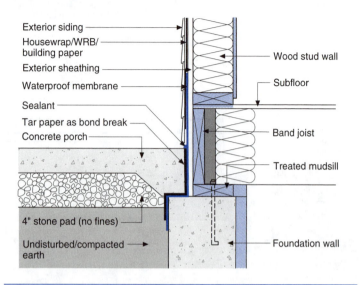

Exterior siding

Housewrap/WRB/
building paper

Exterior sheathing

Waterproof membrane

Sealant

Tar paper as bond break

Concrete porch

Wood stud wall

Subfloor

Band joist

Treated mudsill

4" stone pad (no fines)

Undisturbed/compacted
earth

Foundation wall

Figure 7.44
Porch Slab with Interior Band Joist Insulation

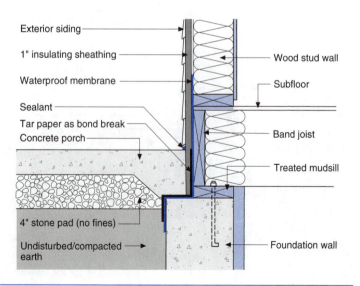

Exterior siding

1" insulating sheathing

Waterproof membrane

Sealant

Tar paper as bond break

Concrete porch

Wood stud wall

Subfloor

Band joist

Treated mudsill

4" stone pad (no fines)

Undisturbed/compacted
earth

Foundation wall

Figure 7.45
Porch Slab with Exterior Insulating Sheathing

Creature and
Dust Control

Infestations of cockroaches, dust mites, mice and rats can all cause allergic reactions. Even after the pests are gone, their skin, hair and feces can remain and can trigger allergic reactions.

Making a home pest-resistant reduces exposure to allergens and asthma triggers released by the pests, and it can reduce the amount of pesticides used by the occupants.

 8

Design and construct the building so it's easy for people to keep pests from colonizing. Take the following steps:

- Make it hard for them to get in by sealing the walls, ceilings, roofs and foundations

- If they do get in, make it hard for them to move around unseen by sealing passages through interior floors, walls and ceilings, kick spaces

- Make it hard for them to find water by: keeping liquid water out, making plumbing easy to inspect and repair, and insulating plumbing pipes to keep them warm (above dewpoint temperatures)

- Make it hard for them to find food using tight food storage, by keeping paper and wood products away from potential moisture sources, and by using pest-resistant materials

- Make directed use of low toxicity pesticides in locations that are heavily infested with problem creatures.

To Know the Critter is to Control The Critter

To actually do the things on the list, you must know the creature. The simplest, safest and most elegant controls are those that work with the creature's natural urges. Creatures that get eaten a lot don't like open spaces. Give them open spaces. No closed-in kick spaces, strips around buildings free of shrubs and organic mulch. Seal around pipes and wires to keep them out of walls.

Keeping Them Out

Keep them out by changing the surrounding landscape and by blocking pest entries and passages. Reduce food and water availability.

Keep bushes and trees at least 3 feet from homes. Bushes and trees near a home provide food, a living place and sheltered passage for pests such as rats, mice, bats, birds, roaches and ants.

Seal utility openings and joints between materials. Use corrosion-proof materials such as copper or stainless steel mesh. Rodents can chew through many materials and squeeze through tiny openings.

Reducing Food and Water

Provide places to store food that are dry and ventilated. Provide a place to store trash and to facilitate recycling.

Design and construct the home to be dry and to dry if and when it gets wet. Absolutely no installed carpet in areas prone to get wet: bathrooms, laundry rooms, kitchens, entryways and damp basements.

In cold climates, dust mites do not generally colonize buildings because buildings are too dry for much of the year. They colonize bedding, stuffed animals and favorite chairs because we humidify these things with our bodies. Control is by washing these items in hot water (greater than 130° Fahrenheit), which kills the mites and washes away allergens.

Pesticides

In the design and construction of new buildings, pesticides have a very limited and targeted role to play. In a neighborhood infested with a difficult species, like roaches or termites, use a limited amount of low toxicity pesticide in targeted locations. In high risk termite areas, exclusion and inspection detailing — plus a combination of treated wooden materials and soil treatment — is useful. For roaches, dusting with boric acid in areas that would be hard to treat later is an effective, low risk strategy. For example, dust with boric acid inside the kick space beneath sink, then seal the kick space as completely as possible.

To assess risk factors associated with a pesticide, look at:
- Registration, classification, use, mode of action
- Specificity, effectiveness, repellency
- Toxicity to humans
- Cautions on label
- Toxicity in the environment
- Resistant populations

Look especially for products like insect growth hormone regulators, which are species-specific, effective and have low toxicity for the applicators, occupants and the environment.

Don't spray pesticides; apply them directly to surfaces to be treated.

Dust

Stop the dust at the door. Vacuum and filter the rest away. And make it easy to clean.

Over two thirds of dust in houses originates outdoors, and is tracked in on feet. House dust is known to contain many hazardous materials. House dust is an asthma trigger.

Entry Control

Pave exterior walks. Use exterior grate track off, interior carpet mat and hard surface floors. Design entries so that there is room to remove and store coats, shoes and boots.

Use a three part track-off approach:

1. Permeable, rugged outdoor mat that collects gritty materials (or a grate over a collection hole is an alternative approach);

2. Rugged indoor mat that collects grit and water and;

3. A hard surface, easily mopped floor to collect very fine particles left by drying foot prints.

Cleanable Surfaces

Whenever possible, replace carpets with smooth flooring which is easy to clean and less likely to retain dust. Use window treatments such as blinds or shades that can be easily wiped. Use hard surfaces rather than textiles. Use semi-gloss latex paints instead of flat or matte finishes. Such surfaces are easier to clean using mild soaps.

Filtration

Construct a tight building enclosure to keep out outside dust and provide filtration. Filters should be MERV 6 – 8 (35 percent or better ASHRAE dust spot efficiency).

Part II

9
10
11
12
13
14
15
16
17
18

General Contractor

All the general contractor has to do is build the building on time and on budget using imperfect materials and imperfect trades, under less than ideal conditions. The client even expects the building to work. There never seems to be enough time or money, but the job still has to get done. In this line of work, Murphy is an optimist.

In the old days, all you needed was good workmanship and good materials. Getting good workmanship today is hard enough given the state of the trades, but it is not enough. Good workmanship cannot compensate for bad design. A general contractor cannot just follow the plans. Plans rarely provide enough information to get the job done, and many times the plans are wrong, in which case, the general contractor has to catch the mistakes. When the plans don't provide enough information, the general contractor has to fill in the gaps. "Not my fault, I was just following the plans" or "Nobody told me I had to do it that way" doesn't work anymore. It's not fair, nor is it right, but it's often the way it is. In order to protect himself or herself, the general contractor has to know everything about everything. Easy, right?

Concerns

Everything is a concern to the general contractor, but of all the concerns, there is one that towers above the rest. Getting the right information to people when they need it in order to do their jobs is more important than anything else. Getting the materials and equipment to the people when they need it comes next. General contracting is all about the flow of information, materials and equipment to the right people at the right time.

Most mistakes happen because of a lack of information, followed by a lack of attention. Attention follows information. Understanding comes from having the right information. Once there is understanding, problems can be caught early and corrected by the people actually doing the

9

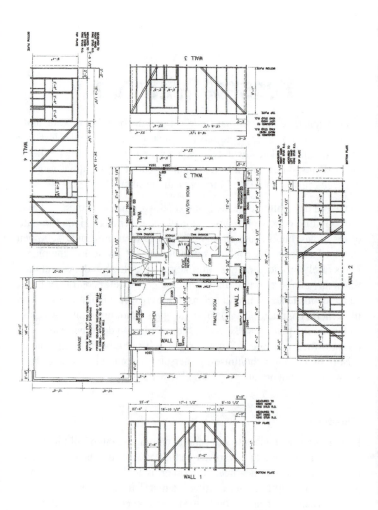

Figure 9.1
Framing Diagram
- Studs, blocking and cross bracing shown on drawings

work. It helps the general contractor to employ workers who understand how their jobs fit into the entire process.

Unfortunately, nowadays it is difficult to find fully trained and knowledgeable trades people. It seems that trades people don't take courses or classes anymore and that apprenticeship programs have fallen by the wayside. The depth of experience in the trades is often missing and the work force is very transient. Even when there has been training, it is often out of date or just plain wrong. In order to get the job done right, the general contractor has to train the trades himself, every day, day in and day out.

The most effective training approach is on-site training, beginning with the first day of work. This should occur during a one-hour period prior to commencing work. "This is what I want and why." "This is how I think you should do it, but I'm open to suggestions on how to do it simpler." "I'm available all day to answer questions."

Training is always easier if you have the right tools. Simple training tools, like visual aids posted on site for easy reference and use, work well. Posters illustrating key details can be developed for framers, electricians, insulators and drywallers.

Detailed framing drawings can be created that illustrate the location of each stud and framing member (Figure 9.1). These drawings can significantly reduce construction time as well as potential confusion. Supervisory time can also be reduced.

The better, the quicker and simpler the training, the better the subcontractors' prices the next time a job is bid.

The detailed framing drawings and the posters developed for the framers, electricians, insulators and drywallers, coupled with checklists and addresses of suppliers, become key elements in developing "competitive pricing" during bid negotiations.

No one likes surprises. The better the information and the better the training, the fewer the surprises. When there are no surprises, you get good prices. Now we're talking.

Materials on the Building Site

Materials should be delivered to the site just before they are to be used. If they are not there, they don't get stolen, rained on, beat-up or destroyed. Of course if materials are not there when you need them, you also have big trouble. "Just-in-time delivery" is more than a slogan, it should be a way of life.

If you have to store materials at the job site, dumping them on the ground in a pond of water in everyone's way is not generally a good idea. Materials should be placed on skids raised off the ground and covered with material to shed water. Materials should be placed in a convenient location. Moving materials more than once is bad planning. A transport trailer makes a good storage shed if you can keep someone from hooking up their truck and pinching it.

Don't get more than you need. If you get more than you need, it's never left over at the end of the job. It somehow disappears. On job sites the law of conservation of mass does not hold. Here's where detailed drawings and detailed take-offs pay for themselves.

Sites should be neat. Sites should be orderly. Sites should be clean. Clean, neat, orderly sites are safe, productive sites. He who makes a mess, cleans the mess. Those of us who were kids once, learned to clean up our toys when we were done. A job site should be run by the same rules.

"Drying in" the building as quickly as possible is a good idea. Roofing felt is relatively inexpensive and can be used to cover everything. Remember that wood loves water. Gypsum board loves water. Mold loves wood and gypsum board when it is wet. Mold is bad.

Materials that come to the job site "flat" and that need to be "flat" when installed should be stored "flat." Roofing shingles should be stored so that the bundles can lie flat without bending. Windows and sliding doors like to be "square." Store them "square." Remember that big piles of building materials can be very heavy. Heavy loads, concentrated on one spot for any appreciable time can be a real problem. Distribute drywall and other heavy loads. Be smart. Think.

Concrete and Excavation

Concrete cracks. Concrete has always cracked. Concrete will always crack. Reinforcing concrete will not prevent it from cracking. It is not possible to build a crack-free concrete slab or foundation wall. However, it is possible to control the cracking process by deliberately cracking the concrete (Figure 10.1). These deliberate cracks are called control joints. Cracks do not naturally occur in straight lines, nor do they happen in predictable places; but control joints cause concrete to crack along straight lines and in predictable locations where builders can better deal with them. Homeowners get annoyed at cracks in concrete, but homeowners don't have a problem with control joints. Cracks are bad. Control joints are good.

Concrete also shrinks, creeps and moves. This is also true for masonry and brick, only more so. Masonry and brick swell when they get wet. Concrete can shrink while brick is swelling. Concrete moves. Masonry moves. Brick moves. Let them move.

When wood gets wet it expands; when it dries it shrinks. Wood is almost always in the process of either getting wet, absorbing moisture or drying. Therefore, wood is almost always moving. Wood has always moved and will always move. Nailing wood, screwing wood and gluing wood will not prevent it from moving. It is not possible to build a wood wall, floor, roof or foundation and not have it move. Let it move.

Mixing wood with concrete, masonry and brick makes building interesting but also makes buildings move. Buildings will always move. It is not possible to build buildings that do not move. Let them move.

Since buildings move, it is not possible to build one without holes. Builders can reduce the number of holes; builders can control the size of holes. Builders can control the types of holes. But make no mistake about it, there will be holes. The trick is to keep the water out, even though you have holes. Fortunately, water is lazy. Water will always

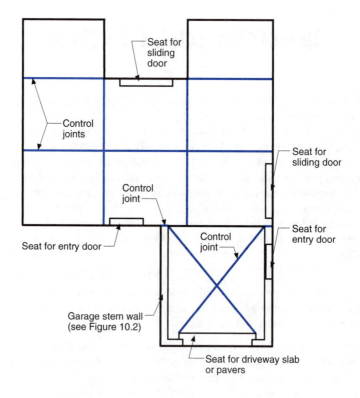

Figure 10.1
Control Joints in Slabs

• Control joint spacing in slab should be a maximum of 12 feet

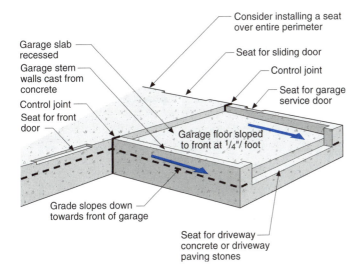

Consider installing a seat over entire perimeter

Garage slab recessed

Garage stem walls cast from concrete

Control joint

Seat for front door

Seat for sliding door

Control joint

Seat for garage service door

Garage floor sloped to front at $1/4$"/ foot

10

Grade slopes down towards front of garage

Seat for driveway concrete or driveway paving stones

Figure 10.2
Garage Foundation Slab

- Garage door sill area recessed to prevent water entry
- Garage service door sill area recessed to prevent water entry
- Stem walls cast from concrete, not constructed from ground contact masonry blocks to prevent capillarity
- Entry door opening in house slab recessed to prevent water entry

Plan View

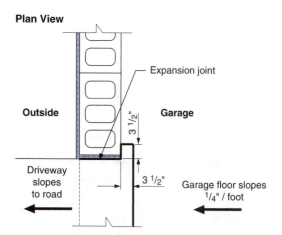

Outside

Garage

Expansion joint

3 1/2"

Driveway
slopes
to road

3 1/2"

Garage floor slopes
1/4" / foot

10

Section View

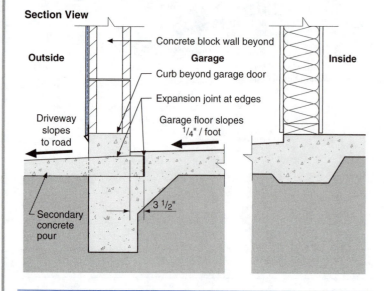

Outside

Garage

Inside

Concrete block wall beyond

Curb beyond garage door

Expansion joint at edges

Driveway
slopes
to road

Garage floor slopes
1/4" / foot

Secondary
concrete
pour

3 1/2"

**Figure 10.3
Garage Floor Door Recess Detail**

Typical Driveway

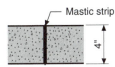

Expansion joint

14'-0"

Control joints @ 4'-0" (max.)

10'-0"

Control joint

Control joint

Control joints @ 6'-0" (max.)

Sidewalk

Apron

Apron

3'-0" Typ.

10

Control Joint

1" deep min.

4"

Expansion Joint

Mastic strip

4"

Figure 10.4
Control Joints and Expansion Joints in Exterior Concrete

- Control joints control shrinkage cracking
- Expansion joints control thermal movement

10

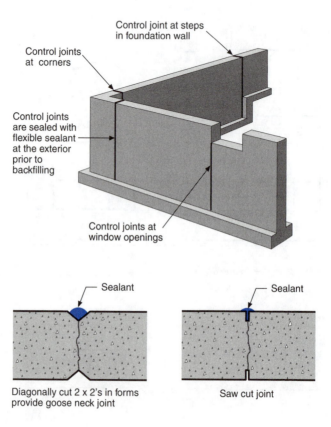

Control joint at steps in foundation wall

Control joints at corners

Control joints are sealed with flexible sealant at the exterior prior to backfilling

Control joints at window openings

Sealant

Sealant

Diagonally cut 2 x 2's in forms provide goose neck joint

Saw cut joint

Figure 10.5
Control Joints in Concrete Foundation Walls

- Control joints should be within 10 feet or corners
- Control joint spacing should be 20 feet maximum

choose the easiest path to travel. If you provide an easy path for water to travel to a foundation drain, it will follow that path rather than a path through a foundation wall, even if the foundation wall has holes. And we know the foundation wall will have holes despite our best efforts.

Polyethylene Under Slabs

A sand layer is sometimes installed over a polyethylene vapor retarder located under a concrete slab. It is thought by some that the sand layer will protect the polyethylene from damage and act as a receptor for excess mix water in the concrete slab when the concrete is cast. This is an extremely bad idea because a reservoir of standing water can be created. If it rains during construction, if a wet cure is used, if excessive irrigation is used or if groundwater rises sufficiently to contact the underside of the polyethylene, water will enter the sand layer and be held in the sand layer by capillary forces, even after the groundwater level drops. Since the polyethylene is between the water-soaked sand and the ground, the only way for the water to get out is up into the building through the concrete slab by diffusion. The wetting of the sand by groundwater can take only minutes, but the drying out may take a decade.

 10

The polyethylene under a concrete slab can function as an effective vapor barrier even if it has holes. It does not need to be protected. The best way to deal with excess mix water is to not have any. Use low water-to-cement ratio concrete with an accelerator or a super-plasticizer. It is faster and easier and, therefore, less expensive. The concrete costs a little more, but the labor is much less.

The rationale for a sand layer is typically as follows:

- Control of cracking by soaking up excess mix water (so use less water and you don't need the sand layer)

- Control of slab curl by soaking up excess mix water (so use less water and you don't need the sand layer)

- Protects poly by reducing punctures (punctures don't matter if air flow doesn't occur — use concrete as an air barrier; so you don't need the sand layer)

- Allows finishing to occur faster by soaking up excess mix water (so use less water and you don't need the sand layer)

We repeat: use less water in concrete, do not use a sand layer, wrap the polyethylene around the slab edge and install a polyethylene capillary break under all plates.

Recommended approach:

- Water-to-cement ratio of 0.45 or less

- Mid-range water reducer (Polyheed, Daracem, Mira)

- Use fly ash; actually required in many regions for sulfate resistance ("Type F" up to 30% may be used)

- Results in lower permeability

- Results in increased sulfate resistance

- Results in reduced shrinkage and cracking

- Results in reduced curl

- Results in increased corrosion resistance

- Strength can be used as surrogate for w/c ratio (4,000 psi) for field verification

You should provide a continuous cure for 72 hours — when you need it, you don't always need it — so pick your spots:

- Necessary for dry windy hot months out west; June through October for CA, AZ, NV

- Dam the slab; pond water on top of slab

- Use burlap and keep it wet; can be covered with polyethylene

- Nothing is better than a continuous wet cure — nothing, nothing, nothing

- Curing compounds don't work; the good ones were banned

- Consider also using control joints or polypropylene mesh to control shrinkage cracking

- To get low water-to-cement ratio, don't just add cement; just adding cement increases shrinkage

- Use combination of mid-range water, reducer, increased cement content and fly ash

- Fly ash replaces cement and doesn't react immediately so that sufficient water is available to finish slab; finishing slabs becomes very difficult with w/c ratios below 0.45; the fly ash kicks in later

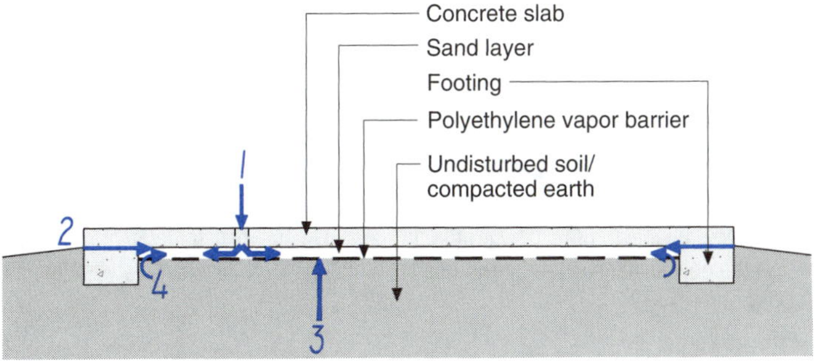

Concrete slab
Sand layer
Footing
Polyethylene vapor barrier
Undisturbed soil/
compacted earth

Figure 10.6
Wetting Mechanisms

- Through top of slab by rain, curing processes, cutting, finishing, cleaning, water testing, etc. (**1**)
- Through slab/footing interface by irrigation water, surface water and groundwater (**2**)
- Through joints, penetrations and punctures in polyethylene vapor barrier by groundwater (**3**)
- Around the edge of polyethylene vapor barrier at footing (**4**)

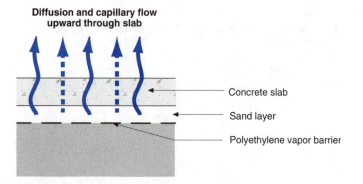

Figure 10.7
Water Distribution in Sand Layers

- Water migrates laterally; pulled by capillary action
- Water is unable to drain downwards through any gaps, penetrations, or punches in the polyethylene vapor barrier due to "blotter paper" effect (capillary forces) of the sand layer

Diffusion and capillary flow upward through slab

Concrete slab

Sand layer

Polyethylene vapor barrier

Figure 10.8
Upward Drying Through Slab

- Sand layer becomes water reservoir to supply water for upward flow through concrete slab by vapor diffusion and capillary transmission
- Drying downward by vapor diffusion or capillary is not possible due to polyethylene vapor barrier

Wood Frame Construction

Three framing approaches are common in residential construction: platform frame, balloon frame, and post and beam. They can use combinations of wood, wood products, steel, masonry and concrete. They can be insulated on the inside, on the outside or in between. However, all have to:

- hold the building up
- keep the rainwater out
- keep the wind out
- keep the water vapor out
- let the water and water vapor out if it gets inside
- keep the heat in during the winter
- keep the heat out during the summer

Concerns

If someone today invented wood, it would never be approved as a building material. It burns, it rots, it has different strength properties depending on its orientation, no two pieces are alike, and most cruelly of all, it expands and contracts based on the relative humidity around it. However, despite all of these problems, wood is the material of choice when building houses. In fact, we can use wood better than we can use steel, masonry and concrete.

We can compensate more easily for wood's poor qualities than for those of steel, masonry and concrete. Steel is worse in a fire than wood because it twists and bends; steel rusts more easily than wood rots; and it is expensive and difficult to compensate for its thermal inefficiencies in an exterior wall. Masonry and concrete can be expensive in many applications to use structurally. In an earthquake, the most dangerous buildings are made of insufficiently reinforced masonry and concrete. Needless to say, it is expensive to reinforce masonry and concrete.

We are a wood dominated industry and will likely remain so. We have learned how to work with wood over the past several hundred years to overcome its inherent deficiencies. However, despite our vast experience, we must use wood better.

We use wood inefficiently. We put too much of it in our buildings in the wrong places and in the wrong ways. And then, in the presence of all this waste, we don't put it where it is needed.

Not enough wood in the right place is obviously a problem. Oops, it fell down. But how can too much wood hurt? Well, wherever you put wood, you can't put insulation. Where you don't have insulation, you have cold spots or hot spots. Cold spots and hot spots always cause trouble. Wood is also expensive; use too much of it and you hurt your wallet.

Frame Movement

Remember, wood always moves. How to attach things to something that is always moving becomes real important. Gypsum board doesn't crack all by itself. It cracks because what it is attached to moves more than the gypsum can. Use too much wood in the wrong places, use too much attachment of gypsum board to wood that is moving in the wrong places, and presto, you have cracks. Lots of them. It's better to use less wood with fewer attachments. One of life's least appreciated ironies is the more you attach gypsum board to wood, the more cracks you get. Longer nails, more nail pops. More nails, more cracks. More about this in the Drywall Chapter (see Chapter 17). In the meantime, in this chapter, we will show how to eliminate as much wood as possible, by making sure we put it only where we really need it — for structure, draftstopping and firestopping.

In the past, framers have used wood much like drunken sailors on leave spend money. They felt they could never run out of wood and they put it everywhere, even where it was not needed. We do still. Look around your job site. Headers can be found in non-load bearing interior and exterior walls. Double plates are everywhere because we have not taken the time to figure out how to line up roof framing with wall and floor framing. Three stud corners to support gypsum board that doesn't need or want support. Cripples under window framing even though we hang windows. Studs on 16 in. centers rather than 24 in. centers to support gypsum board and siding that don't need it. Figures 11.1 through 11.5 and Figure 11.10 describe framing techniques that reduce wood waste, increase structural efficiency, promote thermally efficient walls and help reduce drywall cracking.

But where we really need wood, for draftstopping and firestopping we can't find it.

Rain

Another of life's ironies is that the strategy selected to keep rain out of a building will affect the how it is framed. So, maybe we should decide what the strategy is before we frame. Framers are responsible for installing building paper, sheathing and windows. The rain control strategy will decide whether building papers will be used or not, whether sheathing will be taped or glued or both, and whether window openings will be wrapped or not. Figures 11.59 through 11.62 show important flashing details when using taped insulating sheathing as a drainage plane. Keeping rain out of buildings is discussed in Chapter 2.

Air Barrier

The strategy selected to keep outside and inside air out of the building enclosure will also affect the framing approach. Framers are responsible for installing exterior housewraps, insulating sheathings and maintaining the continuity of polyethylene air barriers (when used as the air barrier system), as well as the draftstops, firestops and framing used in rigid interior air barriers. Figures 11.6 through 11.9 and 11.11 through 11.19 show installation techniques for installing exterior insulating sheathings as exterior air barriers. Figures 11.20 through 11.46 show important air sealing details (draftstopping and firestopping) that the framer must provide. Air barriers are discussed in Chapter 3.

Moisture

The strategy selected to keep water vapor out of the building enclosure yet allow water vapor out of the building enclosure should it get in, will affect the framing approach since the approach selected will specify the type of sheathing used. Sheathings and vapor retarders are discussed in Chapter 4. In addition, if a ventilated roof strategy is to be employed, the framer must install roofing members so that air can in fact flow from soffits to vents (Figures 11.47 through 11.54).

Paint and Trim

The manner in which wood siding and wood trim is installed determines the useful service life of paint and stain coatings. Wood siding and trim should always be coated on all six surfaces and should always be installed over spacers to promote drainage and drying (Figures 11.56 and 11.58). Paints and coatings are discussed in Chapter 18 Painting.

11

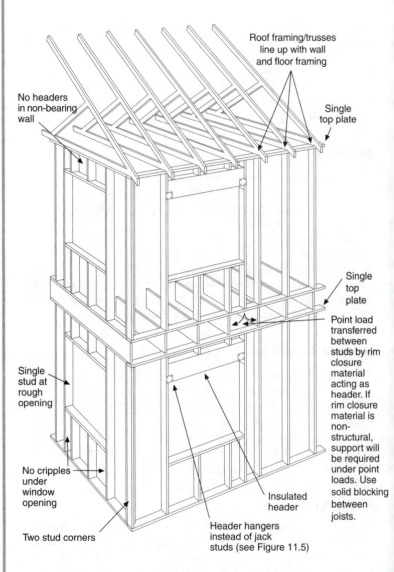

Roof framing/trusses line up with wall and floor framing

No headers in non-bearing wall

Single top plate

Single top plate

Point load transferred between studs by rim closure material acting as header. If rim closure material is non-structural, support will be required under point loads. Use solid blocking between joists.

Single stud at rough opening

No cripples under window opening

Two stud corners

Insulated header

Header hangers instead of jack studs (see Figure 11.5)

Figure 11.1
Stack Framing

- Eliminate headers in non-bearing interior walls
- Headers not needed for openings less than 4'-wide in non-load bearing exterior walls

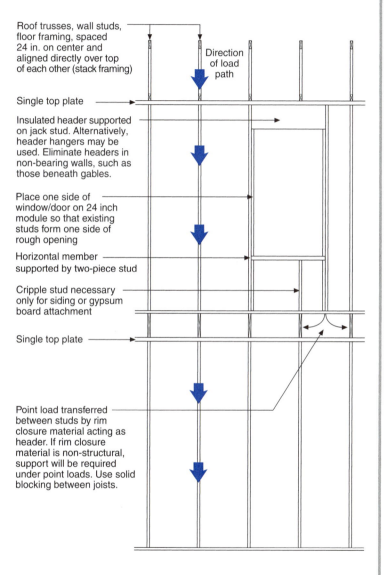

Roof trusses, wall studs, floor framing, spaced 24 in. on center and aligned directly over top of each other (stack framing)

Direction of load path

Single top plate

Insulated header supported on jack stud. Alternatively, header hangers may be used. Eliminate headers in non-bearing walls, such as those beneath gables.

Place one side of window/door on 24 inch module so that existing studs form one side of rough opening

Horizontal member supported by two-piece stud

Cripple stud necessary only for siding or gypsum board attachment

Single top plate

Point load transferred between studs by rim closure material acting as header. If rim closure material is non-structural, support will be required under point loads. Use solid blocking between joists.

Figure 11.2
Stack Framing Elevation View
- Where single plates are used, floor to ceiling heights are affected (97" is standard). Custom cutting dimensional studs to 94" is recommended and results in no impact on gypsum board installation.

Corner framing

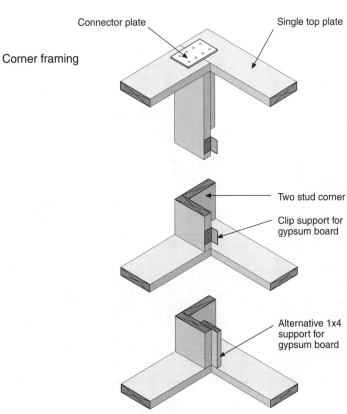

Top plate splice

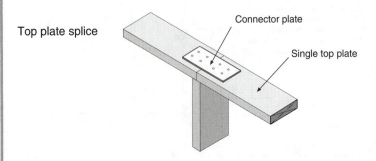

Figure 11.3
Corner Framing and Top Plate Splice

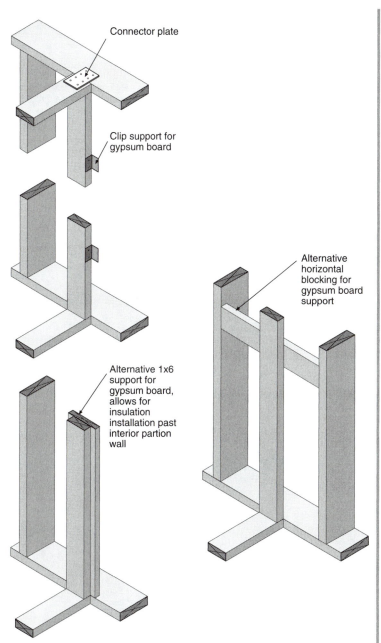

Connector plate

Clip support for gypsum board

Alternative horizontal blocking for gypsum board support

Alternative 1x6 support for gypsum board, allows for insulation installation past interior partion wall

 11

Figure 11.4
Interior Wall at Exterior Wall
• See also Figures 17.3 and 17.7

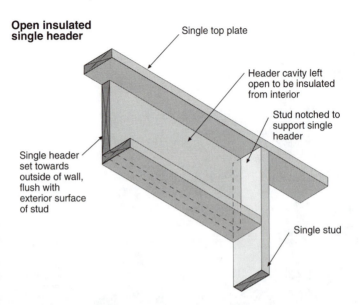

Open insulated single header

Single top plate

Header cavity left open to be insulated from interior

Stud notched to support single header

Single header set towards outside of wall, flush with exterior surface of stud

Single stud

Open insulated double header

Single top plate

Double header with rigid insulation towards inside prevents gypsum board from cracking due to header shrinkage

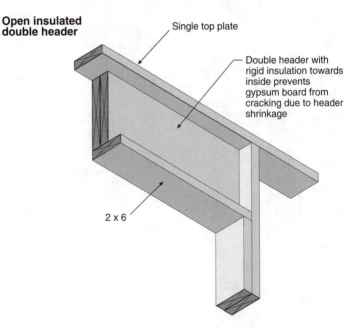

2 x 6

Figure 11.5
Insulated Headers

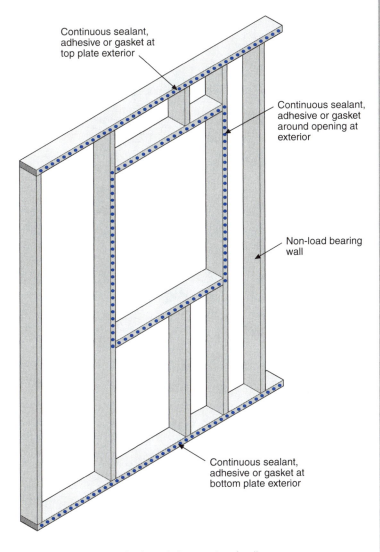

Continuous sealant, adhesive or gasket at top plate exterior

Continuous sealant, adhesive or gasket around opening at exterior

Non-load bearing wall

Continuous sealant, adhesive or gasket at bottom plate exterior

Exterior rigid insulation is sealed to exterior of wall framing at top plates, bottom plates and around openings.

Figure 11.6
Exterior Rigid Insulation Air Sealing on a Non-Load Bearing Wall
• A header is needed above window opening if wall is load bearing; see Figure 11.7.

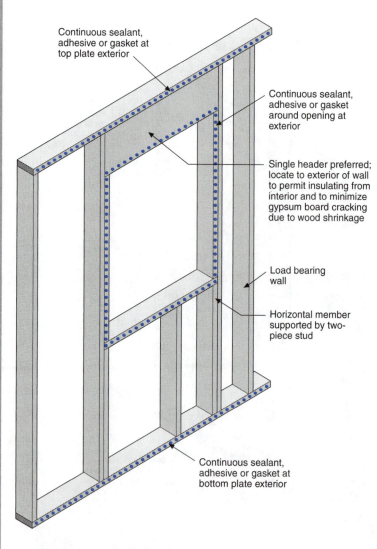

Continuous sealant, adhesive or gasket at top plate exterior

Continuous sealant, adhesive or gasket around opening at exterior

Single header preferred; locate to exterior of wall to permit insulating from interior and to minimize gypsum board cracking due to wood shrinkage

Load bearing wall

Horizontal member supported by two-piece stud

Continuous sealant, adhesive or gasket at bottom plate exterior

Exterior rigid insulation is sealed to exterior of wall framing at top plates, bottom plates and around openings.

Figure 11.7
Exterior Rigid Insulation Air Sealing on a Load Bearing Wall
- A header is needed above window opening if wall is load bearing
- Single header rather than double header preferred, sized appropriately for load

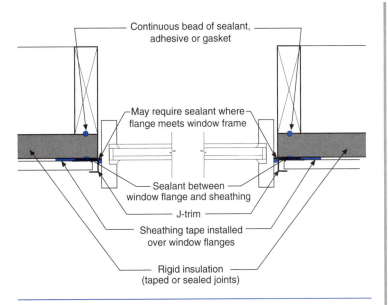

Continuous bead of sealant, adhesive or gasket

May require sealant where flange meets window frame

Sealant between window flange and sheathing

J-trim

Sheathing tape installed over window flanges

Rigid insulation (taped or sealed joints)

**Figure 11.8
Window/Door Jamb Detail for Rigid Insulation**

 11

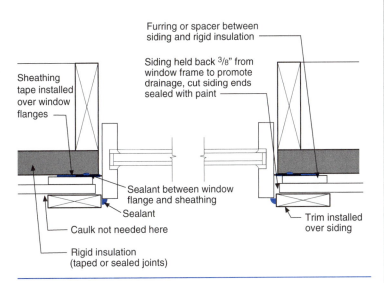

Furring or spacer between siding and rigid insulation

Siding held back $3/8$" from window frame to promote drainage, cut siding ends sealed with paint

Sheathing tape installed over window flanges

Sealant between window flange and sheathing

Sealant

Caulk not needed here

Trim installed over siding

Rigid insulation (taped or sealed joints)

**Figure 11.9
Wood Trim Detail**

• Wood window trim nailed over siding to promote water drainage and drying

253

Each wall should have pairs of cross braces,
crossing from top to bottom in opposite directions.

Cross bracing tied into top and bottom plates

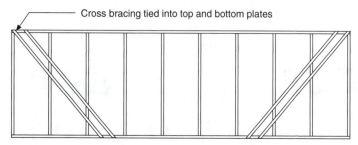

Wrapping metal braces over top of top plates
and under bottom plates and fastening down
into top plate or up into bottom plate
significantly improves shear resistance

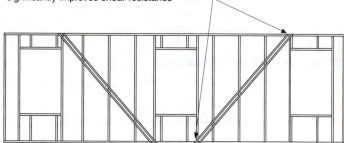

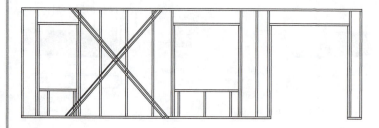

Figure 11.10
Cross Bracing

- Structural requirements and capacity should be determined on a case-by-case basis
- Interior partitions can also be designed to provide shear resistance

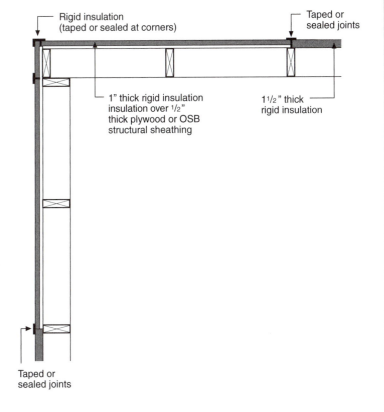

Rigid insulation
(taped or sealed at corners)

Taped or
sealed joints

1" thick rigid insulation
insulation over 1/2"
thick plywood or OSB
structural sheathing

1 1/2" thick
rigid insulation

Taped or
sealed joints

**Figure 11.11
Rigid Insulation Over Structural Sheathing at Corners**

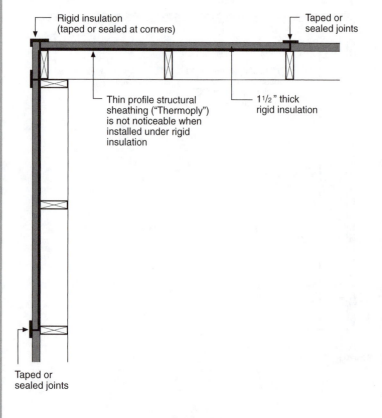

Rigid insulation
(taped or sealed at corners)

Taped or
sealed joints

Thin profile structural
sheathing ("Thermoply")
is not noticeable when
installed under rigid
insulation

1¹/₂" thick
rigid insulation

Taped or
sealed joints

Figure 11.12
Thin Profile Structural Sheathing at Corners

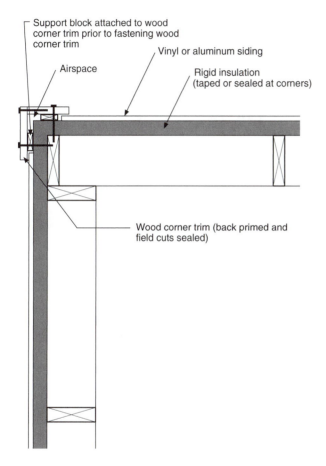

Support block attached to wood corner trim prior to fastening wood corner trim

Airspace

Vinyl or aluminum siding

Rigid insulation (taped or sealed at corners)

Wood corner trim (back primed and field cuts sealed)

 11

Figure 11.13
Exterior Corner Trim Detail for Rigid Insulation — Vinyl or Aluminum Siding with Wood Trim

- Vinyl or aluminum siding
- Trim back primed; all field cut ends in trim sealed with paint

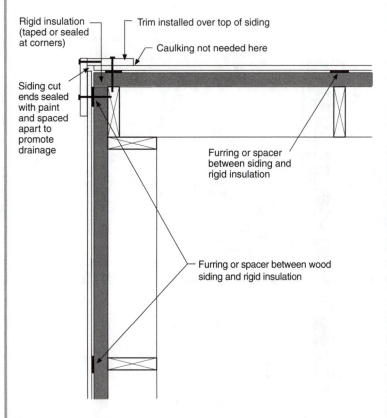

Figure 11.14
Exterior Corner Trim Detail for Rigid Insulation — Wood Siding
- Siding and trim back primed; all field cut ends in siding and trim sealed with paint
- Furring strip can be cut strip of $^3/_8$" pressure treated plywood, wood lath, or 1x4

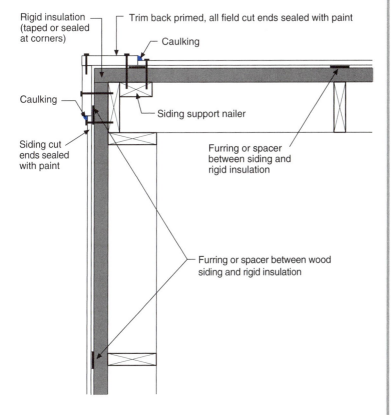

Rigid insulation (taped or sealed at corners)

Trim back primed, all field cut ends sealed with paint

Caulking

Caulking

Siding support nailer

Siding cut ends sealed with paint

Furring or spacer between siding and rigid insulation

Furring or spacer between wood siding and rigid insulation

 11

Figure 11.15
Exterior Corner Trim Detail for Rigid Insulation — Wood Siding
- Siding and trim back primed; all field cut ends in siding and trim sealed with paint
- Furring strip can be cut strip of $^3/_8$" pressure treated plywood, wood lath, or 1x4

11

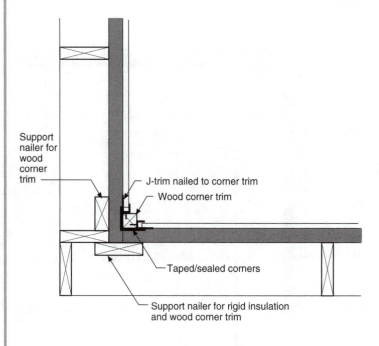

Support nailer for wood corner trim

J-trim nailed to corner trim

Wood corner trim

Taped/sealed corners

Support nailer for rigid insulation and wood corner trim

Figure 11.16
Interior Corner Trim Detail for Rigid Insulation — Vinyl or Aluminum Siding with Wood Trim

- Vinyl or aluminum siding
- Wood trim back primed; all field cut ends in wood trim sealed with paint
- Support nailers not needed if J-trim is attached to wood corner trim

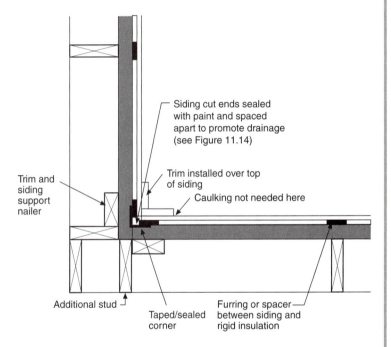

11

Figure 11.17
Interior Corner Trim Detail for Rigid Insulation — Wood Siding
- Rigid insulation installed after wall erection
- Siding and trim back primed; all field cut ends in siding and trim sealed with paint
- Furring strip can be cut strip of $^3/_8$" pressure treated plywood, wood lath, or 1x4

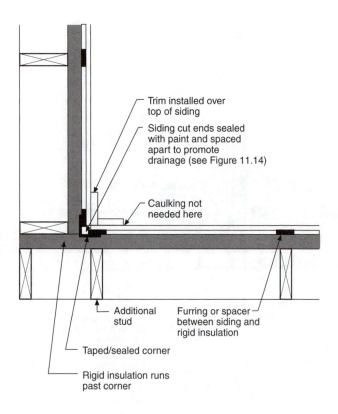

Trim installed over
top of siding

Siding cut ends sealed
with paint and spaced
apart to promote
drainage (see Figure 11.14)

Caulking not
needed here

Additional
stud

Furring or spacer
between siding and
rigid insulation

Taped/sealed corner

Rigid insulation runs
past corner

Figure 11.18
Interior Corner Trim Detail for Rigid Insulation — Wood Siding
- Rigid insulation installed prior to wall erection
- Siding and trim back primed; all field cut ends in siding and trim sealed with paint
- Furring strip can be cut strip of $3/8$" pressure treated plywood, wood lath, or 1x4

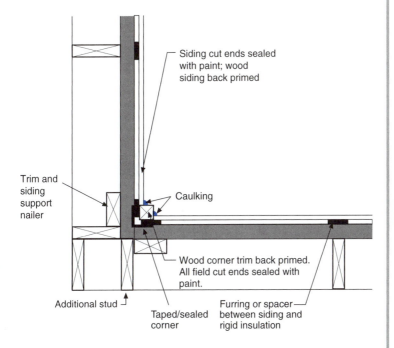

11

Figure 11.19
Interior Corner Trim Detail for Rigid Insulation — Wood Siding

- Siding and trim back primed; all field cut ends in siding and trim sealed with paint
- Furring strip can be cut strip of $^3/_8$" pressure treated plywood, wood lath, or 1x4

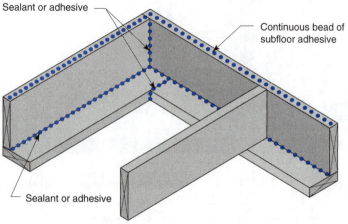

Sealant or adhesive

Continuous bead of
subfloor adhesive

Sealant or adhesive

Figure 11.20
Rim Joist/Band Joist Floor Truss Rim Closure
• Gaskets can be used in place of sealants or adhesives

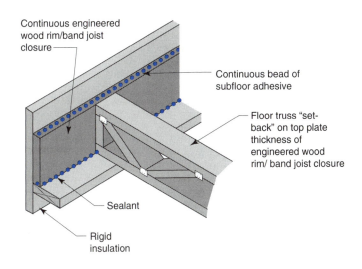

Continuous engineered wood rim/band joist closure

Continuous bead of subfloor adhesive

Floor truss "set-back" on top plate thickness of engineered wood rim/ band joist closure

Sealant

Rigid insulation

Figure 11.21
Floor Truss Rim Closure
- Additional batt insulation can be added to interior

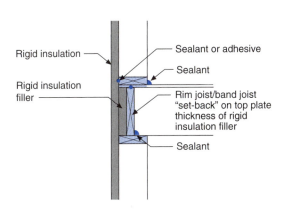

Rigid insulation

Sealant or adhesive

Sealant

Rigid insulation filler

Rim joist/band joist "set-back" on top plate thickness of rigid insulation filler

Sealant

Figure 11.22
Set-Back Rim Joist
- This approach may eliminate the need for additional batt insulation to interior

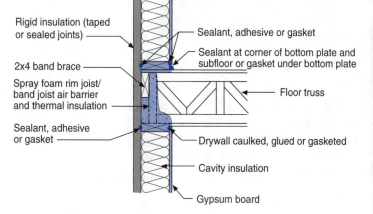

Rigid insulation (taped or sealed joints)

Sealant, adhesive or gasket

Sealant at corner of bottom plate and subfloor or gasket under bottom plate

2x4 band brace

Spray foam rim joist/ band joist air barrier and thermal insulation

Floor truss

Sealant, adhesive or gasket

Drywall caulked, glued or gasketed

Cavity insulation

Gypsum board

Figure 11.23
Band Brace/Floor Truss Closure
• Spray foam is used to provide air barrier continuity across band brace region

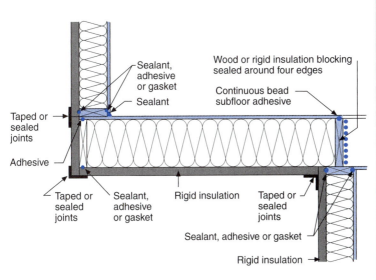

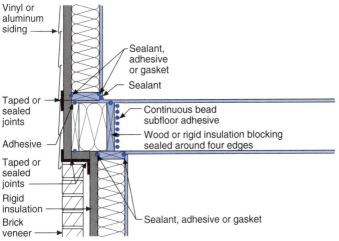

Figure 11.24
Cantilevered Floors

• Floor cavity insulation installed by framers prior to rigid insulation installation

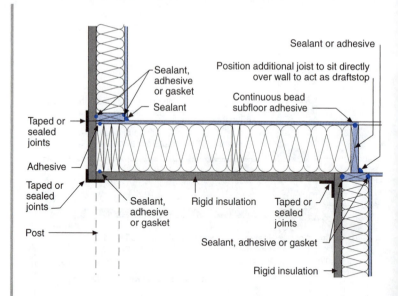

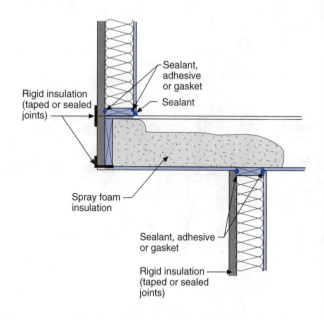

Figure 11.25
Cantilevered Floors
- Floor cavity insulation installed by framers prior to rigid insulation installation
- Alternatively, spray foam insulation can be installed from the interior

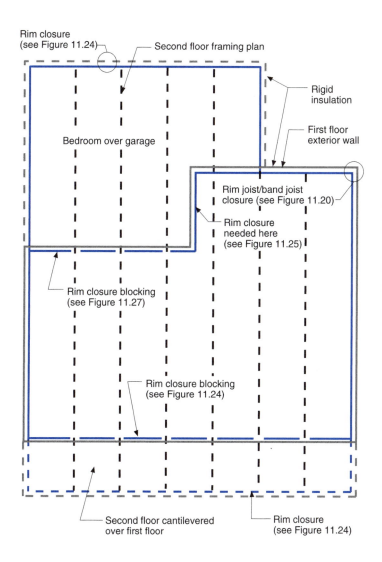

Rim closure
(see Figure 11.24)

Second floor framing plan

Rigid
insulation

First floor
exterior wall

Bedroom over garage

Rim joist/band joist
closure (see Figure 11.20)

Rim closure
needed here
(see Figure 11.25)

Rim closure blocking
(see Figure 11.27)

 11

Rim closure blocking
(see Figure 11.24)

Second floor cantilevered
over first floor

Rim closure
(see Figure 11.24)

Figure 11.26
Critical Rim Joist/Band Joist Closure Areas

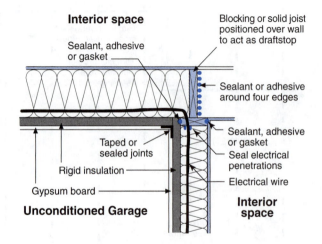

Interior space

Sealant, adhesive or gasket

Blocking or solid joist positioned over wall to act as draftstop

Sealant or adhesive around four edges

Taped or sealed joints

Rigid insulation

Gypsum board

Sealant, adhesive or gasket

Seal electrical penetrations

Electrical wire

Unconditioned Garage

Interior space

Figure 11.27
Floor Over Garage
- Floor cavity insulation installed prior to rigid insulation installation by framers or insulators
- Electrical wires passing to exterior should be sealed with foam sealant or caulking

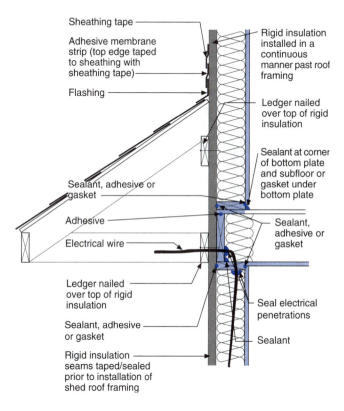

Sheathing tape

Adhesive membrane strip (top edge taped to sheathing with sheathing tape)

Flashing

Sealant, adhesive or gasket

Adhesive

Electrical wire

Ledger nailed over top of rigid insulation

Sealant, adhesive or gasket

Rigid insulation seams taped/sealed prior to installation of shed roof framing

Rigid insulation installed in a continuous manner past roof framing

Ledger nailed over top of rigid insulation

Sealant at corner of bottom plate and subfloor or gasket under bottom plate

Sealant, adhesive or gasket

Seal electrical penetrations

Sealant

Figure 11.28
Shed Roof
- Electrical wires passing to exterior should be sealed with foam sealant or caulking

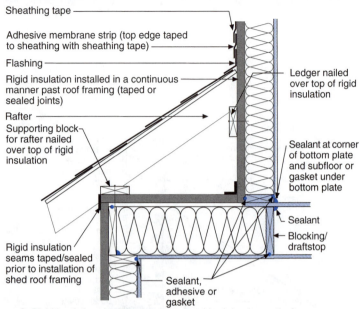

Sheathing tape

Adhesive membrane strip (top edge taped to sheathing with sheathing tape)

Flashing

Rigid insulation installed in a continuous manner past roof framing (taped or sealed joints)

Rafter

Supporting block for rafter nailed over top of rigid insulation

Rigid insulation seams taped/sealed prior to installation of shed roof framing

Ledger nailed over top of rigid insulation

Sealant at corner of bottom plate and subfloor or gasket under bottom plate

Sealant

Blocking/draftstop

Sealant, adhesive or gasket

A: Rigid insulating sheathing is used to provide air barrier continuity

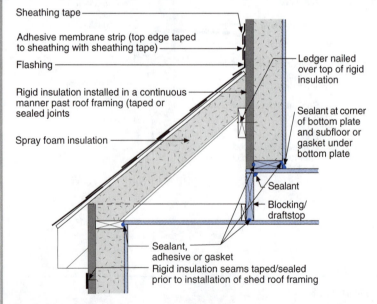

Sheathing tape

Adhesive membrane strip (top edge taped to sheathing with sheathing tape)

Flashing

Rigid insulation installed in a continuous manner past roof framing (taped or sealed joints

Spray foam insulation

Ledger nailed over top of rigid insulation

Sealant at corner of bottom plate and subfloor or gasket under bottom plate

Sealant

Blocking/draftstop

Sealant, adhesive or gasket

Rigid insulation seams taped/sealed prior to installation of shed roof framing

B: Spray foam insulation is used to provide air barrier continuity

Figure 11.29
Set-Back Roof

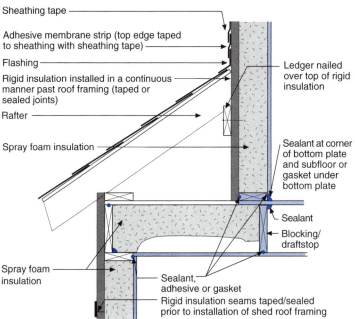

Sheathing tape

Adhesive membrane strip (top edge taped to sheathing with sheathing tape)

Flashing

Rigid insulation installed in a continuous manner past roof framing (taped or sealed joints)

Rafter

Spray foam insulation

Ledger nailed over top of rigid insulation

Sealant at corner of bottom plate and subfloor or gasket under bottom plate

Sealant

Blocking/ draftstop

Spray foam insulation

Sealant, adhesive or gasket

Rigid insulation seams taped/sealed prior to installation of shed roof framing

C: Spray foam insulation is used to provide air barrier continuity

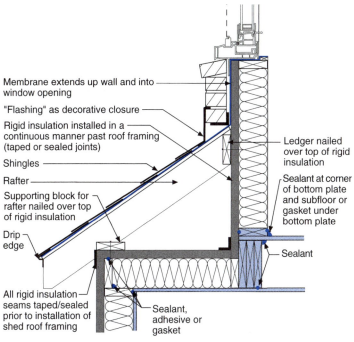

Membrane extends up wall and into window opening

"Flashing" as decorative closure

Rigid insulation installed in a continuous manner past roof framing (taped or sealed joints)

Shingles

Rafter

Supporting block for rafter nailed over top of rigid insulation

Drip edge

All rigid insulation seams taped/sealed prior to installation of shed roof framing

Ledger nailed over top of rigid insulation

Sealant at corner of bottom plate and subfloor or gasket under bottom plate

Sealant

Sealant, adhesive or gasket

D: Membrane extends up wall and into window opening

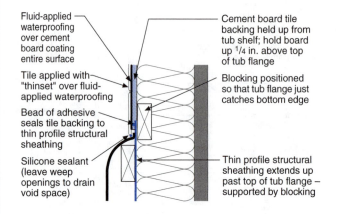

Fluid-applied waterproofing over cement board coating entire surface

Tile applied with "thinset" over fluid-applied waterproofing

Bead of adhesive seals tile backing to thin profile structural sheathing

Silicone sealant (leave weep openings to drain void space)

Cement board tile backing held up from tub shelf; hold board up 1/4 in. above top of tub flange

Blocking positioned so that tub flange just catches bottom edge

Thin profile structural sheathing extends up past top of tub flange – supported by blocking

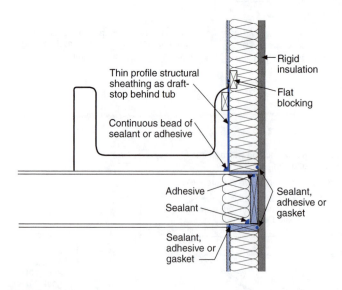

Thin profile structural sheathing as draft-stop behind tub

Continuous bead of sealant or adhesive

Rigid insulation

Flat blocking

Adhesive

Sealant

Sealant, adhesive or gasket

Sealant, adhesive or gasket

Figure 11.30
Tub Framing – Section

- Flat blocking allows cavity insulation to be installed behind tub draftstop
- Cement board tile backing is recommended in place of "green board."
- Cement board is not waterproof, it must be coated with a fluid applied waterproofing
- Installing "tar paper" behind cement board is not recommended as there is no provision to drain penetrating shower water back into the tub enclosure

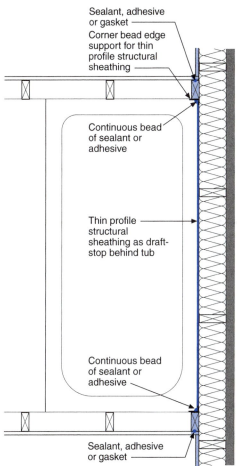

Sealant, adhesive or gasket

Corner bead edge support for thin profile structural sheathing

Continuous bead of sealant or adhesive

Thin profile structural sheathing as draft-stop behind tub

Continuous bead of sealant or adhesive

Sealant, adhesive or gasket

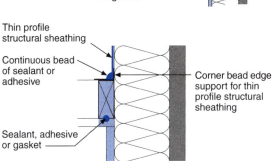

Thin profile structural sheathing

Continuous bead of sealant or adhesive

Corner bead edge support for thin profile structural sheathing

Sealant, adhesive or gasket

Figure 11.31
Tub Framing – Plan

275

Construction

Tub Framing – Interior Elevation

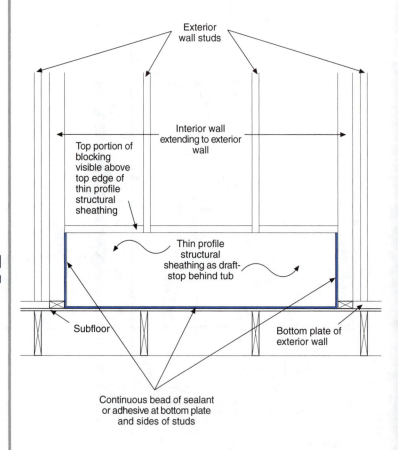

**Figure 11.32
Tub Framing – Interior Elevation**

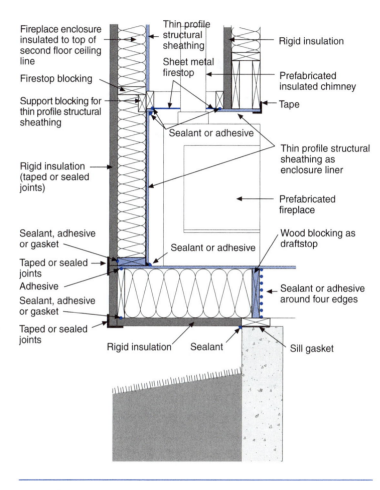

Fireplace enclosure insulated to top of second floor ceiling line

Thin profile structural sheathing

Rigid insulation

Firestop blocking

Sheet metal firestop

Prefabricated insulated chimney

Support blocking for thin profile structural sheathing

Tape

Sealant or adhesive

Thin profile structural sheathing as enclosure liner

Rigid insulation (taped or sealed joints)

Prefabricated fireplace

Sealant, adhesive or gasket

Sealant or adhesive

Wood blocking as draftstop

Taped or sealed joints

Adhesive

Sealant, adhesive or gasket

Sealant or adhesive around four edges

Taped or sealed joints

Rigid insulation   Sealant

Sill gasket

Figure 11.33
Fireplace Section

- Clearances around chimney to be determined by manufacturer's recommendations and local codes.
- Exterior combustion air with a damper should be provided to all fireboxes.
- Ideally, chimneys should be installed within the interior of the building enclosure. Alternatively, chimney enclosures should be insulated full height to keep chimney flue pipes warm to ensure sufficient draft during the "die-down" stages of a fire. Insulated chimney flues are preferred.
- Use of sealed combustion, direct vent gas fireplaces eliminate the need for chimneys

11

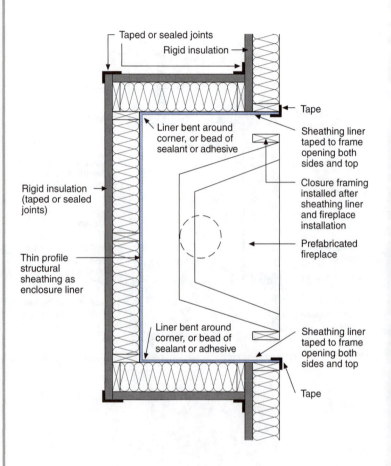

Figure 11.34
Fireplace Section

- Clearances around chimney to be determined by manufacturer's recommendations and local codes.
- Exterior combustion air with a damper should be provided to all fireboxes.
- Ideally, chimneys should be installed within the interior of the building enclosure. Alternatively, chimney enclosures should be insulated full height to keep chimney flue pipes warm to ensure sufficient draft during the "die-down" stages of a fire. Insulated chimney flues are preferred.
- Use of sealed combustion, direct vent gas fireplaces eliminate the need for chimneys

Plan

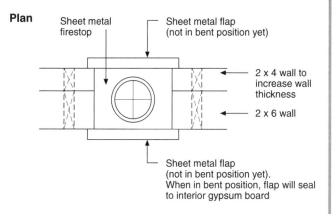

Sheet metal firestop

Sheet metal flap (not in bent position yet)

2 x 4 wall to increase wall thickness

2 x 6 wall

Sheet metal flap (not in bent position yet). When in bent position, flap will seal to interior gypsum board

Section

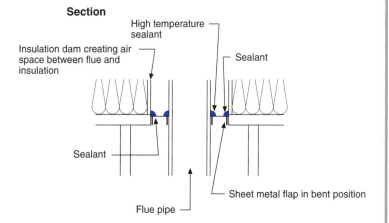

High temperature sealant

Insulation dam creating air space between flue and insulation

Sealant

Sealant

Flue pipe

Sheet metal flap in bent position

Figure 11.35
Flue Closure
 • Interior gypsum board sealed with adhesive to sheet metal flap or firestop

11

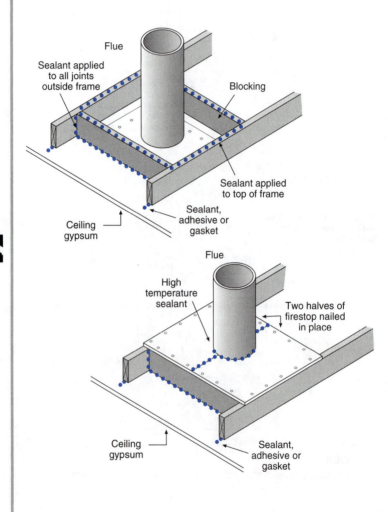

Flue

Sealant applied
to all joints
outside frame

Blocking

Sealant applied
to top of frame

Sealant,
adhesive or
gasket

Ceiling
gypsum

Flue

High
temperature
sealant

Two halves of
firestop nailed
in place

Ceiling
gypsum

Sealant,
adhesive or
gasket

Figure 11.36
Alternative Flue Closure

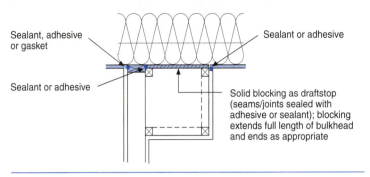

Sealant, adhesive
or gasket

Sealant or adhesive

Sealant or adhesive

Solid blocking as draftstop
(seams/joints sealed with
adhesive or sealant); blocking
extends full length of bulkhead
and ends as appropriate

**Figure 11.37
Interior Soffit**

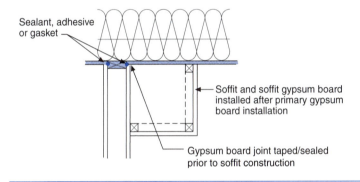

Sealant, adhesive
or gasket

Soffit and soffit gypsum board
installed after primary gypsum
board installation

Gypsum board joint taped/sealed
prior to soffit construction

**Figure 11.38
Interior Soffit Constructed After Gypsum Board Installed**

• Soffit framed after gypsum board installed in order to provide air barrier
continuity

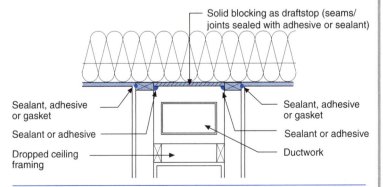

Solid blocking as draftstop (seams/
joints sealed with adhesive or sealant)

Sealant, adhesive
or gasket

Sealant, adhesive
or gasket

Sealant or adhesive

Sealant or adhesive

Dropped ceiling
framing

Ductwork

**Figure 11.39
Dropped Ceiling**

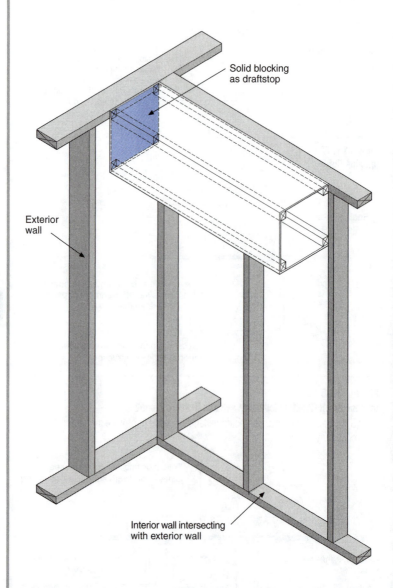

Solid blocking as draftstop

Exterior wall

Interior wall intersecting with exterior wall

Figure 11.40
Interior Soffit Footprint Against Exterior Wall
- Alternatively, interior soffit can be constructed after gypsum board installation (see Figure 11.38)

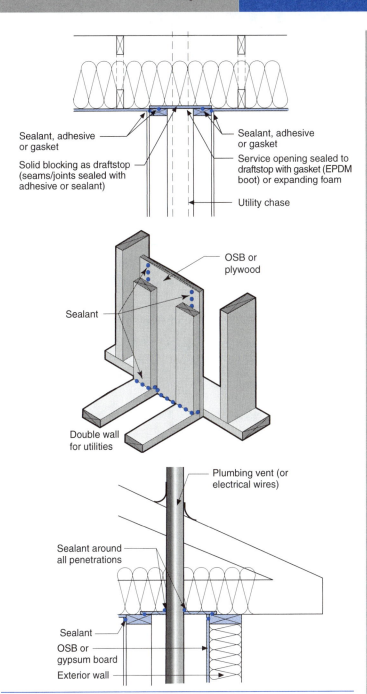

Sealant, adhesive or gasket

Solid blocking as draftstop (seams/joints sealed with adhesive or sealant)

Sealant, adhesive or gasket

Service opening sealed to draftstop with gasket (EPDM boot) or expanding foam

Utility chase

OSB or plywood

Sealant

Double wall for utilities

Plumbing vent (or electrical wires)

Sealant around all penetrations

Sealant

OSB or gypsum board

Exterior wall

Figure 11.41
Utility Chase

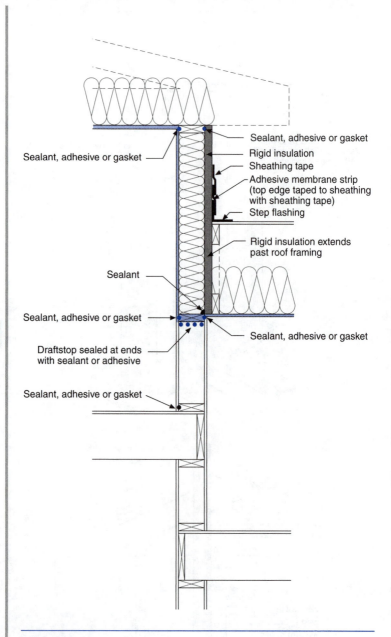

Sealant, adhesive or gasket

Sealant, adhesive or gasket

Rigid insulation

Sheathing tape

Adhesive membrane strip (top edge taped to sheathing with sheathing tape)

Step flashing

Rigid insulation extends past roof framing

Sealant

Sealant, adhesive or gasket

Sealant, adhesive or gasket

Draftstop sealed at ends with sealant or adhesive

Sealant, adhesive or gasket

Figure 11.42
Split Level

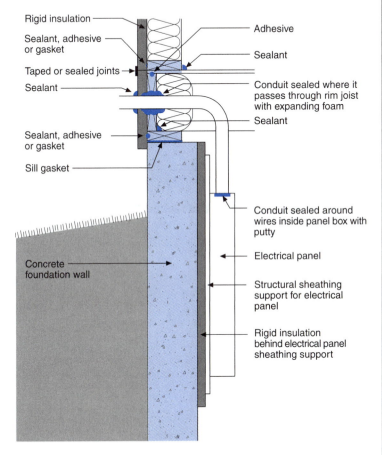

Rigid insulation

Sealant, adhesive or gasket

Taped or sealed joints

Sealant

Sealant, adhesive or gasket

Sill gasket

Concrete foundation wall

Adhesive

Sealant

Conduit sealed where it passes through rim joist with expanding foam

Sealant

Conduit sealed around wires inside panel box with putty

Electrical panel

Structural sheathing support for electrical panel

Rigid insulation behind electrical panel sheathing support

Figure 11.43
Electrical Panel
- Rigid insulation installed under electrical panel as a thermal break and to provide continuity for interior basement insulation

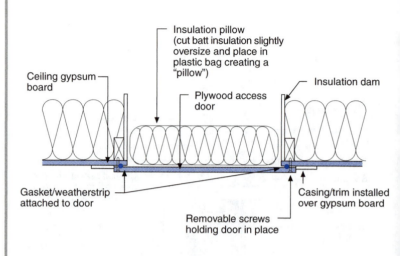

Figure 11.44
Attic Access or Removable Cover for Whole House Ventilation Fan

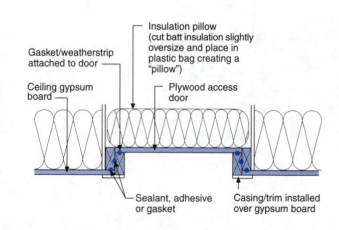

Figure 11.45
Attic Access via Scuttlehole

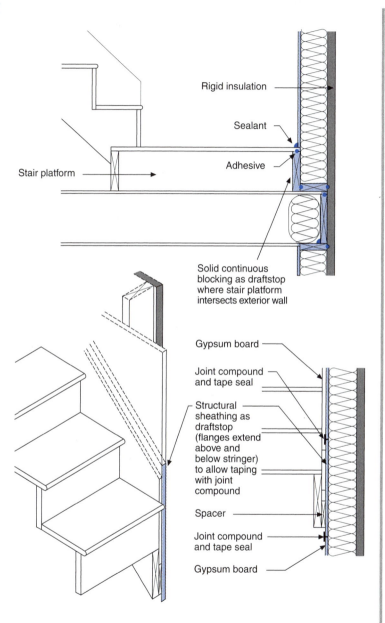

Figure 11.46
Stairs at Exterior Wall or Garage Wall
- Alternatively, stairs can be held away from wall framing to allow gypsum board to be installed in continuous manner between stairs and wall framing

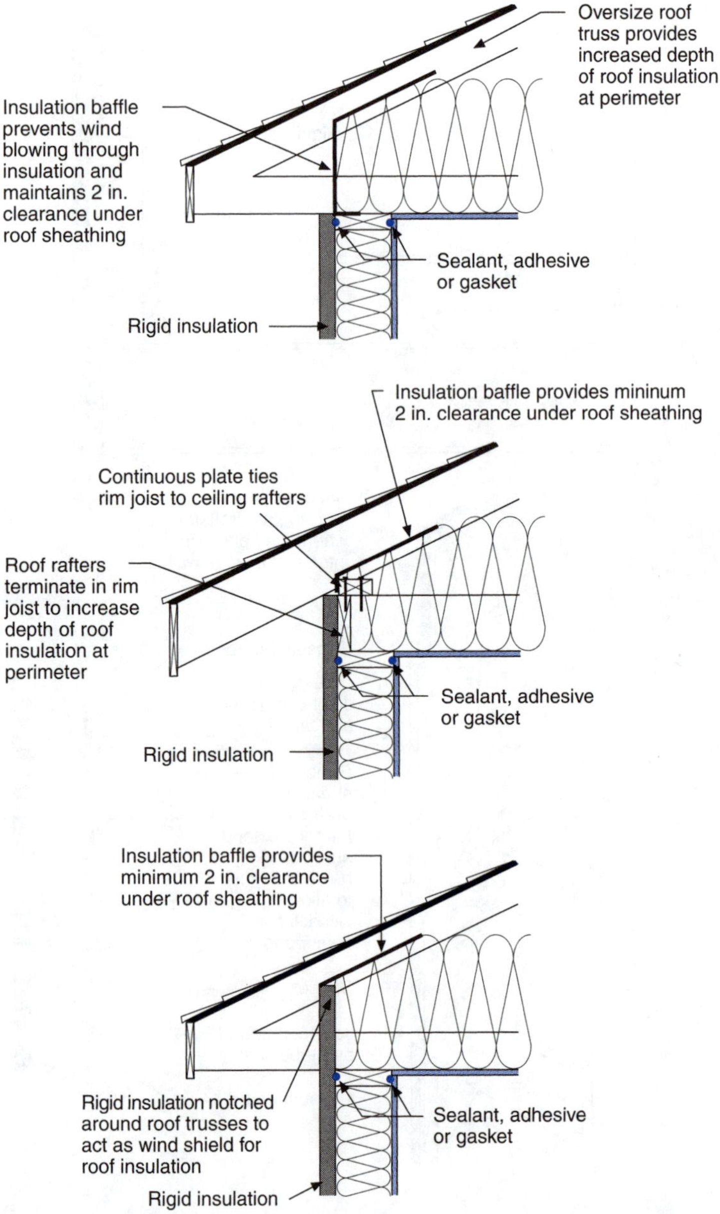

Oversize roof truss provides increased depth of roof insulation at perimeter

Insulation baffle prevents wind blowing through insulation and maintains 2 in. clearance under roof sheathing

Sealant, adhesive or gasket

Rigid insulation

Insulation baffle provides minimum 2 in. clearance under roof sheathing

Continuous plate ties rim joist to ceiling rafters

Roof rafters terminate in rim joist to increase depth of roof insulation at perimeter

Sealant, adhesive or gasket

Rigid insulation

Insulation baffle provides minimum 2 in. clearance under roof sheathing

Rigid insulation notched around roof trusses to act as wind shield for roof insulation

Sealant, adhesive or gasket

Rigid insulation

Figure 11.47
Roof Framing

- Roof insulation thermal resistance (depth) at truss heel (roof perimeter) should be equal or greater than thermal resistance of exterior wall

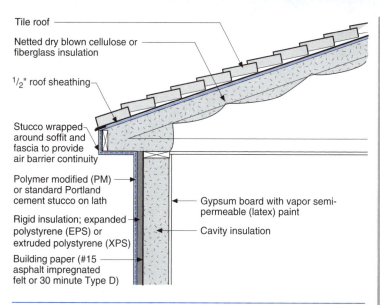

Tile roof

Netted dry blown cellulose or fiberglass insulation

$1/2$" roof sheathing

Stucco wrapped around soffit and fascia to provide air barrier continuity

Polymer modified (PM) or standard Portland cement stucco on lath

Rigid insulation; expanded polystyrene (EPS) or extruded polystyrene (XPS)

Building paper (#15 asphalt impregnated felt or 30 minute Type D)

Gypsum board with vapor semi-permeable (latex) paint

Cavity insulation

Figure 11.48
Unvented Roof — Stucco Closure
- Hot-dry climate only

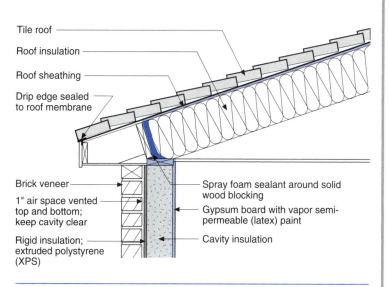

Tile roof

Roof insulation

Roof sheathing

Drip edge sealed to roof membrane

Brick veneer

1" air space vented top and bottom; keep cavity clear

Rigid insulation; extruded polystyrene (XPS)

Spray foam sealant around solid wood blocking

Gypsum board with vapor semi-permeable (latex) paint

Cavity insulation

Figure 11.49
Unvented Roof — Blocking Closure
- Hot-dry climate only

11

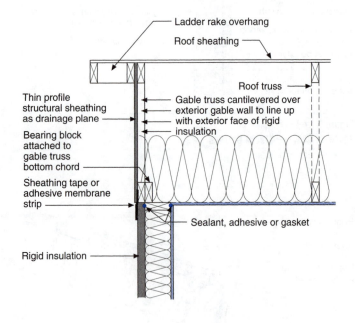

Ladder rake overhang

Roof sheathing

Thin profile structural sheathing as drainage plane

Roof truss

Gable truss cantilevered over exterior gable wall to line up with exterior face of rigid insulation

Bearing block attached to gable truss bottom chord

Sheathing tape or adhesive membrane strip

Sealant, adhesive or gasket

Rigid insulation

Figure 11.50
Aligning Gable Truss Over Wall with Insulating Sheathing

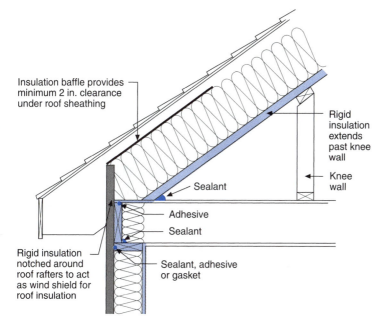

Insulation baffle provides minimum 2 in. clearance under roof sheathing

Rigid insulation extends past knee wall

Knee wall

Sealant

Adhesive

Sealant

Rigid insulation notched around roof rafters to act as wind shield for roof insulation

Sealant, adhesive or gasket

Figure 11.51
Roof Knee Wall
• Knee wall installed after rigid insulation

11

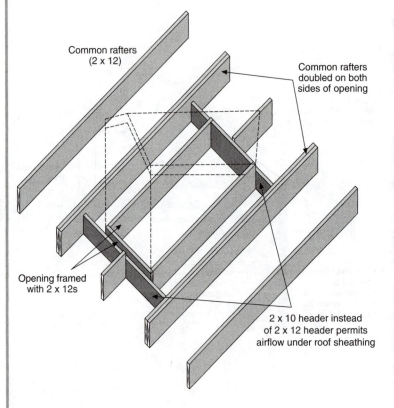

Common rafters
(2 x 12)

Common rafters
doubled on both
sides of opening

Opening framed
with 2 x 12s

2 x 10 header instead
of 2 x 12 header permits
airflow under roof sheathing

Figure 11.52
Skylight and Dormer Venting
- Reducing size of headers below that of rafters allows roof ventilation past opening

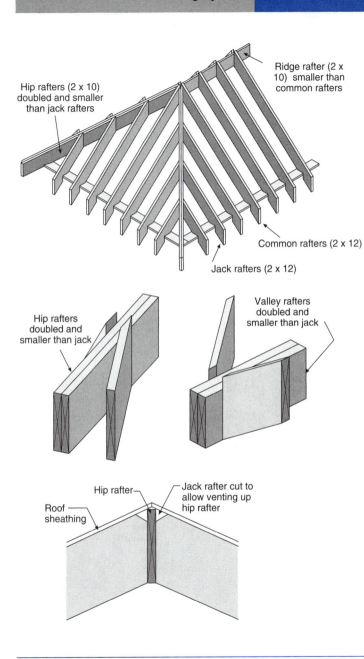

Hip rafters (2 x 10) doubled and smaller than jack rafters

Ridge rafter (2 x 10) smaller than common rafters

Hip rafters doubled and smaller than jack

Valley rafters doubled and smaller than jack

Common rafters (2 x 12)

Jack rafters (2 x 12)

Roof sheathing

Hip rafter

Jack rafter cut to allow venting up hip rafter

Figure 11.53
Venting Hip Roofs

- Smaller hip, ridge and valley rafters permit venting of hip roofs and valleys to ridge locations

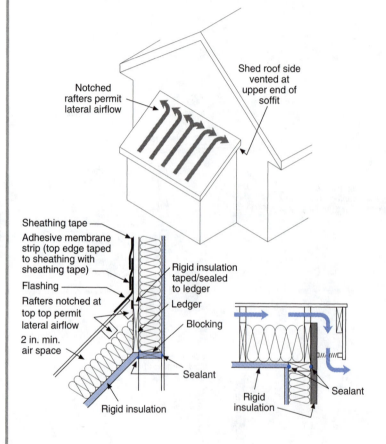

Notched rafters permit lateral airflow

Shed roof side vented at upper end of soffit

Sheathing tape

Adhesive membrane strip (top edge taped to sheathing with sheathing tape)

Flashing

Rafters notched at top top permit lateral airflow

2 in. min. air space

Rigid insulation

Rigid insulation taped/sealed to ledger

Ledger

Blocking

Sealant

Rigid insulation

Sealant

Figure 11.54
Venting Shed Roofs

- Allowance for shear strength reduction must be provided due to notching of rafters

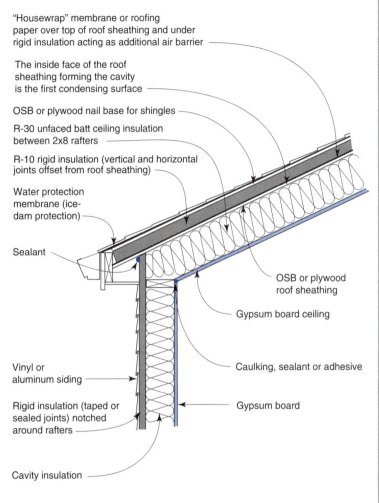

"Housewrap" membrane or roofing paper over top of roof sheathing and under rigid insulation acting as additional air barrier

The inside face of the roof sheathing forming the cavity is the first condensing surface

OSB or plywood nail base for shingles

R-30 unfaced batt ceiling insulation between 2x8 rafters

R-10 rigid insulation (vertical and horizontal joints offset from roof sheathing)

Water protection membrane (ice-dam protection)

Sealant

OSB or plywood roof sheathing

Gypsum board ceiling

Caulking, sealant or adhesive

Vinyl or aluminum siding

Rigid insulation (taped or sealed joints) notched around rafters

Gypsum board

Cavity insulation

 11

Figure 11.55
Hot Roof
- Rigid insulation raises dew point temperature of the first condensing surface
- This roof assembly is intended to be site constructed; prefabricated panels are not intended to be used in this assembly unless they are installed in a manner that provides air barrier continuity
- Offsetting joints in sheathing, rigid insulation and nail base creates air barrier continuity
- See Chapter 6 Roof Design

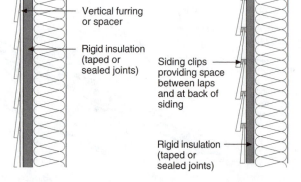

Vertical furring
or spacer

Rigid insulation
(taped or
sealed joints)

Siding clips
providing space
between laps
and at back of
siding

Rigid insulation
(taped or
sealed joints)

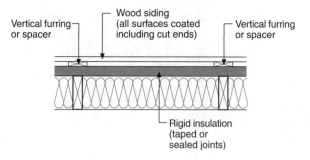

Vertical furring
or spacer

Wood siding
(all surfaces coated
including cut ends)

Vertical furring
or spacer

Rigid insulation
(taped or
sealed joints)

Figure 11.56
Wood Siding Installation

• Furring or spacer can be made by ripping strips of $^3/_8$" thick pressure treated plywood

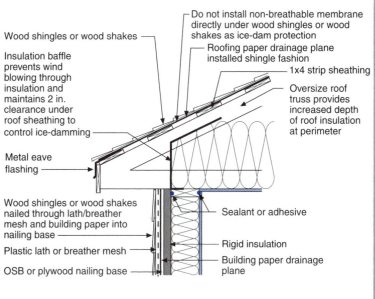

Do not install non-breathable membrane directly under wood shingles or wood shakes as ice-dam protection

Wood shingles or wood shakes

Roofing paper drainage plane installed shingle fashion

1x4 strip sheathing

Insulation baffle prevents wind blowing through insulation and maintains 2 in. clearance under roof sheathing to control ice-damming

Oversize roof truss provides increased depth of roof insulation at perimeter

Metal eave flashing

Wood shingles or wood shakes nailed through lath/breather mesh and building paper into nailing base

Sealant or adhesive

Plastic lath or breather mesh

Rigid insulation

OSB or plywood nailing base

Building paper drainage plane

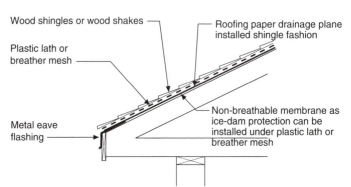

Wood shingles or wood shakes

Roofing paper drainage plane installed shingle fashion

Plastic lath or breather mesh

Metal eave flashing

Non-breathable membrane as ice-dam protection can be installed under plastic lath or breather mesh

Figure 11.57
Wood Shingles or Wood Shake Walls and Roofs
- Wood shingle or wood shakes on roofs and walls should be installed so they can "breathe" to back side (install over strip sheathing, plastic lath or breather mesh)
- Continuous soffit roof venting is more effective as ice-dam control than water protection membrane

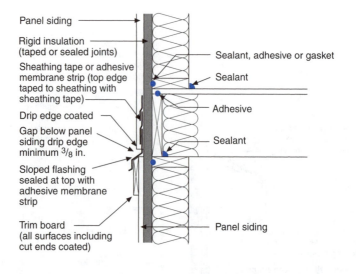

Panel siding

Rigid insulation
(taped or sealed joints)

Sheathing tape or adhesive
membrane strip (top edge
taped to sheathing with
sheathing tape)

Drip edge coated

Gap below panel
siding drip edge
minimum $3/8$ in.

Sloped flashing
sealed at top with
adhesive membrane
strip

Trim board
(all surfaces including
cut ends coated)

Sealant, adhesive or gasket

Sealant

Adhesive

Sealant

Panel siding

Figure 11.58
Panel Trim

11

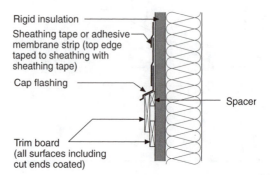

Rigid insulation

Sheathing tape or adhesive
membrane strip (top edge
taped to sheathing with
sheathing tape)

Cap flashing

Spacer

Trim board
(all surfaces including
cut ends coated)

Figure 11.59
Flashing Installed Over Padded Horizontal Trim

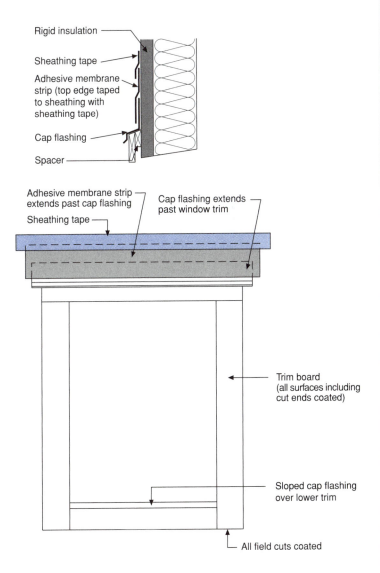

Rigid insulation

Sheathing tape

Adhesive membrane strip (top edge taped to sheathing with sheathing tape)

Cap flashing

Spacer

Adhesive membrane strip extends past cap flashing

Cap flashing extends past window trim

Sheathing tape

Trim board (all surfaces including cut ends coated)

Sloped cap flashing over lower trim

All field cuts coated

Figure 11.60
Flashing Over and Under Window Trim
- Back-ventilate trim materials so that absorbed water can evaporate and be vented to the exterior

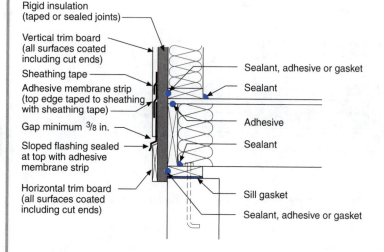

Rigid insulation
(taped or sealed joints)

Vertical trim board
(all surfaces coated
including cut ends)

Sheathing tape

Adhesive membrane strip
(top edge taped to sheathing
with sheathing tape)

Gap minimum $3/8$ in.

Sloped flashing sealed
at top with adhesive
membrane strip

Horizontal trim board
(all surfaces coated
including cut ends)

Sealant, adhesive or gasket

Sealant

Adhesive

Sealant

Sill gasket

Sealant, adhesive or gasket

Figure 11.61
Vertical Trim and Flashing Installed Over Horizontal Trim

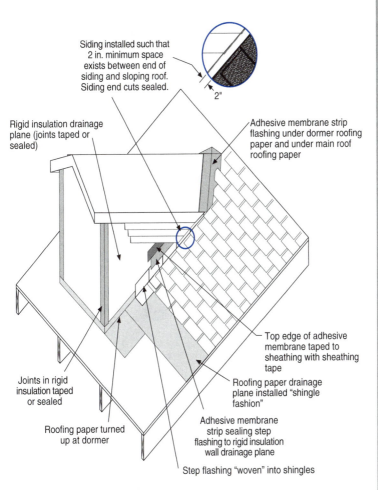

Siding installed such that 2 in. minimum space exists between end of siding and sloping roof. Siding end cuts sealed.

2"

Rigid insulation drainage plane (joints taped or sealed)

Adhesive membrane strip flashing under dormer roofing paper and under main roof roofing paper

Top edge of adhesive membrane taped to sheathing with sheathing tape

Joints in rigid insulation taped or sealed

Roofing paper drainage plane installed "shingle fashion"

Roofing paper turned up at dormer

Adhesive membrane strip sealing step flashing to rigid insulation wall drainage plane

Step flashing "woven" into shingles

Note: Layering cut away in this figure shown for clarity, not as recommendation for installation sequencing

Figure 11.62
Dormer Siding Installation
• See also Figure 2.22

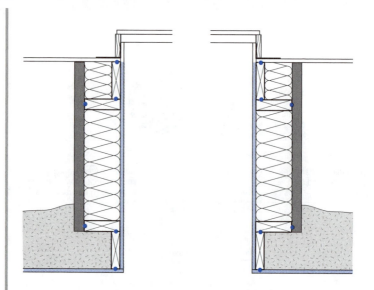

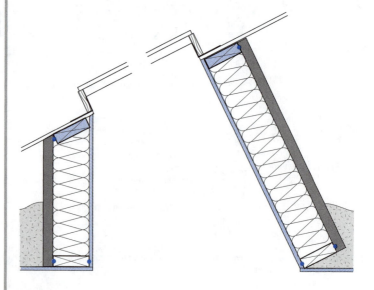

Figure 11.61
Skylight Well

• Skylight wells should be constructed and insulated similar to exterior walls

Masonry Construction

Masonry construction is often used in conjunction with wood framing and steel framing in hot-humid climates. It is common to construct structures where the first floor is masonry, the second floor is wood frame and the interior partitions on the first and second floor are steel.

The drivers for selecting the combination of approaches are typically cost and the availability of trades. In some locations it is more cost effective to construct both stories of a typical residence out of masonry, and in other locations it is more cost effective to mix masonry with wood frame on the upper level.

Masonry construction can be insulated on the inside or the outside. Whatever the approach, it should

- hold the building up
- keep the rain water out
- keep the wind out
- keep the water vapor out
- let the water and water vapor out if they get inside
- keep the heat out during the summer
- keep the heat in during the winter

Concerns

Masonry walls in hot-humid climates are typically constructed as barrier walls from the perspective of rain control. That is they usually have a stucco rendering applied directly on their exterior surface. In effect the exterior surface becomes the drainage plane. With such an approach it is expected that some rain water penetration through the stucco rendering and absorption by the masonry wall will occur. The key to the performance of these walls is reducing the rainwater absorption and penetration of the stucco rendering and the ability of these walls to harmlessly store this penetrating moisture and then release it to the interior and exterior in a controlled manner that does not damage either interior or exterior surfaces.

Accordingly, impermeable interior finishes or insulation systems should be avoided including vinyl wall coverings, reflective foil radiant barriers, foil-faced rigid insulations, polypropylene faced rigid insulations, and faced fiberglass batt cavity insulations. These type of materials interfere with the ability of the masonry wall to dry towards the interior (Figure 12.1).

Note that stucco is not a waterproof layer — even when painted. Some rainwater will enter the masonry. For this reason a "seat" is recommended to facilitate rainwater storage and provide drainage to the exterior (Figure 12.2 and Figure 12.3).

Vapor semi-permeable or permeable interior finishes or insulation systems are recommended such as latex paint, unfaced expanded or extruded polystyrene rigid insulation, glass or paper-faced isocyanurate rigid insulation, unfaced fiberglass rigid or batt cavity insulation or spray cellulose.

On exterior stucco surfaces, water repellant vapor permeable and vapor semi-permeable coating or paint systems are recommended. Recall that water vapor flow occurs from both a higher concentration to a lower concentration and from the warm side of an assembly to the cold side of an assembly. A rain wetted stucco covered masonry wall that is heated by solar radiation will be warmer and wetter than both the interior and the exterior air. Drying initially will be both to the outside and to the inside. A heavy coat of impermeable paint over exterior stucco will blister under such conditions (see Chapter 18: Painting). Premium "elastomeric" paints are typically vapor semi-permeable and work well over exterior stucco surfaces in hot-humid climates.

Many building enclosures in hot-humid climates mix rain control strategies between stories and elevations. For example, a wood frame water managed stucco wall is often constructed as a second story enclosure over a stucco covered masonry barrier wall first story. The intersection of these different strategies and assemblies is important. The second story must be flashed and drained where it meets the first story in this instance.

Where steel studs and steel framing are used for interior partitions, they should not be used to separate garage spaces from interior spaces due to their high conductivity and poor thermal performance. Wood framing should be used at this location. Steel framing or steel studs should not be used in any exterior insulated assemblies including their use as interior furring on masonry walls.

Experience shows that most windows, doors and sliding door assemblies are prone to rain water leakage. Prudent installation should take this into account and installation should be done using assembly details that provide drainage to the exterior through the use of pan flashings, "seats" in concrete slabs that act as pan flashings, or masonry sill details that are designed to drain to the exterior.

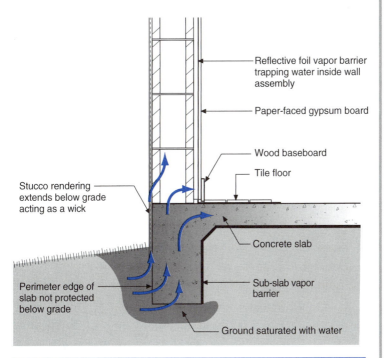

Reflective foil vapor barrier trapping water inside wall assembly

Paper-faced gypsum board

Wood baseboard

Tile floor

Stucco rendering extends below grade acting as a wick

Concrete slab

Sub-slab vapor barrier

Perimeter edge of slab not protected below grade

Ground saturated with water

Figure 12.1
Impermeable Interior Materials
• Reflective foil vapor barrier traps water in assembly

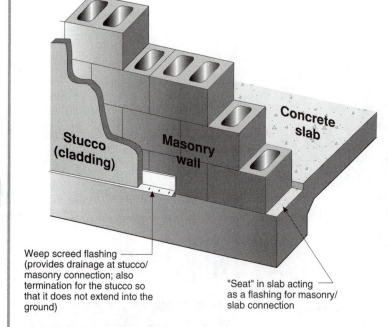

Concrete slab

Stucco (cladding)

Masonry wall

Weep screed flashing (provides drainage at stucco/ masonry connection; also termination for the stucco so that it does not extend into the ground)

"Seat" in slab acting as a flashing for masonry/ slab connection

Figure 12.2
Seat in Slab Promotes Drainage and Rain Water Storage

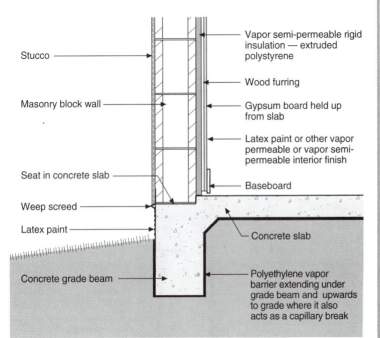

Stucco

Vapor semi-permeable rigid insulation — extruded polystyrene

Wood furring

Masonry block wall

Gypsum board held up from slab

Latex paint or other vapor permeable or vapor semi-permeable interior finish

Seat in concrete slab

Baseboard

Weep screed

Latex paint

Concrete slab

 12

Concrete grade beam

Polyethylene vapor barrier extending under grade beam and upwards to grade where it also acts as a capillary break

Figure 12.3
Seat in Slab

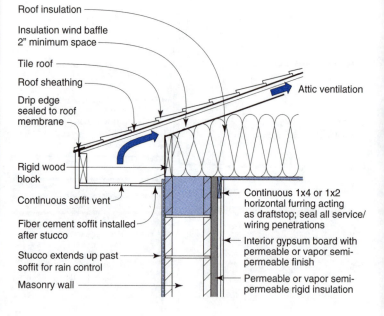

Roof insulation

Insulation wind baffle
2" minimum space

Tile roof

Roof sheathing

Drip edge
sealed to roof
membrane

Attic ventilation

Rigid wood
block

Continuous soffit vent

Fiber cement soffit installed
after stucco

Stucco extends up past
soffit for rain control

Masonry wall

Continuous 1x4 or 1x2
horizontal furring acting
as draftstop; seal all service/
wiring penetrations

Interior gypsum board with
permeable or vapor semi-
permeable finish

Permeable or vapor semi-
permeable rigid insulation

Figure 12.4
Masonry Wall — Interior Rigid Insulation — Stucco —Vented Roof
- Horizontal furring at top of wall acts as draftstop; must be continuous
- Stucco extends up past soffit for rain control

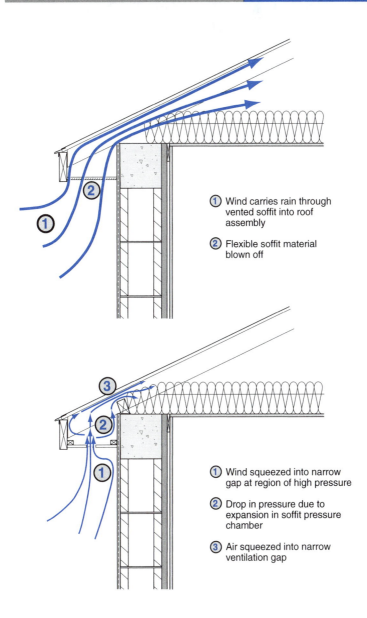

(1) Wind carries rain through vented soffit into roof assembly

(2) Flexible soffit material blown off

(1) Wind squeezed into narrow gap at region of high pressure

(2) Drop in pressure due to expansion in soffit pressure chamber

(3) Air squeezed into narrow ventilation gap

Figure 12.5
Pressure Equalized Soffit
• Non-rigid soffits blow off leading to rain entry and roof pressurization

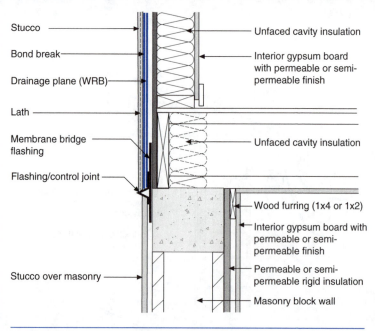

Stucco

Bond break

Drainage plane (WRB)

Lath

Membrane bridge flashing

Flashing/control joint

Stucco over masonry

Unfaced cavity insulation

Interior gypsum board with permeable or semi-permeable finish

Unfaced cavity insulation

Wood furring (1x4 or 1x2)

Interior gypsum board with permeable or semi-permeable finish

Permeable or semi-permeable rigid insulation

Masonry block wall

Figure 12.6
Masonry Wall — Wood Frame Intersection — Stucco
  • Control joint flashing attached to lower edge of wood sheathing

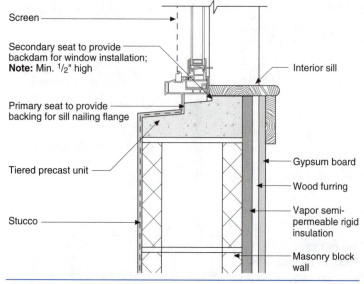

Screen

Secondary seat to provide backdam for window installation;
Note: Min. $1/2$" high

Primary seat to provide backing for sill nailing flange

Tiered precast unit

Stucco

Interior sill

Gypsum board

Wood furring

Vapor semi-permeable rigid insulation

Masonry block wall

Figure 12.7
Window Sill Drainage Detail — Section

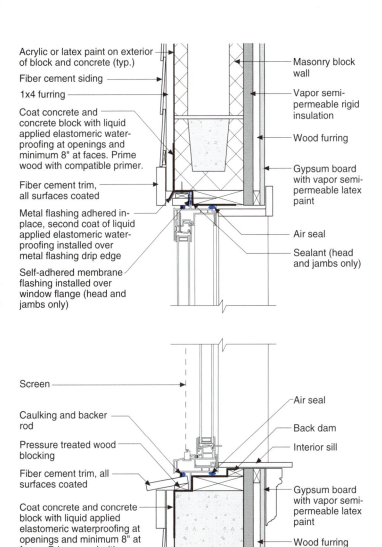

Acrylic or latex paint on exterior of block and concrete (typ.)

Fiber cement siding

1x4 furring

Coat concrete and concrete block with liquid applied elastomeric water-proofing at openings and minimum 8" at faces. Prime wood with compatible primer.

Fiber cement trim, all surfaces coated

Metal flashing adhered in-place, second coat of liquid applied elastomeric water-proofing installed over metal flashing drip edge

Self-adhered membrane flashing installed over window flange (head and jambs only)

Masonry block wall

Vapor semi-permeable rigid insulation

Wood furring

Gypsum board with vapor semi-permeable latex paint

Air seal

Sealant (head and jambs only)

Screen

Caulking and backer rod

Pressure treated wood blocking

Fiber cement trim, all surfaces coated

Coat concrete and concrete block with liquid applied elastomeric waterproofing at openings and minimum 8" at faces. Prime wood with compatible primer.

Acrylic or latex paint on exterior of block and concrete (typ.)

Fiber cement siding

1x4 furring

Air seal

Back dam

Interior sill

Gypsum board with vapor semi-permeable latex paint

Wood furring

Vapor semi-permeable rigid insulation

Masonry block wall

**Figure 12.8
Window Head and Sill Drainage Detail — Section with Siding**

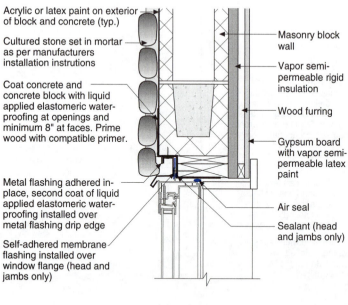

Acrylic or latex paint on exterior of block and concrete (typ.)

Cultured stone set in mortar as per manufacturers installation instrutions

Coat concrete and concrete block with liquid applied elastomeric waterproofing at openings and minimum 8" at faces. Prime wood with compatible primer.

Metal flashing adhered inplace, second coat of liquid applied elastomeric waterproofing installed over metal flashing drip edge

Self-adhered membrane flashing installed over window flange (head and jambs only)

Masonry block wall

Vapor semi-permeable rigid insulation

Wood furring

Gypsum board with vapor semi-permeable latex paint

Air seal

Sealant (head and jambs only)

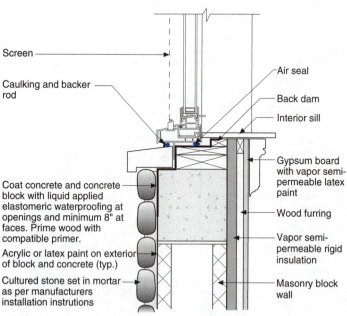

Screen

Caulking and backer rod

Air seal

Back dam

Interior sill

Gypsum board with vapor semi-permeable latex paint

Wood furring

Vapor semi-permeable rigid insulation

Masonry block wall

Coat concrete and concrete block with liquid applied elastomeric waterproofing at openings and minimum 8" at faces. Prime wood with compatible primer.

Acrylic or latex paint on exterior of block and concrete (typ.)

Cultured stone set in mortar as per manufacturers installation instrutions

Figure 12.9
Window Head and Sill Drainage Detail — Section with Stone Veneer

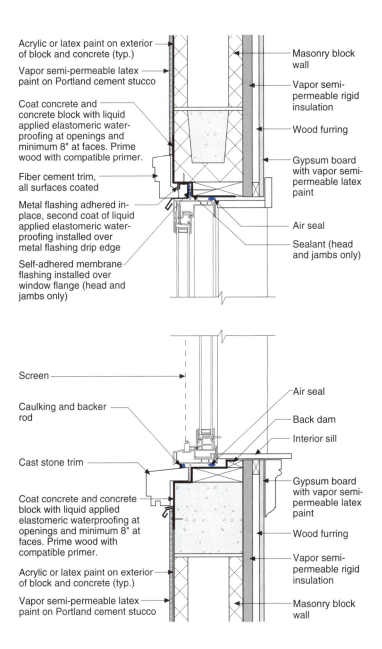

Acrylic or latex paint on exterior of block and concrete (typ.)

Vapor semi-permeable latex paint on Portland cement stucco

Coat concrete and concrete block with liquid applied elastomeric waterproofing at openings and minimum 8" at faces. Prime wood with compatible primer.

Fiber cement trim, all surfaces coated

Metal flashing adhered in-place, second coat of liquid applied elastomeric waterproofing installed over metal flashing drip edge

Self-adhered membrane flashing installed over window flange (head and jambs only)

Masonry block wall

Vapor semi-permeable rigid insulation

Wood furring

Gypsum board with vapor semi-permeable latex paint

Air seal

Sealant (head and jambs only)

Screen

Caulking and backer rod

Cast stone trim

Coat concrete and concrete block with liquid applied elastomeric waterproofing at openings and minimum 8" at faces. Prime wood with compatible primer.

Acrylic or latex paint on exterior of block and concrete (typ.)

Vapor semi-permeable latex paint on Portland cement stucco

Air seal

Back dam

Interior sill

Gypsum board with vapor semi-permeable latex paint

Wood furring

Vapor semi-permeable rigid insulation

Masonry block wall

Figure 12.10
Window Head and Sill Drainage Detail — Section with Stucco

Step 1
Concrete block wall with concrete
header

Step 2
Install 2-tier precast or cast-in-
place sill

Figure 12.11
Installing Window in Concrete Block Wall in Nine Steps

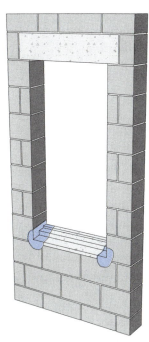

Step 3
Liquid applied elastic membrane
installed at sill corners

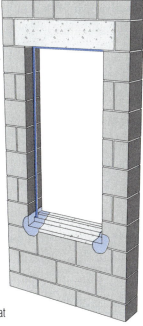

Step 4
Apply sealant for wood buck at
jambs and head

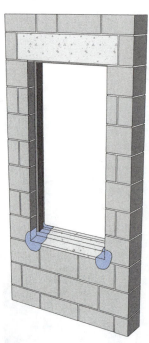

12

Step 5
Install wood buck over sealant

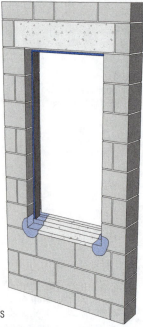

Step 6
Apply window sealant at jambs and head over wood buck

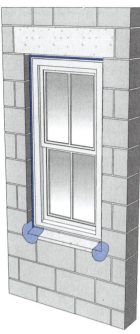

Step 7

Install window plumb, level and square per manufacturer's instructions; apply sealant over exposed wood buck at jambs and head

Step 8

Apply stucco finish

Construction

Step 9

Complete stucco finish with continuous bead of sealant at jambs and head sealing stucco finish to window unit

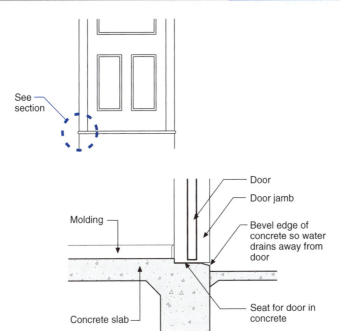

Figure 12.12
Door Threshold Detail — Section
- Prevents damage to bottom of casings/leaks at front door

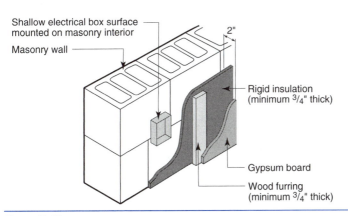

Figure 12.13
Electrical Box Installation — Masonry Wall
- Use of rigid insulation on interior of masonry wall eases the installation of interior electrical boxes
- Electrical boxes can be surface mounted eliminating chiseling/chipping of masonry

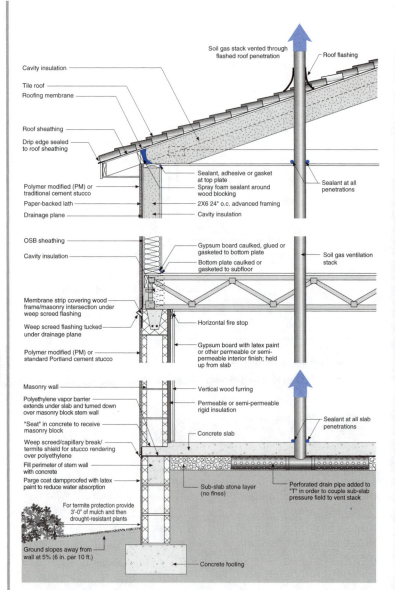

Soil gas stack vented through flashed roof penetration

Roof flashing

Cavity insulation

Tile roof

Roofing membrane

Roof sheathing

Drip edge sealed to roof sheathing

Sealant, adhesive or gasket at top plate

Spray foam sealant around wood blocking

Polymer modified (PM) or traditional cement stucco

Paper-backed lath

Drainage plane

2X6 24" o.c. advanced framing

Cavity insulation

Sealant at all penetrations

OSB sheathing

Cavity insulation

Gypsum board caulked, glued or gasketed to bottom plate

Bottom plate caulked or gasketed to subfloor

Soil gas ventilation stack

Membrane strip covering wood frame/masonry intersection under weep screed flashing

Weep screed flashing tucked under drainage plane

Polymer modified (PM) or standard Portland cement stucco

Horizontal fire stop

Gypsum board with latex paint or other permeable or semi-permeable interior finish; held up from slab

Masonry wall

Polyethylene vapor barrier extends under slab and turned down over masonry block stem wall

"Seat" in concrete to receive masonry block

Weep screed/capillary break/ termite shield for stucco rendering over polyethylene

Fill perimeter of stem wall with concrete

Parge coat dampproofed with latex paint to reduce water absorption

For termite protection provide 3'-0" of mulch and then drought-resistant plants

Ground slopes away from wall at 5% (6 in. per 10 ft.)

Vertical wood furring

Permeable or semi-permeable rigid insulation

Concrete slab

Sealant at all slab penetrations

Sub-slab stone layer (no fines)

Perforated drain pipe added to "T" in order to couple sub-slab pressure field to vent stack

Concrete footing

Figure 12.14
Building Section

HVAC

Two heating and cooling approaches are common in residential construction: those with forced air and those without. Three controlled ventilation approaches are common in residential construction: exhaust systems, supply systems and balanced systems (Figures 13.1, 13.2, 13.3). When mixed and matched all these systems have to:

- heat when it's cold
- cool when it's hot
- humidify when it's dry
- dehumidify when it's humid
- bring in outside air
- distribute outside air
- exhaust strong pollutant and water vapor point sources
- filter the air
- do all this when needed without noise, vibration, drafts and odors

Concerns

As window systems have improved and as roof and wall insulation levels have increased, the solar gain or heat load on buildings has been significantly reduced. This should lead to a reduction in the size of air conditioning equipment. Often, this is not the case. This "oversizing" leads to an inability of the air conditioning system to control interior humidity. Interior moisture is removed by air conditioning equipment only when air conditioning is activated by the thermostat. If the air conditioning system is not running, it is not dehumidifying. Oversized equipment runs for shorter periods such that the coils don't get cold enough to remove much moisture before the thermostat is satisfied, especially under part load conditions.

In many cases, units small enough to match the loads are not commonly available. This is particularly acute in townhouse and condominium construction where the smallest commonly available units are 2-Ton and the typical loads are 1 to 1 $\frac{1}{2}$ Ton. These are full load values, meaning that under most cooling periods, part load conditions are the norm and even correctly sized equipment doesn't run often enough or long enough to control interior humidity.

The problems are further exacerbated by air leakage at equipment cabinets and ductwork located external to the conditioned space. This leakage and the associated air pressure differences lead to infiltration of exterior humid air. The latent (moisture) load associated with this infiltrating air dwarfs the sensible (temperature) load. Comfort and interior humidity problems are likely.

By constructing a high performance building enclosure that minimizes heat gain and by providing outside air, it is not likely that even correctly sized air conditioning equipment will be able to control interior humidity levels under part load conditions in this climate. Supplemental dehumidification will be required. A dehumidifier or ventilating dehumidifier will be necessary.

The key steps in providing a comfortable, healthy, durable interior environment in hot-humid climates are as follows:

- minimize heat gain through the use of spectrally selective glazing (Low E^2) in non-conductive (vinyl or wood) frames
- minimize heat gain through the use of attic insulation and tile roofs, white colored roofing or metal roofing on furring
- reduce latent and sensible loads by locating ductwork and air handlers within building conditioned spaces eliminating leakage to the exterior
- reduce latent and sensible loads by balancing interior pressures by transfer grilles eliminating induced infiltration
- provide outside air only as needed and in the quantities needed
- properly size equipment
- provide supplemental dehumidification

Choices

The first choice a builder makes is which type of energy source to use: combustion, electricity or the sun (if we split hairs, the sun is the source of all energy, except nuclear generated electricity - where we split atoms). Choosing between combustion, electricity or the sun requires the wisdom of Solomon, the intelligence of Newton and the wit of Wilde (the reason we have trouble with this choice is that none of these people are alive to actually ask). Wars have been fought over less. Sometimes no choice is possible. We will assume that some type of rational choice will be made and avoid this messy discussion.

If a combustion energy source is selected, combustion appliances should not be subject to backdrafting or spillage of combustion products. If electricity is selected, resistance heating should be avoided except in cases where building enclosures are so ultra-efficient that conventional thinking does not apply. In all cases, utilizing the sun for at least a partial contribution should be considered.

The second choice a builder makes is to use forced air or not. If air conditioning is desired (and by air conditioning we mean cooling), forced air systems usually become a necessity.

Forced air heating and cooling systems include an air handler with a heating and cooling coil (such as a furnace with an A/C coil, a heat pump, or a fan-coil unit with an A/C coil and a heating coil coupled to a hot water heater). The air handler is typically connected to a supply and return ductwork system.

The third choice a builder makes is how to ventilate. Ventilation? Remember, HVAC stands for heating, ventilating and air conditioning. Most builders seem to ignore the ventilating part. When builders think about it at all many seem to think they can handle ventilation by not building too tight and installing operable windows. Building leaky buildings or telling homeowners to open windows somehow seems acceptable. How do you build a leaky building? Do you rely on lousy workmanship? Do you deliberately put holes in the building? How many holes do you put in and where do you put them? How often do you open window and how open do you keep them? When you think about it, building leaky buildings or relying on operable windows for ventilation doesn't make sense.

Why do we need to ventilate? We need to ventilate to protect building occupants and the building materials and finishes. Ventilation controls odors and airborne contaminants. Ventilation also can control interior moisture levels when outside air dewpoint temperatures are low (typically below 55°F).

Ventilation Requirements

Adding controlled ventilation in hot-humid climates can seem at first to be a contradiction. Supplying outside air increases interior moisture levels and the latent load. Isn't this bad? Yes, but people also need outside air and outside air is required to dilute interior contaminants emitted from furnishings, the building structure and occupant activities. The key is to only supply the amount of outside air that is needed — and not more. And then address the resulting latent (interior humidity) load.

The contradiction gets worse when you consider the effect of duct leakage. Most ductwork and air handlers leak and are located outside of the conditioned space in vented attics, vented crawl spaces or garages. This results in excessive air change. That's right, most houses in humid climates are over-ventilated when the air handler operates. So you'd have to be crazy to add controlled ventilation? Well, no. The key is to get rid of the duct leakage induced air change which is uncontrolled and then add controlled ventilation. This approach typically results in a lower latent load. It is not unusual to have more than 150 to 200 cfm of duct leakage induced air change. This is replaced by 50 to 60 cfm of controlled ventilation after the duct leakage is eliminated.

Relying on duct leakage-induced air change to provide ventilation causes other problems. The outside air that is induced to enter often enters from the attic, crawl space or garage — locations that are not conducive to the highest quality of outside air by virtue of temperature, humidity or contaminants.

The ideal approach to ventilation in hot-humid climates is to properly size equipment and minimize the need for outside air. The air should be obtained in a controlled manner (mechanically with a fan). The air should be conditioned where it comes into the building. It should be dehumidified by cooling it below its dewpoint (typically below 55° F) and used to maintain the building enclosure at a slight positive pressure relative to the exterior. By doing so, it can be used to control the infiltration of exterior hot, humid air. Furthermore, the building enclosure should be built in a manner that aids in the pressurization of the building. Tight construction with duct work and air handling located within conditioned spaces is recommended.

Building tight is right. Buildings can never be built too tight. However, they can be under-ventilated.

The best way to go is to build a tight building envelope and install controlled ventilation using a fan or several fans which operate when people are present. Why the tight building enclosure? Because before you can control air you must first enclose air. Why controlled ventilation? Because you don't want to over-ventilate during cold weather and under-ventilate during warm weather.

How much air do you need? ASHRAE Standard 62.2 provides guidance. Somewhere between 10 cfm and 20 cfm per person when the building is occupied. If a building doesn't have strong interior pollutant sources, as low as 10 cfm per person will work. If a building has strong interior pollutant sources, not even 20 cfm per person will be enough. What are strong interior pollutant sources? Smokers. Damp basements. Pets that are not house trained. Unvented gas fireplaces or space heaters. Unusual hobby activities. Gas ovens and gas cooktops. Generally, if you keep the water out of a building, vent combustion appliances to the exterior, don't smoke, and don't have unusual habits or an uncommon lifestyle, 10 cfm per person will be just fine.

How do you decide how many people live in a house? A good rule of thumb is to take the number of bedrooms and add 1. This assumes two people in the master bedroom and one person in each additional bedroom. The following ventilation requirements result when you follow ASHRAE Standard 62.2.

- 7.5 cfm/person plus 0.01 cfm/ft^2 of conditioned floor area
- For a 2,000 ft^2 three bedroom house with 4 occupants:
  4 x 7.5 cfm = 30 cfm (people load)
  2,000 ft2 x 0.01 cfm/ft2 = 20 cfm (house/furnishings load)
  30 cfm (people load) + 20 cfm (house/furnishings load) = 50 cfm

Ventilation air should be provided when the building is occupied. Why ventilate when no one is in the building? Ventilation air should also be distributed (circulated) when the building is occupied.

Indoor Humidity and Airborne Pollutants

Indoor humidity and airborne pollutants are both controlled by ventilation. There are two kinds of ventilation: spot ventilation ("point source exhaust") and dilution ventilation. Both are necessary. Spot ventilation deals with point sources of pollution such as bathrooms and kitchens. Dilution ventilation deals with low-level pollutants throughout the home.

This ventilation is in addition to the use of operable windows.

Every home needs to have exhaust from kitchens and from bathrooms. In kitchens, recirculating fans should be avoided because they become breeding grounds for biologicals, a major source of odors, and in all cases allow grease vapors to coat surfaces throughout the home. Kitchen range hoods must be exhausted to the outside to remove moisture, odors and other pollutants.

Bathroom fans must exhaust to the exterior — even bathrooms with operable windows. No exceptions. Low sone fans (less than 3 sones) are recommended because they are quiet (so they are more likely to be used) and more durable (in order to make them quiet they must be made durable).

Dilution ventilation can be provided three ways: exhaust (Figure 13.1), supply (Figure 13.2) or balanced (Figure 13.3). In all cases, it should be continuous and fan powered.

The key to dilution ventilation is good distribution. Outside air should be provided throughout the house. Forced air duct systems can be excellent distribution systems (either by directly providing outside air or by providing mixing of interior air). Where duct distribution systems do not exist, multiport exhaust strategies can be used.

Formaldehyde and other emissions form particleboard can be harmful. To reduce emissions from particleboard surfaces, reduce the amount of particleboard. Use wire shelving in closets. Wire shelving is easy to clean and permits air circulation. With kitchen and bathroom cabinets constructed from particleboard, the exposed particleboard sources can be sealed with 100 percent acrylic paint or clear sealant.

Since most ventilation system airflow across the building enclosure is in the 50 cfm or less range, the effect on building pressures is typically negligible. However, in general, exhaust ventilation systems have a slight depressurization effect on building enclosures; supply ventilation systems have a slight pressurization effect on building enclosures, and balanced ventilation systems have no effect on building air pressures. See Figures 13.4 through 13.10 for more detail on different types of ventilation systems.

In hot-humid climates it is desirable to have supply ventilation or balanced ventilation systems. Note that the figures show side wall locations for the exit or entrance point of ventilation ducts because venting through the ceiling plane is not desirable due to the difficulties in air sealing fan housings and rainproofing roof penetrations.

Furnace fans or air handler fans should not run continuously unless at reduced speed using electrically commutated motors (ECM). These types of motors are typically found only on premium units.

Even with efficient electric motors and blowers HVAC units should not run continuously if outside air is ducted to the air handling system. Dehumidification of outside air ducted into the air handling system will only occur when the coils are energized. If blowers are run continuously, re-evaporation of condensate off the cooling coil will occur and condensation of humidity contained in ducted outside air can occur on cold ductwork (i.e. ductwork cooled by cold air coming off energized coils). This can be avoided by cycling the air handler such that once coils are de-energized, blowers are also shutdown for 15 minutes to allow ductwork to warm-up and condensate to drain from wet coils and drain pans.

A flow controller with a motorized damper can be used in a system where outside air is ducted to the return side of the air handler. When the air handler is operating and the motorized damper is open, outside air is brought into the building and distributed. The controller will only operate the blower with the damper open if the coils are energized or with a time-delay where the blower is shutdown for a given period after coils are de-energized to allow for system condensate drainage and duct system warming. This approach will also allow the blower to cycle several minutes each hour even if heating or cooling is not required. This will bring in outside air throughout the year in cases where the outside air supply is provided by the blower independent of thermostat control (Figures 13.7 and 13.8). The motorized damper prevents overventilation during long blower duty cycles.

These problems can be avoided by installing a stand-alone supply ventilation system such as a ventilating dehumidifier (Figure 13.4) where outside air is brought in, dehumidified and distributed throughout the house. Such a system can also have an operable outside damper that closes when the house is unoccupied and recirculation of interior air can occur through the dehumidifier. This is ideal for houses, apartments or condominiums that are used or occupied only seasonally. Air conditioning systems can be shutdown and the dehumidifier operated on a recirculation setting (i.e. no outside air). Humidity will be controlled without the energy waste of running the A/C system. The space will be warm or hot, but the humidity will be low.

Exhaust fans extracting less than 50 cfm will typically not significantly increase interior latent loads, radon ingress, soil gas ingress or backdrafting problems with fireplaces or wood stoves due to their negligible effect on building air pressures. However, larger exhaust air flows may lead to unacceptably high negative air pressures (above 5 Pascals negative is considered unacceptable by some codes; 3 Pascals or less negative is a recommended maximum allowable design metric or design standard in hot-humid and other climates) and result in interior humidity and moisture problems in building enclosure assemblies. Exhaust fans should not run continuously in hot-humid climates. Timers or some other control approach is recommended.

Dehumidification

It is common to attempt to use dilution (controlled air change by an exhaust, supply or balanced ventilation system) during heating periods (cold, dry winter months) to limit/control interior moisture levels. During cooling periods (hot, sometimes humid, summer months) the dehumidification characteristics of air conditioning (mechanical cooling) systems are used to reduce interior moisture levels.

However, dilution ventilation during the winter and air conditioning during the summer will not likely be able to control interior moisture levels in hot-humid climates without supplemental dehumidification. Furthermore spring or fall conditions, when neither heating nor cooling occurs and exterior humidity is high will lead to even more difficult interior moisture control problems. In hot-humid climates, the winters are particularly mild and the exterior air during the winter period is humid. Dilution (or air change) will not remove much moisture under these conditions since the incoming air is humid. Air change is still required to remove/dilute interior pollutants. However, air change will not remove moisture. Under such conditions a dehumidification system is needed. A stand alone dehumidifier or a ventilating dehumidifier is recommended.

Similarly, under part load conditions in the summer or especially during the spring or fall (when the outside air is humid, but at the same temperature or slightly higher than the interior) when the air conditioner does not operate much, dehumidification may also be needed. Air conditioning will only remove moisture from the interior air when the air conditioning system is cooling the interior air. If there is not much need (demand) for cooling, dehumidification by the air conditioning system will not occur. Again, under such conditions a dehumidification system is needed and can be supplied by a stand alone dehumidifier or a ventilating dehumidifier.

Dehumidifiers or ventilating dehumidifiers should be selected with a minimum capacity or 40 to 60 pints per day. Interior relative humidities should always be maintained below 60%. In homes with carpets over slab-on-grade, interior RH should be kept below 50% to control dust mites. Dehumidifier capacity should be rated at 75°F and 50% RH.

It is recommended that dehumidifiers with enhanced recuperative dehumidification cycles be used. These types of dehumidifiers add a recuperative heat exchanger to the conventional dehumidification cycle. They typically require only $1/3$ of the electrical power and operating cost as conventional dehumidifiers.

Combustion Appliances

Spillage or backdrafting of combustion appliances is unacceptable. Only sealed combustion, direct vented, power vented or induced draft combustion appliances should be installed inside conditioned spaces for space conditioning or for domestic hot water. Traditional gas water heaters with draft hoods are prone to spillage and backdrafting. They should be avoided. Gas ovens, gas stoves or gas cooktops should only be installed with an exhaust range hood directly vented to the exterior. Wood-burning fireplaces or gas-burning fireplaces should be supplied with glass doors and exterior combustion air ducted to the firebox. Wood stoves should have a direct ducted supply of combustion air. Unvented (ventless) gas fireplaces or gas space heaters should never be installed. Sealed combustion direct vent gas fireplaces are an acceptable alternative. Portable kerosene heaters should never be used indoors. Figures 13.27 through 13.31 describe several different systems for safely installing gas fired furnaces and hot water tanks.

Garages

Ideally, garages should not be connected to a home. Discrete, separate garages constructed away from homes are preferred. If garages are connected to a home, they should be ventilated to the exterior with a passive vent stack (a "chimney" to the outside — 6-inch duct). Air handling devices such as furnaces or air conditioners should never be located in garages. Nor should forced air ductwork. Weatherstrip the door between the garage and the home and air seal the common wall.

When ductwork passes through a chase or a floor above a garage or adjacent to an exterior wall bordering a garage, it is important that the ductwork be sealed airtight against the migration of pollutants from the garage to inside the home.

Smoke

Smoking should not occur in homes. If you must smoke, smoke outside. Candles and incense produce soot as do fireplaces and wood stoves. Soot can be unhealthy.

Recirculating Fans

Recirculating range hoods and recirculating bathroom fans should be avoided due to health concerns. If recirculating range hood filters are not regularly replaced and units not regularly cleaned, they become a breeding ground for biologicals and a major source of odors.

Whole House Fans

Whole house fans are commonly used to provide cooling and comfort as an alternative to air conditioning (mechanical cooling) or as a means to displace some air conditioning load. They are typically sized to exhaust approximately 10 air changes per hour.

When the exterior air is humid, whole house ventilation air conditioning energy savings are illusionary, due to the large latent (moisture) load typically associated with whole house ventilation.

Whole house fans should only be operated when the exterior moisture levels (vapor pressure) are lower than the interior moisture levels (vapor pressure). Enthalpy (moisture) controllers are recommended for use with whole house fans. Interior surfaces and furnishing are hygroscopic (absorb/adsorb or hold moisture) and act as moisture reservoirs or batteries which are "charged" during cool, humid evenings when whole house fans are typically operated. This stored moisture is "discharged" into the interior air when the air conditioning system is fi-

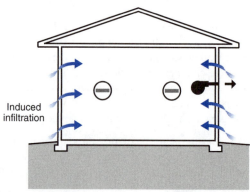

Figure 13.1
Exhaust Ventilation System
- Not recommended for hot-humid climates

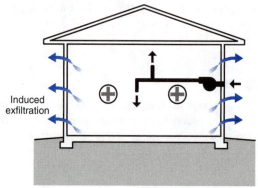

Figure 13.2
Supply Ventilation System

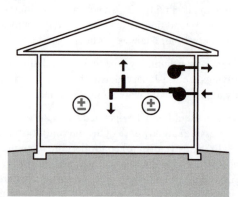

Figure 13.3
Balanced Ventilation System

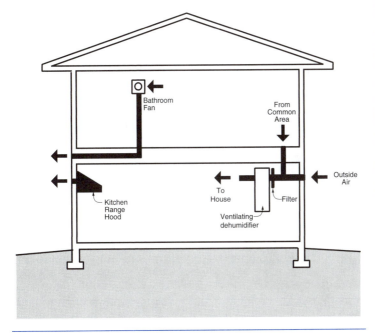

Figure 13.4
Supply Ventilation System with Dehumidification and Point Source Exhaust

- Ventilating dehumidifier brings in outside air and mixes it with air pulled from a common area (living room, hallway) to provide circulation and tempering prior to supplying to common area.
- Run time is based on time of occupancy or interior humidity.
- In supply ventilation systems, pre-filtration is recommended as debris can affect duct and fan performance reducing air supply.
- Kitchen range hood and bathroom fans provide point source exhaust as needed.
- Exhaust fans should not run continuously.
- Outside air supply can be controlled by a damper to allow dehumidifier to dehumidify without providing outside air during unoccupied periods. A/C can be also shutdown during this period allowing dehumidifier only to control interior RH.
- Outside air supply should have a closeable damper for periods when outside air is poor, i.e. smoke from fires.

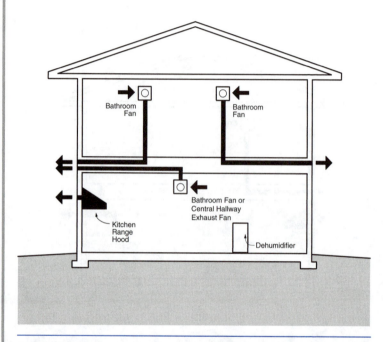

Figure 13.5
Exhaust Ventilation System with Point Source Exhaust

- Individual exhaust fans pull interior air out of bathrooms. One of these fans is selected to also serve as the exhaust ventilation fan for the entire building with a run time based on time of occupancy. Alternatively, an additional centrally located (hallway) exhaust fan can be installed.
- Replacement air is drawn into bathrooms from hallways and bedrooms providing circulation and inducing controlled infiltration of outside air.
- Kitchen range hood provides point source exhaust as needed.
- Exhaust ventilation system and A/C can be shut down during unoccupied periods and only dehumidifier operated to control interior humidity.

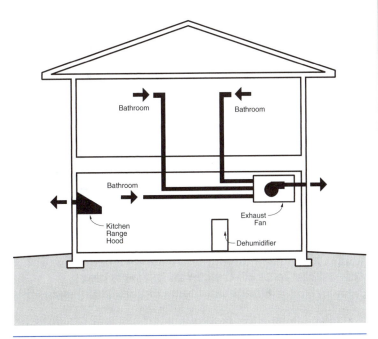

Figure 13.6
Central Exhaust Ventilation System

- A single exhaust fan is ducted to individual bathrooms pulls interior air out of bathrooms
- Replacement air is drawn into bathrooms from hallways and bedrooms providing circulation and inducing controlled infiltration of outside air.
- Run time is based on time of occupancy.
- Individual bathroom fans are eliminated.
- Kitchen range hood provides point source exhaust as needed.
- Exhaust ventilation system and A/C can be shut down during unoccupied periods and only dehumidifier operated to control interior humidity.

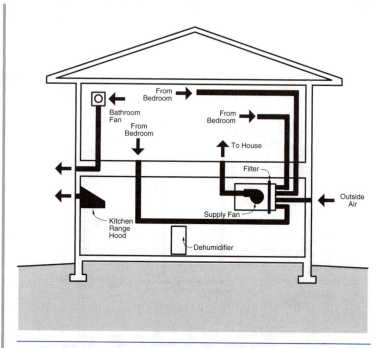

Figure 13.7
Supply Ventilation System with Circulation and Point Source Exhaust

- Supply fan brings in outside air and mixes it with air pulled from bedrooms to provide circulation and tempering prior to supplying to common area.
- Run time is based on time of occupancy.
- In supply ventilation systems, pre-filtration is recommended as debris can affect duct and fan performance reducing air supply.
- Kitchen range hood and bathroom fans provide point source exhaust as needed.
- Exhaust fans should not run continuously.
- Supply ventilation system and A/C can be shutdown during unoccupied periods and only dehumidifier operated to control interior humidity
- Outside air supply should have a closeable damper for periods when outside air is poor, i.e. smoke from fires.
- Supply ventilation system and A/C can be shut down during unoccupied periods and only dehumidifier operated to control interior humidity.

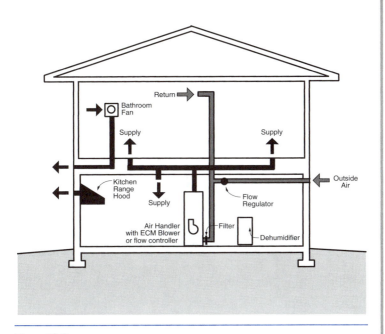

Figure 13.8
Supply Ventilation System Integrated with Heating and A/C

- Air handler operated based on time of occupancy by a flow controller pulling outside air into the return system.
- Flow controller should not allow air handler to run for 15 minutes after coil becomes de-energized to prevent re-evaporation of condensate from coils and drain pan.
- A flow regulator provides fixed outside air supply quantities independent of air handler blower speed.
- House forced air duct system provides circulation and tempering.
- Point source exhaust is provided by individual bathroom fans and a kitchen range hood.
- In supply ventilation systems, pre-filtration is recommended as debris can affect duct and fan performance reducing air supply.
- Kitchen range hood and bathroom fans provide point source exhaust as needed.
- Exhaust fans should not run continuously.
- Outside air supply can be controlled by a damper. Closing the outside air damper during unoccupied periods will allow the flow controller to periodically mix the interior air without bringing in outside air helping the dehumidifier control interior RH — humid air is brought to the dehumidifier. The cooling function of the A/C can also be shutdown during this time (i.e. A/C on blower only operation).
- Outside air supply should have a closeable damper for periods when outside air is poor, i.e. smoke from fires.

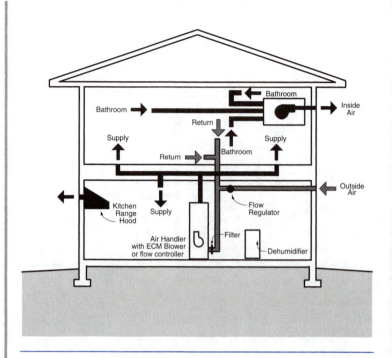

Figure 13.9
Balanced Ventilation System Using a Supply Ventilation System Integrated with Heating and A/C with a Stand Alone Central Exhaust

- The supply system integrated with heating and A/C as in Figure 13.8.
- A central exhaust system is added as in Figure 13.6. Both systems are operated simultaneously (i.e. interlocked).
- Run time is based on time of occupancy.
- Air handler operated based on time of occupancy by a flow controller pulling outside air into the return system.
- In supply ventilation systems pre-filtration is recommended as debris can affect duct and fan performance reducing air supply.
- Flow controller should not allow air handler to run for 15-minutes after coil becomes de-energized to prevent re-evaporation of condensate from coils and drain pan.
- Kitchen range hood provides point source exhaust as needed.
- Outside air supply can be controlled by a damper. Closing the outside air damper during unoccupied periods will allow the flow controller to periodically mix the interior air without bringing in outside air helping the dehumidifier control interior RH — humid air is brought to the dehumidifier. The cooling function of the A/C can also be shutdown during this time (i.e. A/C on blower only operation).
- Outside air supply should have a closeable damper for periods when outside air is poor, i.e. smoke from fires.

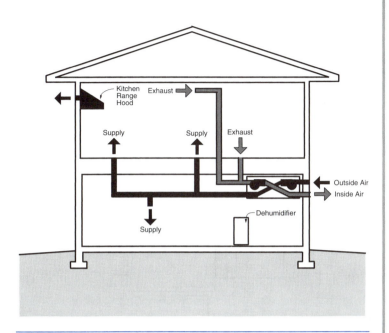

Figure 13.10
Balanced Ventilation System with Energy Recovery via an Energy Recovery Ventilator (ERV)

- The ventilation system has a separate duct system not integrated with the heating and A/C system.
- Run time is based on time of occupancy.
- Exhausts are typically from bathrooms and supplies are typically to bedrooms.
- In supply ventilation systems, and with energy recovery ventilation, pre-filtration is recommended as debris can affect duct, heat exchanger and fan performance reducing air supply.
- Energy recovery allows transfer of moisture across incoming and outgoing air flows.
- May require supplemental dehumidification.
- Outside air supply should have a closeable damper for periods when outside air is poor, i.e. smoke from fires.

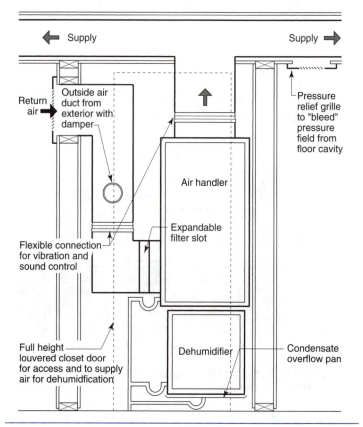

Figure 13.11
Mechanical Closet Configuration

- Air handler located and accessed within interior conditioned space
- Dehumidifier and condensate overflow pan supported by metal stand
- Dehumidifier drain plumbed to air handler condensate system
- Return ducted with two offsets to reduce sound and vibration
- Outside air provided to return side of system with damper control
- Flow controller should not allow air handler to run for 15 minutes after cooling is de-energized to prevent re-evaporation of condensate from coils and drain pan
- Outside air supply can be controlled by a damper. Closing the outside air damper during unoccupied periods will allow the flow controller to periodically mix the interior air without bringing in outside air helping the dehumidifier control interior RH — humid air is brought to the dehumidifier. The cooling function of the A/C can also be shutdown during this time (i.e. A/C on blower only operation).
- Pressure relief grille provided to "bleed" accidental/incidental interstitial pressure field created by duct leakage

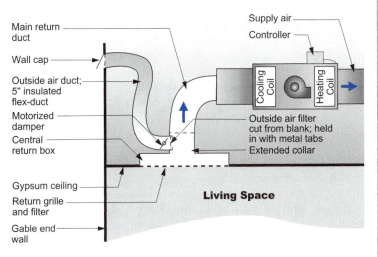

Conditioned Attic

Main return duct

Wall cap

Outside air duct; 5" insulated flex-duct

Motorized damper

Central return box

Supply air

Controller

Cooling Coil

Heating Coil

Outside air filter cut from blank; held in with metal tabs

Extended collar

Gypsum ceiling

Return grille and filter

Gable end wall

Living Space

13

Figure 13.12
Supply Ventilation System with Horizontal AHU in Attic

- Air handler operation draws air from the conditioned space (main return), as well as from the outside, via the outside air duct
- Balancing damper adjusted to provide required flow
- Outside air duct should have filter in the return (provides easiest access to change outside air filter), as well as insect screen at wall cap
- Outside air should be drawn from either a sidewall, the gable end, or a porch soffit
- Drawing from the roof is not recommended because the air can be hotter and of poor quality (i.e. off-gassing of asphalt shingles and air from exhausts and vents).
- Controller can be mounted on the air handler (as shown) or in the main space near the thermostat
- Motorized damper allows control of ventilation air duty cycle separate from air handler duty cycle

nally turned on. The air conditioning system now has to work substantially harder/longer to cool the interior.

In general, whole house fans should be both thermostatically and humidistatically controlled. Whole house fans should not operate when the exterior temperature is greater than 75°F or when the exterior relative humidity is greater than 60 percent with an exterior temperature greater than 75°F.

Whole house fans can also significantly depressurize building enclosures. Combustion appliance spillage, radon and soil gas infiltration concerns should be addressed by providing sub-slab ventilation systems, and by not using combustion appliances, such as fireplaces, when whole house fans are operating. Windows need to be open to provide make-up/relief air when whole house fans are operating. Additionally, adequate roof venting must be provided to prevent over-pressurization of attic/roof assemblies when exhaust air is dumped into the attic.

Ceiling Fans

Ceiling fans can provide significant comfort benefits while simultaneously displacing air conditioning energy. With ceiling fans and the associated air flow, people can feel more comfortable at higher thermostat settings (thereby reducing A/C run times). As a general rule, thermostats can be set approximately 4°F higher without affecting comfort if the ceiling fan provides air movement at 100 to 150 feet per minute. The table below provides information on sizing ceiling fans.

Largest Room Dimension	Minimum Fan Diameter (inches)
12 feet or less	36
12 - 16 feet	48
16 - 17.5 feet	52
17.5 - 18.5 feet	56
18.5 or more feet	2 fans

System Sizing

Equipment should be sized correctly and return air flow paths should be planned. If a similar floor plan is constructed several times in a subdivision and sited with different orientations, heat gain and heat loss calculations should be done for each orientation. Equipment should be specifically selected for each orientation.

Incorrectly sized equipment can lead to operational and cost problems. Oversizing or undersizing heat pump systems with resistance strip back-up (supplemental) heat can alter the thermal balance point and increase resistance strip heat use and, therefore, operating costs.

Oversizing air conditioning and heat pump systems can increase cycling losses, induce high wear and lead to loss of comfort control. During cooling periods, the dehumidification capabilities of the air conditioning system are used to control interior humidity. Oversizing of air conditioning equipment can lead to high interior humidity problems in humid climates since oversized equipment will not run for extended periods of time, and, therefore, will dehumidify less than properly sized equipment.

Incorrectly sized ductwork that is improperly laid out can also lead to operational and cost problems. Heat pumps and air conditioning systems lose significant operating system efficiencies if air flow volumes across the coils are reduced as a result of improperly sized ducts, duct leakage or blockages due to layout, improper installation or a lack of servicing. If equipment is not located so it is accessible, dirty coils and dirty filters will occur from a lack of servicing and result in a reduction of air flow. Too little air across the indoor coil can potentially lower the coil temperature to the point of ice formation and create serious damage.

Correct heat gain and heat loss determination is necessary in order to size and select equipment and systems. The Air Conditioning Contractors Association (ACCA) provides a recognized standard procedure in the publication, *Manual J*. Size of ductwork and distribution system calculation procedures are outlined in a second publication, *Manual D*. However, users of these procedures are cautioned against using default values or rules of thumb for air leakage inputs.

Exterior coils of air conditioning and heat pump systems located under decks or adjacent to vegetation will experience recirculation of air, resulting in greater operating losses.

Incorrectly selected controls, or controls adjusted incorrectly, can result in significant operational losses. Strip heat should be kept off with an outdoor thermostat that only allows operation when outdoor temperature is below the calculated balance point (e.g. 20°F). Heat strips should be installed in banks of 5 kW or less and each bank should have its own outdoor thermostat.

Thermostats for forced air systems are low voltage and typically have a built-in anticipation circuit, and a manual changeover switch or a heating/cooling lockout to prevent cross-cycling ("dueling") between heating and cooling modes. If thermostats have setback settings, they should have a ramped recovery or intelligent recovery feature that limits use of heat strips (supplementary heat) during the recovery period.

Improperly charged systems can significantly affect efficiency. Overcharging or undercharging refrigerants by 10 or 15 percent reduces equipment efficiency by 10 or 15 percent.

Incorrectly selected blower speeds can also result in significant operational losses. Blower speeds are usually different for cooling modes and heating modes. Approximately 425 to 450 cfm of air flow per ton of cooling is typically required over dry coils (400 cfm per ton for wet coils). Significantly less air flow is usually required through the same system during heating periods, and may vary further under different heating modes. Three to five blower speed settings are common and are usually set manually. Some units use two speeds, one for cooling, and a lower speed for heating. Blowers that automatically adjust for changes in duct resistance are also available.

Air Handlers and Ductwork

Furnaces, air handlers and ductwork should always be located within conditioned spaces and allow for easy access to facilitate servicing, filter replacement, drain pan cleaning, and future replacement as technology improves. Furnaces, air handlers and ductwork should not be located in vented attics, vented crawlspaces or garages. Ductwork should not be located in exterior walls or in concrete floor slabs. See Figure 13.40 for suggested conceptual ductwork layouts. In designs where ducts are unavoidably located in an unconditioned space, they should be sealed airtight and insulated (Figures 13.13 through 13.18).

Proper location of supply registers and return grilles is important to good system performance. The best approach involves either high supply registers and low return grilles, or low supply registers and high return grilles. Care should be taken to locate supply registers so conditioned air is not blown directly on people occupying the space.

Locate air conditioning supply registers so that cold air is not blown directly at or across wall and ceiling surfaces. Improper placement can potentially chill these surfaces below dew point temperatures and cause mold growth and damage to interior finishes.

Ductwork, furnaces and air handlers should be sealed against air leakage. The only place air should be able to leave the supply duct system and the furnace or air handling unit is at the supply registers. The only place air should be able to enter the return duct system and the furnace or air handling unit is at the return grilles. A forced air system should be able to be pressure tested the way a plumber pressure tests a plumbing system for leaks. Builders don't accept leaky plumbing systems, they should not accept leaky duct systems.

Supply systems should be sealed with mastic in order to be airtight. All openings (except supply registers), penetrations, holes and cracks should be sealed with mastic or fiberglass mesh and mastic. Tape, especially duct tape, does not work and should not be used. Sealing of the supply system includes sealing the supply plenum, its attachment to the air handler or furnace, and the air handler or furnace itself. Joints, seams and openings on the air handler, furnace or ductwork near the air handler or furnace should be sealed with both fiberglass mesh and mastic due to greater local vibration and flexure. See Figures 13.13 through 13.18 for suggested ways to air seal your ductwork.

Return systems should be "hard" ducted and sealed with mastic in order to be airtight. Building cavities should never be used as return ducts. Stud bays or cavities should not be used for returns. Panned floor joists should not be used. Panning floor joists and using stud cavities as returns leads to leaky returns and the creation of negative pressure fields within interstitial spaces. Carpet dust marking at baseboards, odor problems, mold problems and pollutant transport problems typically occur when building cavities are used as return ducts.

The return side of the air handler or furnace also must be sealed, especially the filter access. The filter access should be easy to get at and have a gasketed airtight fitting door. Gaskets also need to be used around the filters in order to avoid bypass of air around the filters.

Sheet Metal Duct Sealing

The longitudinal seams and transverse joints in sheet metal ducts should be sealed (Figures 13.14 and 13.19). The inner liner of insulated plastic flex duct should be sealed where flex ducts are connected to other ducts, plenums, junction boxes and boots/registers (Figure 13.15 through Figure 13.18).

Flex Duct Sealing

In flex duct installation, the outer liner and insulation should be pulled back and the inner liner attached to the collar with a tie. Fiberglass mesh tape (fabric) should be installed over the inner liner and collar such that at least 1 inch of fiberglass mesh tape covers the exposed collar. Mastic is then applied over the fiberglass mesh tape. The insulation and the outer liner is then pulled back over the connection and sealed with a second tie (Figure 13.18). When flex ducts are used, care must be taken to prevent restricting air flow by "pinching" ducts.

Sealing Boots and Grilles

Connections between grilles, registers and ducts at ceilings, floors or knee walls typically leak where the boot does not seal tightly to the grille or gypsum board. Air from the attic, basement, or crawlspace can

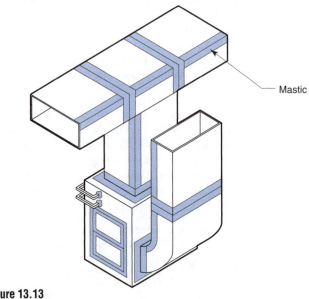

Mastic

Figure 13.13
Air Handler Air Sealing

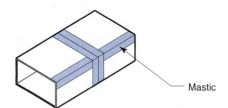

Mastic

Figure 13.14
Rigid Duct Air Sealing

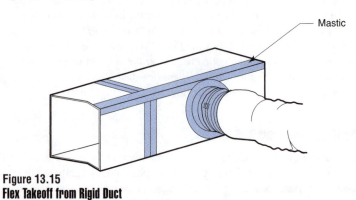

Mastic

Figure 13.15
Flex Takeoff from Rigid Duct

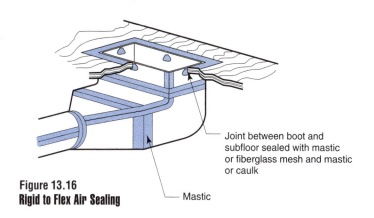

Joint between boot and subfloor sealed with mastic or fiberglass mesh and mastic or caulk

Mastic

Figure 13.16
Rigid to Flex Air Sealing

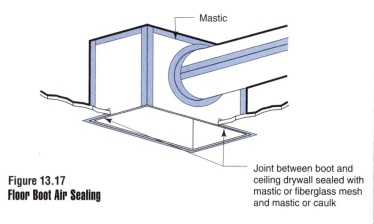

Mastic

Joint between boot and ceiling drywall sealed with mastic or fiberglass mesh and mastic or caulk

Figure 13.17
Floor Boot Air Sealing

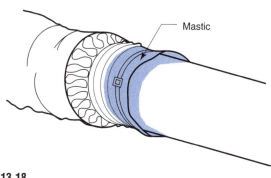

Mastic

Figure 13.18
Ceiling Boot Air Sealing

be drawn into the return. Leaks can also exist within the boot and where the ducts connect to the boot. Several examples of recommended installations are given in Figures 13.19 through 13.22.

If the gap between boots and gypsum board opening or subfloor openings is kept to less than $^3/_8$", a bead of sealant or mastic may be used to seal the gap. Where gaps are larger than $^3/_8$", fabric and mastic should both be used. The optimum approach is to keep the gaps to less than $^3/_8$" and use a bead of sealant. This requires careful coordination with the drywall contractor to make sure that the rough openings for the boots are cut no more than $^3/_8$" bigger than the actual boot size on all sides.

Plenums

When a return plenum draws directly through a wall, the wall cavity may inadvertently become a duct. If the penetration through the wall is not blocked and sealed, return leaks can occur with air drawn from the wall cavity. The wall cavity should be isolated from the return (Figures 13.25 and 13.26).

Sometimes, return plenums leak through the floor. In floors with crawlspaces or basements, the plenum floor may not be airtight, allowing air from those zones to be drawn into the return.

In some slab homes, a 4"-diameter chase pipe enters the plenum. The chase pipe carries the refrigerant lines, condensate piping, and control wiring which connect the indoor and outdoor units. This chase pipe is frequently unsealed, allowing unconditioned air or soil gases (radon, pesticides, herbicides, moisture) to be drawn into the return. Chases should never terminate inside the return air stream.

Return plenums are sometimes formed by the enclosed space below the air handler support platform. This plenum will leak to adjacent walls and directly to the space in which it is located. A return plenum in an air handler closet may have no gypsum board separating it from an adjacent tub enclosure. As a result, air may be drawn from the attic. The adjacent walls often have plumbing and wiring in them that either comes from the attic, crawlspace, garage, basement, outside, or some other interior space. Many of these platforms are lined with insulation or fibrous duct board because of fire codes and soundproofing requirements. This lining is not an air barrier and leakage will occur if the joints and penetrations are not sealed.

In designs where ducts are unavoidably located in an unconditioned space, they should be sealed airtight and insulated. Penetrations of ductwork through building enclosures should also be sealed (Figure 13.39).

Leakage can also occur at the connection between the air handler and the support platform. All sides of the air handler must be sealed to the support platform. Supply plenums also leak at the seams, particularly sleeved plenums. The preferred approach to seal supply and return plenums and gaps at air handlers is to use both fabric and mastic due to the greater flexure and vibration typically present near air handlers.

Return Path

All supply registers should have clear access to a return grille in order to prevent the pressurization of bedrooms and the depressurization of common areas. Bedrooms should either have a direct-ducted return or a transfer grille. Undercutting of bedroom doors rarely works and should not be relied upon to relieve bedroom pressurization. A central "hard" ducted return that is airtight and coupled with transfer grilles to relieve bedroom pressurization significantly outperforms a return system with leaky ducted returns in every room, stud bays used as return ducts and panned floor joists. See Figure 13.34 for an effective transfer grille and Figure 13.35 for a jump duct detail.

The design of HVAC ductwork is important as to whether extensive mold growth will occur or not. If debris can readily collect on the internal surface of ductwork, mold growth is more likely. On the other hand, if the internal surface is smooth like bare galvanized metal, debris is much less likely to collect and mold to grow. For most residential homes, the use of internal duct liners in the HVAC system is not necessary or desired. Porous liners (fibrous glass) should never be used for sound attenuation. Normally liners are only necessary near fan discharge and inlet plenums. If liner is used, only "Tough" liner (smooth coated fibrous glass) should be specified and then only for short distances from the fan (normally less than ten feet). Internal liners should never be used in fresh air intakes or return air ductwork. Internal porous liners should never be used near moisture sources, only a smooth surface closed cell foam liners should be used. Air handling units, if insulation is necessary, should be insulated on the exterior or double walled (metal insulation metal). If perforated metal is used adjacent to the airstream, the fibrous glass inside the liner needs to be protected by plastic film.

Cooling Coils and Drain Pans

The most common mold amplification sites identified are the cooling coils and adjacent areas of the HVAC system. The cooling coils discussed here are not the condenser coils located outside your house, these are located inside ductwork inside the house. There are three potential growth sites in the cooling coil area:

- The cooling coils — Debris can build up on the cooling coils (especially on the upstream side of the coils) and act as a food source for the mold. This area is a high humidity area during the cooling season.

- The cooling coil drainage area — Debris can build up in the drainage pans and act as a food source for the mold to grow in this area. In addition, the debris can block the drainage from the cooling coils and cause water to be sprayed or leaked onto the ductwork downstream from the cooling coils. Drainage systems which are not properly engineered will not drain and will allow standing water to accumulate in the drainage pans and/or adjacent ductwork. Double-sloped smooth surface drainage pans are recommended. Drainage pans must slope to the middle of the pan and the middle should slope to one end and a trap installed. Pitch of $1/4$-inch per foot is acceptable but this slope needs to be in both directions. The depth of the trap should be 1 x the static pressure present in the cooling coil drainage pan location. Normally a 3-inch trap is needed for most residential homes. The drainage pan must drain to leave puddles no larger than 2-inches in diameter and no more than $1/8$-inch deep.

- Ductwork downstream and adjacent to cooling coils — Internal ductwork surfaces within 10 feet up and downstream of the cooling coils are the most critical. Internal ductwork surfaces need to have a smooth washable surface to avoid the collection of debris. Internal ductwork surfaces which have a rough surface will collect debris from the air being circulated. Mold grows on the debris which has collected on the internal surfaces. Rough, porous internal liners (fibrous glass) should not be used in high humidity areas of the HVAC system because these materials readily collect debris and are impossible to clean. The goal in the operation of the HVAC system is to control the population of mold. All cooling coil areas need to have easy accessible "gasketed clean-out doors" upstream and downstream of cooling coils. A single large access door, which allows access to both the up and downstream areas of the cooling coils is adequate in many cases. The cooling coil area needs to be cleaned with a cleaning solution in the early spring or late fall of each year. This cleaning needs to include ductwork three feet upstream and downstream of the cooling coils.

Air Filtration

One of the best methods of controlling dirt and debris in the air conveyance system is using good quality filters to remove particulates from the airstream. In the past, it has been common to use one-thick low efficiency "boulder catcher" filters. These are lo longer considered adequate. Air filters are primarily tested using the American Society of Heating, Refrigeration and Air Conditioning Engineers Society (ASHRAE) Standard 52.2. Filters are rated from a MERV 1 (typical furnace filter) to MERV 16 (typical HEPA filter). A practical filter which does an excellent job is a MERV 8 filter. A big problem with better filters is the pressure drop across the filter (how hard it is to move air through the filter). The higher the pressure drop, the harder it is to move air through the filter.

Large Exhaust Fans

Large exhaust fans and appliances such as whole house fans, attic ventilation fans, indoor grills, clothes dryers, fireplaces and kitchen exhaust range hoods can significantly depressurize buildings.

Whole house fans should only be used with windows open in order to relieve building pressures.

Attic ventilation fans should never be installed. A correctly constructed attic makes attic ventilation fans unnecessary. An incorrectly constructed attic should be repaired, not saddled with an energy wasting and problem-creating attic ventilation fan.

Kitchen exhaust range hoods should not be oversized. Anything larger than 100 cfm exhaust capacity for a kitchen exhaust range hood should be carefully integrated into the entire design of the building. Indoor barbecue grills should only be installed with a provision for make-up air.

Clothes Dryers

Nothing practical can be done with the large unbalanced air flows created by clothes dryers except to reduce the impact of the negative effects by installing only sealed combustion, direct vent, induced draft or power vented combustion furnaces and water heaters. By the way, never duct a clothes dryer into the house. The resulting moisture load is more difficult to deal with than an intermittent pressure imbalance. Additionally, lint and other particulates are known to aggravate allergies and contribute to dustmarking on interior finishes.

It is important that dryers be located so that exhaust ducting can be installed within the manufacturers' limitations on length and number of elbows. If long lengths and excessive elbows are unavoidable a booster fan designed for dryer use should be provided.

Central Vacuum Systems

These units should discharge to the outside of the house due to indoor air quality concerns. Their impact on house pressure is minimal because of run time.

1

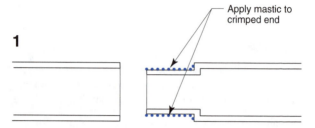

Apply mastic to
crimped end

2

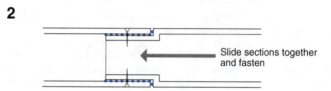

Slide sections together
and fasten

3

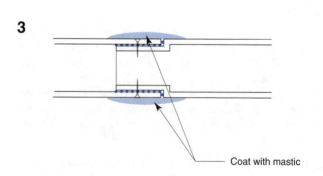

Coat with mastic

Figure 13.19
Sheet Metal Duct Sealing

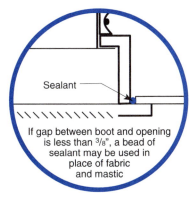

Sealant

If gap between boot and opening is less than ³/₈", a bead of sealant may be used in place of fabric and mastic

Alternative Detail

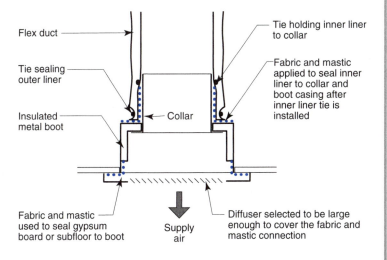

Flex duct

Tie sealing outer liner

Insulated metal boot

Collar

Tie holding inner liner to collar

Fabric and mastic applied to seal inner liner to collar and boot casing after inner liner tie is installed

Fabric and mastic used to seal gypsum board or subfloor to boot

Supply air

Diffuser selected to be large enough to cover the fabric and mastic connection

Figure 13.20
Grille and Boot Connections at Ceiling

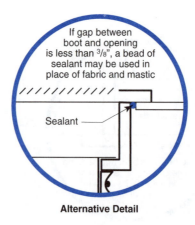

If gap between boot and opening is less than 3/8", a bead of sealant may be used in place of fabric and mastic

Sealant

Alternative Detail

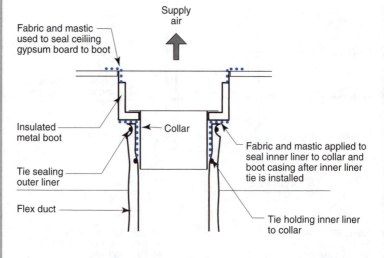

Supply air

Fabric and mastic used to seal ceiliing gypsum board to boot

Insulated metal boot

Collar

Fabric and mastic applied to seal inner liner to collar and boot casing after inner liner tie is installed

Tie sealing outer liner

Flex duct

Tie holding inner liner to collar

Figure 13.21
Grille and Boot Connections at Floor

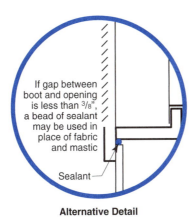

If gap between boot and opening is less than $3/8$", a bead of sealant may be used in place of fabric and mastic

Sealant

Alternative Detail

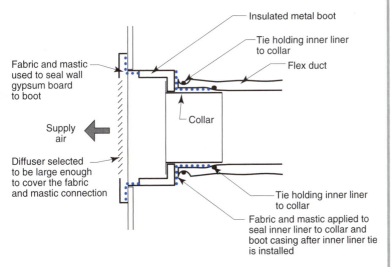

Insulated metal boot

Tie holding inner liner to collar

Flex duct

Fabric and mastic used to seal wall gypsum board to boot

Collar

Supply air

Diffuser selected to be large enough to cover the fabric and mastic connection

Tie holding inner liner to collar

Fabric and mastic applied to seal inner liner to collar and boot casing after inner liner tie is installed

Figure 13.22
Grille and Boot Connections at Wall

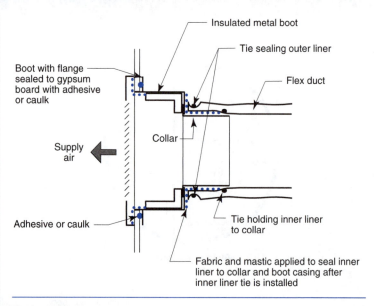

**Figure 13.23
Boot Connection with Flange**

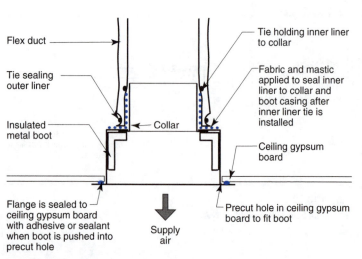

Note: Boot installed after gypsum board

**Figure 13.24
Ceiling Boot Connection**

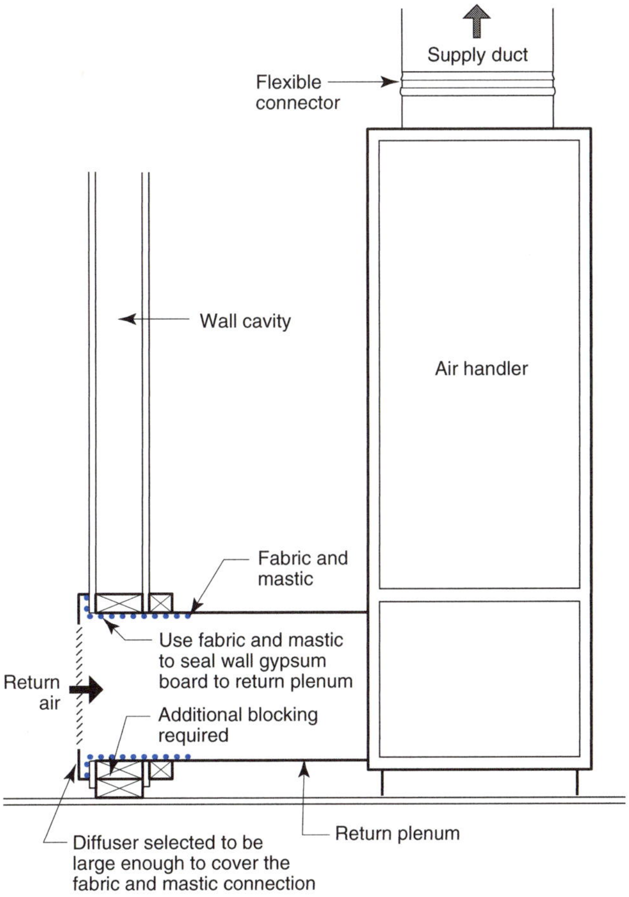

Supply duct

Flexible
connector

Wall cavity

Air handler

Fabric and
mastic

Use fabric and mastic
to seal wall gypsum
board to return plenum

Return
air

Additional blocking
required

Diffuser selected to be
large enough to cover the
fabric and mastic connection

Return plenum

Figure 13.25
Sealing Return Plenums

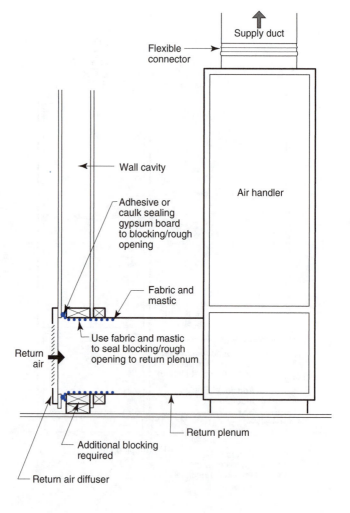

Supply duct

Flexible connector

Wall cavity

Adhesive or caulk sealing gypsum board to blocking/rough opening

Air handler

Fabric and mastic

Use fabric and mastic to seal blocking/rough opening to return plenum

Return air

Additional blocking required

Return air diffuser

Return plenum

Figure 13.26
Sealing Return Plenum Alternative

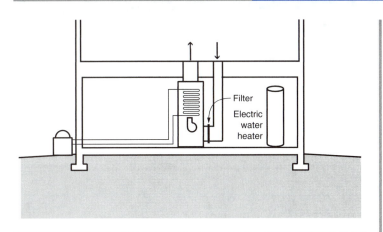

Figure 13.27
Air-to-Air Heat Pump
- Heating and cooling provided by an electrically driven heat pump with exterior air used as a heat source/sink

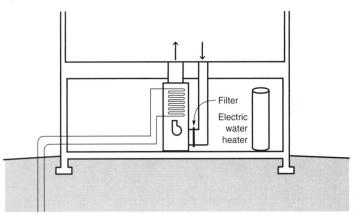

To underground loop

Figure 13.28
Ground Source Heat Pump
- Heating and cooling provided by an electrically driven heat pump with ground used as a heat source/sink

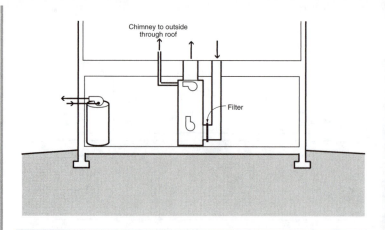

Figure 13.29
Sealed Combustion Power Vented Gas Water Heater, Induced Draft Gas Furnace
- Water heater combustion air supplied directly to water heater from exterior via duct; product os combustion exhausted directly to exterior also via duct
- Furnace flue gases exhausted to the exterior using a fan to induce draft; combustion air taken from the interior

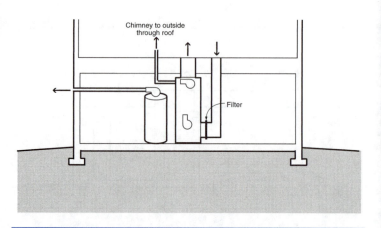

Figure 13.30
Power Vented Gas Water Heater, Induced Draft Gas Furnace
- Water heater flue gases exhausted to exterior using a fan to maintain draft; combustion air taken from the interior
- Furnace flue gases exhausted to the exterior using a fan to induce draft; combustion air taken from the interior

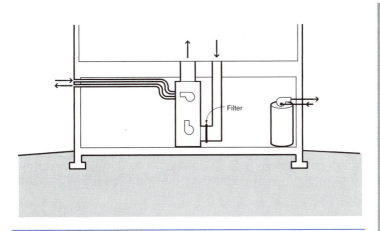

Figure 13.31
Sealed Combustion Power Vented Gas Water Heater, Sealed Combustion Power Vented Furnace

- Water heater combustion air supplied directly to water heater from exterior via duct; products of combustion exhausted directly to exterior also via duct
- Furnace flue gases exhausted to the exterior using a fan; combustion air supplied directly to furnace from exterior via duct

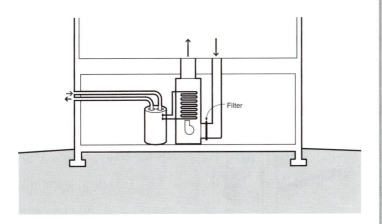

Figure 13.32
Sealed Combustion Power Vented Gas Water Heater

- Water heater flue gases exhausted to the exterior using a fan; combustion air supplied directly to water heater from exterior via duct
- No furnace; heat provided by hot water pumped through a water-to-air-heat exchanger (fan-coil)

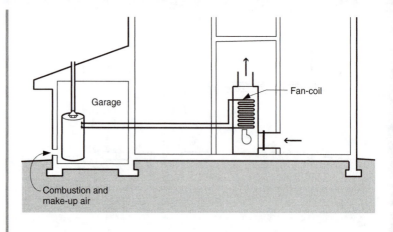

Figure 13.33
Traditional Gas Water Heater with a Water-to-Air Heat Exchanger
- Standard/traditional gas water heater with a draft hood is located out of the conditioned space (exterior to the building enclosure "pressure boundary") in a garage supplied with combustion and make-up air.
- No furnace; heat is provided by hot water pumped through a water-to-air heat exchanger (fan-coil).

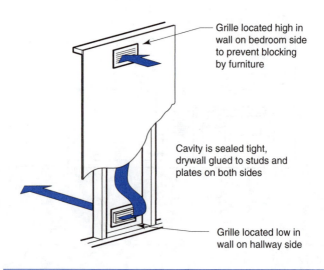

Figure 13.34
Transfer Grille — Offset
- Relieves pressure differences between spaces
- Typically 8x16 or 10x16
- Door undercut of 1" minimum still required

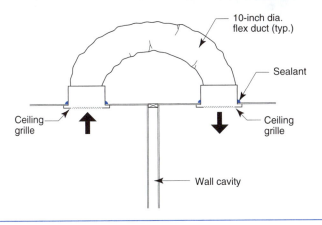

**Figure 13.35
Jump Duct**
- Relieves pressure differences between spaces
- Door undercut of 1" minimum still required

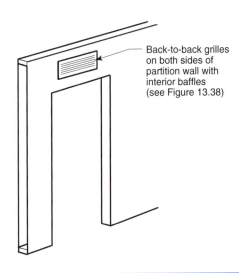

**Figure 13.36
Transfer Grille — Over Door Opening**
- Relieves pressure differences between spaces
- Interior baffles control sound and light transfer
- Door undercut of 1" minimum still required

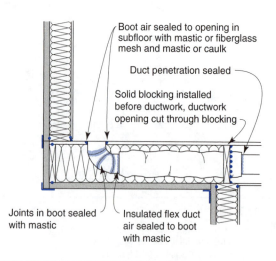

Boot air sealed to opening in subfloor with mastic or fiberglass mesh and mastic or caulk

Duct penetration sealed

Solid blocking installed before ductwork, ductwork opening cut through blocking

Joints in boot sealed with mastic

Insulated flex duct air sealed to boot with mastic

Figure 13.37
Transfer Grille — Construction

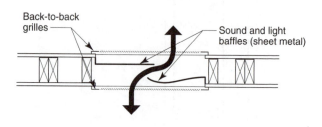

Back-to-back grilles

Sound and light baffles (sheet metal)

Figure 13.38
Transfer Grille — Section
• Typically 6x20

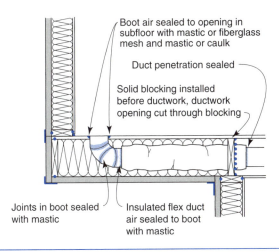

Boot air sealed to opening in subfloor with mastic or fiberglass mesh and mastic or caulk

Duct penetration sealed

Solid blocking installed before ductwork, ductwork opening cut through blocking

Joints in boot sealed with mastic

Insulated flex duct air sealed to boot with mastic

 13

Figure 13.39
Ducts in Cantilevered Spaces or Serving Bedrooms Over Garages

- Avoid whenever possible
- Use insulated ducts
- Air seal all joints with mastic including boot penetrations through subfloors and duct penetrations through draftstops
- Consider spraying boot exterior with foam insulation

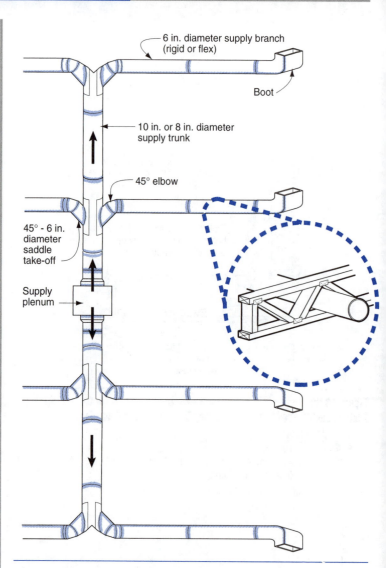

Figure 13.40
Supply Duct System
- Must be sized/designed on a case-by-case basis
- Supply duct system sized to fit within 14"-deep or 12"-deep open webbed floor trusses.
- Trunk ducts can be 10"-diameter for 14"-deep floor trusses or 8"-diameter for 12"-deep floor trusses depending on air flows and heat losses
- Branch supply ducts can be insulated flex
- Rounded ducts, 45° takeoffs and 45° elbows reduce air flow resistance so ducts can be made smaller to fit in floor system

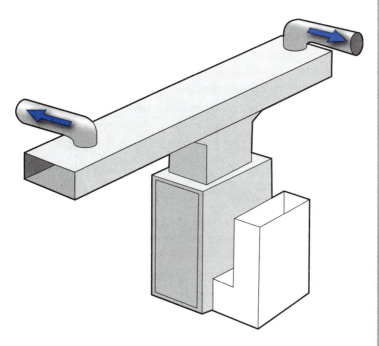

Figure 13.41
Supply Duct System
- Supply ducts taken off top of supply trunk duct
- AHU located at midpoint of supply trunk duct
- Note "dropped tee" supply plenum (see Figure 13.44)

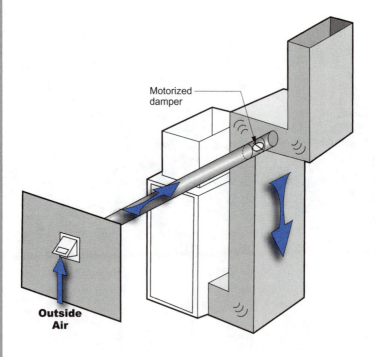

Motorized damper

Outside
Air

Figure 13.42
Return Duct System

- Outside air cut connection to return located far enough from AHU to allow mixing
- Motorized damper regulates outside air

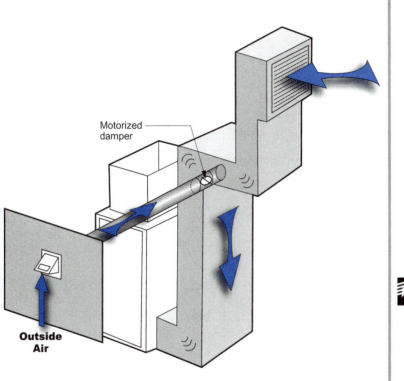

Motorized damper

**Outside
Air**

13

Figure 13.43
Return Duct System

- Outside air duct connection to return located far enough from central return in region of high negative pressure to facilitate flow (suction) in outside air duct
- Motorized damper regulates outside air

Supply plenum located at midpoint of supply trunk

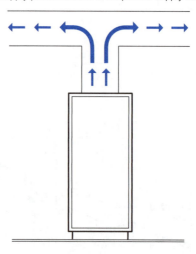

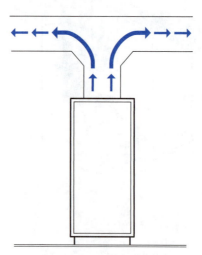

"Dropped tee" supply plenum reduces resistance to flow and noise

Figure 13.44
Supply Plenum
• Significant reduction in pressure and increase in efficiency

Return plenum is offset for noise control and turning vanes to reduce resistance and noise

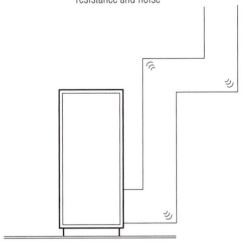

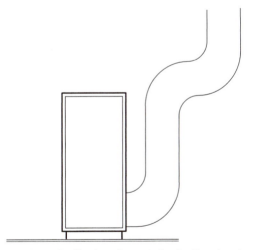

Radius return plenum is offset for noise control and radius return to reduce resistance and noise

Figure 13.45
Return Plenum
- Every 90° change in direction is equal to 15 ft. of equivalent floor resistance
- Radius sections reduce resistance by almost half

Construction

Plumbing

Plumbing systems have to:

- supply cold water

- supply hot water

- remove gray water and solid wastes

- not leak water, odors or air

Concerns

Plumbing system penetrations can be a major source of air leakage. Don't put plumbing in outside walls. Let us repeat that for those that may not get it. Don't put plumbing in outside walls. Holes in outside walls cause drafts; outside walls get cold. Pipes freeze. Owners get annoyed. Get it?

Plumbing penetrations through rim joists should be sealed with expanding foam or caulk. Vent stacks penetrating into attics should be sealed with flexible seals to handle expansion of pipes. See Figures 14.1 through 14.3 for details.

Openings in concrete slabs for tubs and showers must be sealed with concrete; avoid "earth" or "dirt" surfaces.

Tubs and Shower Stalls

Tubs, shower stalls, and one-piece tub-shower enclosures installed on exterior walls can be one of the single largest sources of air leakage across a building enclosure. It is essential that rigid draftstopping material is installed prior to tub and shower stall installation. With one-piece tub-shower enclosures, the entire height of the interior surface of the exterior wall should be insulated and sheathed prior to tub-shower enclosure installation. See Figures 11.30 through 11.32 for details.

Water Consumption

Low-flow toilets and shower heads should be installed to minimize water consumption. Pressure balanced shower controls should be used to reduce the dangers of scalding.

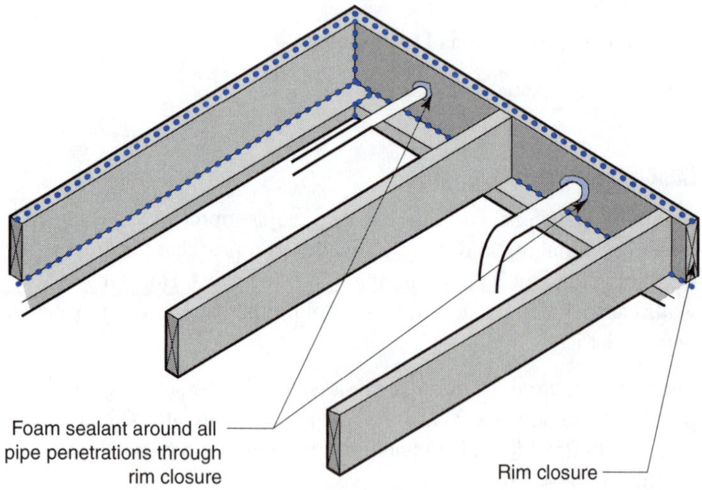

Foam sealant around all
pipe penetrations through
rim closure

Rim closure

Figure 14.1
Rim Penetrations

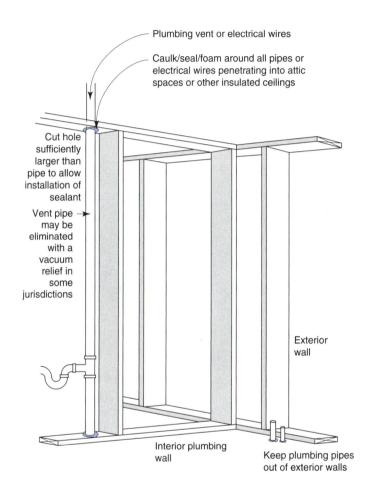

Plumbing vent or electrical wires

Caulk/seal/foam around all pipes or electrical wires penetrating into attic spaces or other insulated ceilings

Cut hole sufficiently larger than pipe to allow installation of sealant

Vent pipe → may be eliminated with a vacuum relief in some jurisdictions

Exterior wall

Interior plumbing wall

Keep plumbing pipes out of exterior walls

Figure 14.2
Locating Plumbing Pipes
- Where connections are made to toilets and pedestal sinks and exposed piping, chrome plated or other surface finished materials should be considered for aesthetic reasons
- Sealants should be flexible, non-hardening

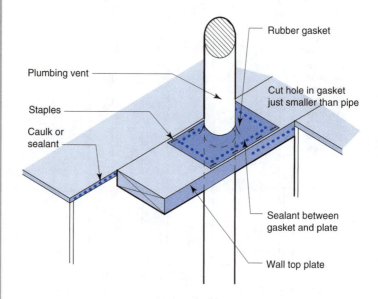

Rubber gasket

Plumbing vent

Cut hole in gasket just smaller than pipe

Staples

Caulk or sealant

Sealant between gasket and plate

Wall top plate

Figure 14.3
Vent Stack Penetration to Attic

- Prefabricated roof vent pipe flashings can be adapted to use as air sealing gaskets

Electrical

Electrical systems have to:

- supply electricity
- supply communication and control signals
- not leak air

Concerns

Electrical system penetrations through the building enclosure can be a major source of air leakage.

Airtight outlet boxes should be installed in exterior walls and insulated ceilings. Specialized boxes are available. Alternatively, sealants can be used to seal penetrations in standard outlet boxes. See Figure 15.2 for details.

Electrical penetrations through rim joists should be sealed with expanding foam or caulk. Wires penetrating into attics, and through top and bottom plates in exterior walls should be sealed with expanding foam or caulk. Air can also leak through service penetrations in studs where interior walls intersect exterior walls. These penetrations should also be sealed (Figure 15.3).

Recessed light fixtures in insulated ceilings should be insulation cover (IC) rated fixtures which are airtight and can be covered with insulation (Figure 15.4). Recessed light fixtures installed in dropped ceilings or soffits need to be draftstopped (Figure 15.5).

Where electrical panels are installed on exterior walls, air sealing of all penetrations is necessary (Figure 15.1).

Wires should be located along plates or against studs rather than through the center of insulated cavities to minimize insulation compression where batt insulation is used (Figure 15.4).

15

Lighting fixtures, locations and approaches should be selected in conjunction with daylighting design. Energy efficient lighting fixtures, bulbs and controls should be specified.

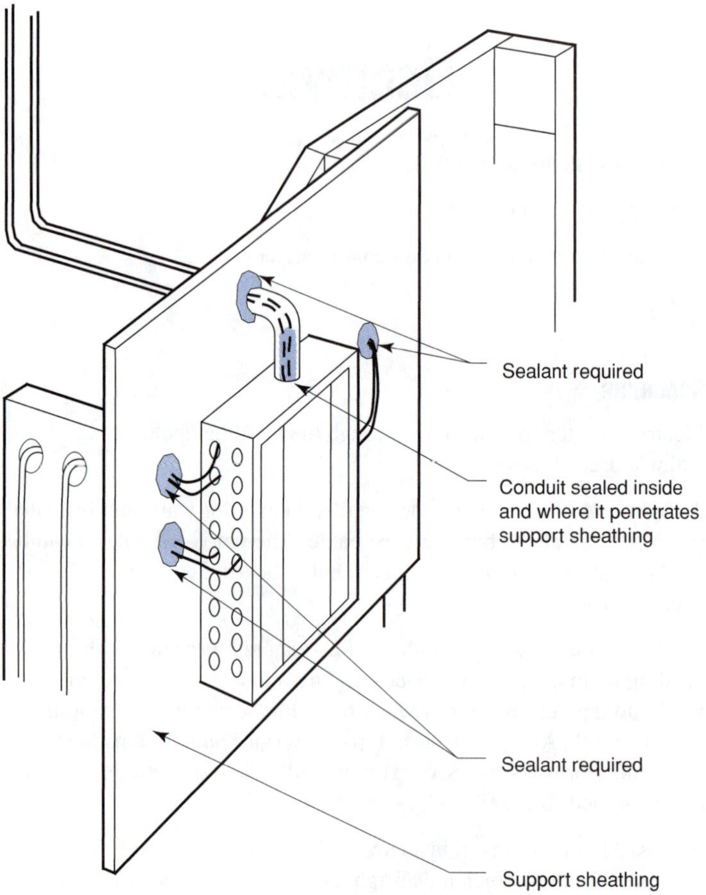

Sealant required

Conduit sealed inside and where it penetrates support sheathing

Sealant required

Support sheathing

Figure 15.1
Electrical Panels

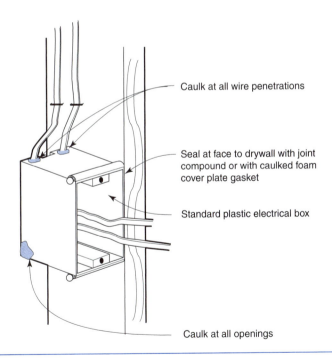

Built-in seal at wire entrance

Flange for sealing to drywall air barrier

Gasket built into box

Special air-sealing box

Nailing flange

Caulk at all wire penetrations

Seal at face to drywall with joint compound or with caulked foam cover plate gasket

Standard plastic electrical box

Caulk at all openings

Figure 15.2
Electrical Boxes

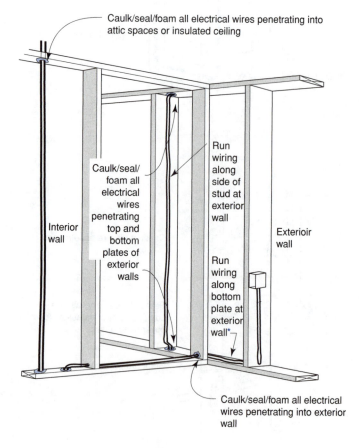

Caulk/seal/foam all electrical wires penetrating into attic spaces or insulated ceiling

Caulk/seal/ foam all electrical wires penetrating top and bottom plates of exterior walls

Interior wall

Run wiring along side of stud at exterior wall

Run wiring along bottom plate at exterior wall*

Exterioir wall

Caulk/seal/foam all electrical wires penetrating into exterior wall

15

Figure 15.3
Sealing Electrical Wires

- Run low voltage wires in plastic conduit to allow for future upgrade or service
* Some codes require wires to be held up from bottom plates 6" to 8" to protect wires from future drilling of holes through plates

Avoid placing recessed lights in insulated ceilings unless they are specifically designed to be airtight. Install IC-rated fixtures that have passed the ASTM E-283 test for air leakage.

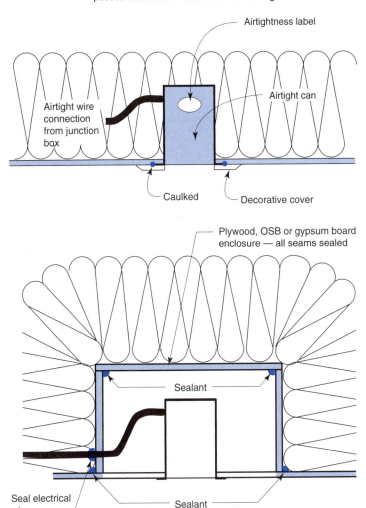

Alternate Recessed Light Box Detail

Figure 15.4
Airtight Recessed Light Box

15

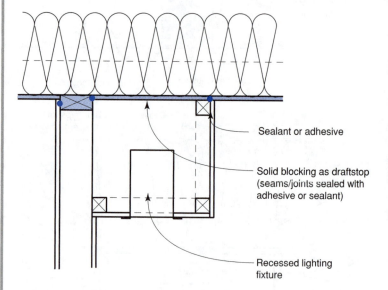

Sealant or adhesive

Solid blocking as draftstop
(seams/joints sealed with
adhesive or sealant)

Recessed lighting
fixture

Figure 15.5
Blocking and Sealing Around Recessed Lights
- Alternatively, interior soffit can be constructed after gypsum board installation
  (see Figure 11.38)

Insulation

Cavity insulation combined with insulating sheathings are common in residential wall construction. Cavity insulations are typically fiberglass batt, damp spray cellulose, dry spray cellulose, fiberglass, rock or slag wool supported by netting or reinforced polyethylene and spray foam. Insulating sheathings are typically extruded and expanded polystyrenes, foil and fiber faced isocyanurates, and rigid fiberglass. Roof insulations are typically blown fiberglass, blown cellulose, fiberglass batt, and spray foam. All have to:

• keep the heat in during the winter

• keep the heat out during the summer

Concerns

Fiberglass batt, damp spray cellulose, dry spray cellulose, blown fiberglass, rock or slag wool and blown cellulose cavity and roof insulations are not air barriers. They should be used in conjunction with air barriers. Spray foams and rigid insulations (when their joints are sealed) are air barriers in their own right.

16

Just so that we are all clear, blowing a cathedral ceiling "solid" with cellulose will not eliminate the need for an interior air barrier or the need for roof ventilation if the temperature of the first condensing surface (underside of roof deck) is not sufficiently warm by virtue of the local climate or controlled by the use of rigid insulation. Packing fiberglass around window frames or around plumbing stacks to reduce air flow, while better than nothing at all, does not effectively stop air flow. Foam sealants or caulks provide effective air sealing in these areas.

Let's also point out that properly treated cellulose is not, and we repeat is not, more of a fire hazard than fiberglass insulation. Furthermore, regarding health related risks associated with different insulation types, loose fibers and particulates of many sizes and many materials (fiberglass, rock or slag wool, cellulose) have been known to irritate many people. As long as proper precautions are taken during installation and proper containment and air sealing of these insulations is made, the health risk is negligible compared to the benefits provided by the energy savings.

To be perfectly clear: insulation is good. All insulation is good. More insulation is better. All insulation is environmentally friendly, even the rigid foams because of the sheer quantity of energy (barrels of oil not burned, pounds of carbon not dumped into the atmosphere) they save over their useful service lives. The amount of energy used to make all insulation (the embodied energy, even in the rigid foams) is trivial compared to the energy they save when used in buildings that last (don't rot and fall down) at least one mortgage period. So the key to environmentally friendly construction is durable building enclosures that are extremely well insulated.

Fiberglass

In wall cavities, fiberglass batt insulation should be cut to fit and carefully installed to completely fill the cavity. Batts should not be cut short or cut long and forced/compressed into small areas. Batts should be fluffed to full thickness and split around plumbing and wiring (Figure 16.2). Even better, move the wires so that insulation does not have to be split. Where kraft paper faced batts are used it is preferable to face staple the fiberglass batts as inset stapling can affect performance

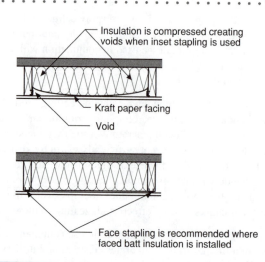

Figure 16.1
Face Stapling vs. Inset Stapling

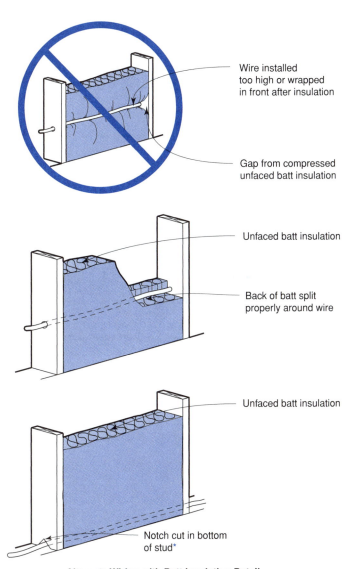

Wire installed
too high or wrapped
in front after insulation

Gap from compressed
unfaced batt insulation

Unfaced batt insulation

Back of batt split
properly around wire

Unfaced batt insulation

Notch cut in bottom
of stud*

Alternate Wiring with Batt Insulation Detail

Figure 16.2
Installing Batt Insulation in Cavity with Electrical Wiring
  * Some codes require wires to be held up from bottom plates 6" to 8" to protect
  wires from future drilling of holes through plates

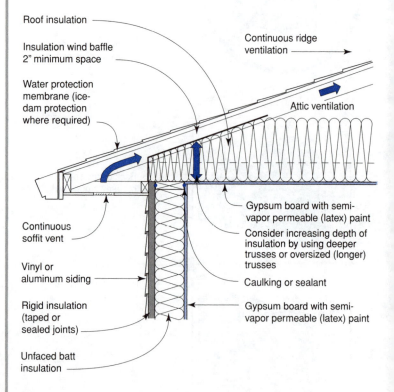

Roof insulation

Insulation wind baffle 2" minimum space

Water protection membrane (ice-dam protection where required)

Continuous ridge ventilation

Attic ventilation

Gypsum board with semi-vapor permeable (latex) paint

Consider increasing depth of insulation by using deeper trusses or oversized (longer) trusses

Caulking or sealant

Gypsum board with semi-vapor permeable (latex) paint

Continuous soffit vent

Vinyl or aluminum siding

Rigid insulation (taped or sealed joints)

Unfaced batt insulation

Figure 16.3
Baffle Installation

- Roof insulation thermal resistance (depth) at truss heel (roof perimeter) should be equal or greater to thermal resistance of exterior wall

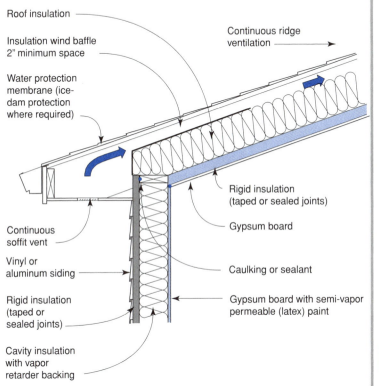

Roof insulation

Insulation wind baffle
2" minimum space

Water protection
membrane (ice-
dam protection
where required)

Continuous ridge
ventilation

Rigid insulation
(taped or sealed joints)

Gypsum board

Continuous
soffit vent

Vinyl or
aluminum siding

Rigid insulation
(taped or
sealed joints)

Cavity insulation
with vapor
retarder backing

Caulking or sealant

Gypsum board with semi-vapor
permeable (latex) paint

Figure 16.4
Baffle Installation in a Cathedral Ceiling
- Under hot, humid conditions, ventilation air brought into the roof assembly
  from the exterior will not condense on the upper surface of the rigid insulation
  (if the rigid insulation is a vapor diffusion retarder) if the R-value of the
  insulation is sufficient; R-5 or higher is recommended for mixed climates or
  hot, humid climates.

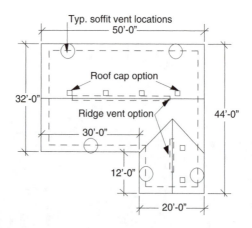

Obtain Local Code Requirement; E.G. 1:300*

1. Total sq. ft. attic
   (32 x 50) + (12 x 20) = 1840 sq. ft.

2. Total free vent required
   1850/300 = 6 sq. ft.

3. Location of vents
   50% (3 sq. ft.) at ridge
   50% (3 sq. ft.) at eave

4. Total vents at ridge (cap type)
   3 sq. ft./net free area per vent
   3 sq. ft./0.6** = 5 vents required

5. Alt. vent at ridge (strip ridge vent)
   3 sq. ft./net free area per lin. ft. of vent
   3 sq. ft./0.125** = 24 linear ft. of vent

6. Total vents at eave
   3 sq. ft./net free area per vent
   3 sq. ft./0.9** = 3 or 4 vents required

Continuous ridge combined with continuous soffit vents typically exceed code requirements for ventilated roof assemblies

Figure 16.5
Determining Attic Space Vent Area Requirements

 * 1:300 requires 1 sq. ft. of vent area for every 300 sq. ft. of ceiling area divided between the ridge soffit

 ** See vent manufacturer's literature for net free vent area of particular vent

(Figure 16.1). Yes, face stapled batts can interfere with drywall installation. If you now need a vapor retarder because you are not using faced batts, use paint. The use of higher density batts will typically improve the quality of installation.

Cellulose

Damp spray cellulose insulation should only be used in wall assemblies that are able to dry towards the interior or toward the exterior. If a vapor impermeable sheathing is installed on the exterior of a wall assembly, a vapor retarder should not be installed on the interior. Dry spray cellulose can be used in wall cavities with netting and can be used with any type of sheathing. Cellulose is not a vapor retarder. That means if you need a vapor retarder on the interior of a building assembly, you should use a vapor retarder paint. Keep in mind that polyethylene on the inside of building assemblies in mixed-humid, mixed-dry, marine, cold, hot-humid and hot-dry climates is not a good idea.

Roofs

In all truss and roof assemblies, baffles should be installed at roof perimeters to prevent the wind washing of thermal insulation and to prevent insulation from blocking soffit vents in vented roof assemblies (Figure 16.3 and 16.4).

Ventilation can effectively be used to remove moisture from roof assemblies and to control ice-damming during winter months in cold climates, mixed-humid and mixed-dry climates. Attic ventilation is most effective when half the vent area is near the ridge and half is near the eave. Typical practice requires 1 square foot of net free vent area for every 300 square feet of ceiling area (Figure 16.5).

In mixed-humid, mixed-dry, hot-humid and hot-dry climates it is not recommended that a vapor retarder be installed in vented roof assemblies. During hot, humid weather hot, humid ventilation air is brought into vented roof assemblies. Under such conditions roof assembly vapor retarders are typically on the "wrong" side of the assembly (i.e. towards the interior). It is better to not have a ceiling vapor retarder to permit drying to the interior. Under heating conditions, roof ventilation will flush moisture migrating from the interior out of vented roof assemblies.

Crawlspaces

Crawlspaces should not be vented. Crawlspaces should be constructed as mini-basements. They should be part of the conditioned space of the house. See Chapter 7.

Spray Foam

Spray foam insulations provide excellent air sealing characteristics and can be used to provide air barrier continuity at difficult details such as across rim joist/band joist assemblies. Use of low density foams result in flexible installations, forgiving of movement. Higher density foams are more abuse-resistant but are not as tolerant of movement.

Drywall

Drywall (gypsum board) has to:

- provide rigidity

- provide aesthetics

- provide fire protection

- not leak air

Concerns

Wood moves. Drywall does not move. Interesting problem. The more you attach drywall to wood, the more cracks you have. Easy, attach the drywall to less wood, and, in a manner that allows the wood to move.

Nail pops happen because as wood dries, it shrinks. Nails do not shrink. Actually, nails do not pop. The wood shrinks away from the back face of the drywall as it dries. How about getting dry wood? Sure. Better to use shorter nails. Even better, use glue. With glue, as the wood shrinks, it pulls the drywall inwards with it. But, you can't only use glue, you've got to use something until the glue begins to work. Now, shorter nails don't hold very well, and we don't want to use more of them, so use shorter screws and glue.

Wood is weird. When it shrinks, it shrinks differently along the grain than perpendicular to the grain. It shrinks much more at right angles to the grain, than along the grain. Studs don't get shorter, but they get thinner in thickness and in width (Figure 17.1 and 17.2). When we attach drywall, we need to keep this in mind especially when we box in built-up beams made out of 2x10s and 2x12s. What's nice about engineered wood is it doesn't shrink. Drywall likes engineered wood, especially above windows as header material. Drywall doesn't like big pieces of real wood.

Stairwells provide an interesting problem. With real wood floor joists (2x10s), you get more shrinkage in the $9\frac{1}{4}$ inches of floor framing (remember this is at right angles to the grain of the floor joists) than in the 8 feet of wall framing (along the grain of the studs) above and below. Old timers used to balloon frame two story-spaces for this reason, or provide control joints between the floors in the plaster or drywall. Better to use floor trusses or engineered wood joists, they don't shrink.

Truss Uplift

What can we say about truss uplift? You can't prevent it. Truss uplift occurs because of moisture content differences between the upper and lower chords of wood trusses. Moisture content differences are inevitable if one member is cold and the other member is warm. If you insulate a wood roof truss, the lower chords will be warm and the upper ones will be cold. Remember, truss uplift is not truss uplift if the owner can't see it. Let the trusses move. Floating corners for drywall attachment is the way to go. The truss moves, the drywall bends, no crack, end of story (Figure 17.4 and Figure 17.5). This is also the same principle to use at corners of exterior and interior walls. Why use three stud corners? If we attach the drywall to the wood on both sides of the corner, when the wood shrinks, the drywall cracks. Two stud corners are better. Don't attach the drywall (Figure 17.3). Let the wood move. If you are going to use a three stud corner, at least don't attach the drywall to one of the sides, just support the drywall until it is taped. Let the tape hold the corner together.

Air Barriers

One of the really nice things about drywall is that air doesn't leak through it. It leaks around it, but not through it. Tape the joints of drywall together, glue it to top and bottom plates and around window openings and presto - you have an air barrier (Figures 17.6 through 17.8). Now paint it and shazzam - you have a vapor retarder. Add draftstops and firestops out of rigid material to the framing package and we have an airtight building.

Now drywall is not the only air barrier. Polyethylene installed in a continuous fashion (seams taped or sealed) works as an air barrier ("danger, danger"; remember, only in very cold climates). So does spray foam. If you are using polyethylene, it is very important that the installation of the drywall does not cause rips and tears in the poly.

Ceramic Tile Tub and Shower Enclosures

Treated gypsum board ("green board") is typically used as a base for ceramic tile enclosures around bathtubs and showers. Don't use it. The gypsum board goes to mush (green board included, it just takes longer). Moisture gets in because grout joints are permeable to moisture. Cement boards or cement and wire lath should be used in place of gypsum board and coated with a fluid-applied waterproofing.

Winter Construction

Drywall under insulated ceilings should be $^5/_8$-inch thick. Interior and exterior walls should be framed on 24-inch centers and standard $^1/_2$ - inch drywall is fine. Don't tape under insulated ceilings in the winter if you haven't insulated. Big mistake. Try it and you will learn all about condensation and how expensive it is to put up new drywall. Yes, but I really like blown insulation and I can't blow until my drywall is up. Well, you can board first, then blow and then tape. Or, you can use a batt and blow strategy. Batt the ceiling with a little insulation when you do the walls, and then blow the loose stuff on top later.

Winter construction is always fun. Propane heaters release lots and lots of moisture. Interesting problem. You want to dry out your building while you are humidifying it. Okay, so we open some windows. Let's now heat a building with open windows. Propane heaters also release lots and lots of carbon dioxide. Joint compound does not like carbon dioxide. Bad things happen with lots of carbon dioxide, moisture and joint compound; carbonation. Carbonation is bad. Better to hook your

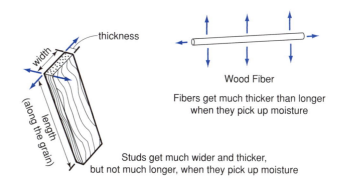

Wood Fiber

Fibers get much thicker than longer when they pick up moisture

Studs get much wider and thicker, but not much longer, when they pick up moisture

**Figure 17.1
Wood Shrinkage**

heating system up. If you have gas heat, make sure you have a chimney. The alternative is to install temporary heat, properly vented. You still need air change to flush out the moisture, but not as much as before. Moisture is bad, carbon dioxide is bad, heat and ventilation are good. Use a setting type compound with humid or cool conditions. These compounds can be selected for a range of specific properties.

Floor joist
2 x 10

Do not attach
close to beam

Do not attach
close to beam

Float corners

Shrinkage

Float corners

Nail or screw
only in center
of beam

3 - 2 x 10 built-up
beam

Shrinkage

Crimp corner
bead into dry
wall - do not nail
or screw corner
bead into beam

Shrinkage gap can
occur without
causing drywall
to crack

Figure 17.2
Built-up Beam Shrinkage
- Float corners and crimp corner bead without nails to allow for beam shrinkage

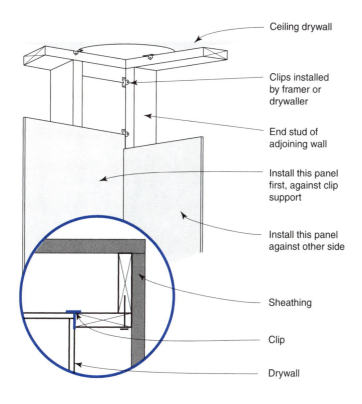

Ceiling drywall

Clips installed
by framer or
drywaller

End stud of
adjoining wall

Install this panel
first, against clip
support

Install this panel
against other side

Sheathing

Clip

Drywall

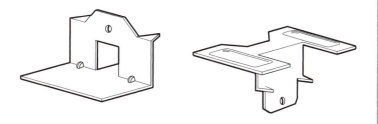

Figure 17.3
Typical Clip Support for Gypsum Board
- Use of clip support for gypsum board results in floating corners and
  significantly less drywall cracking

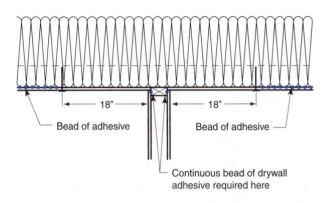

18"   18"

Bead of adhesive   Bead of adhesive

Continuous bead of drywall
adhesive required here

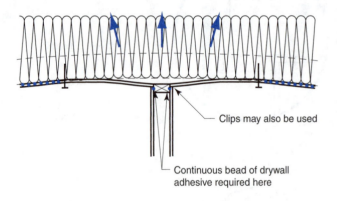

Clips may also be used

Continuous bead of drywall
adhesive required here

Figure 17.4
Truss Uplift

- Do not install ceiling drywall adhesive or ceiling drywall screws/nails closer than 18" to interior partition top plates in order to control drywall cracking from truss uplift.
- "Floating corners" of ceiling drywall allow truss movement without drywall cracking. Note that a continuous bead of drywall adhesive is required along both sides of top plates for wall drywall to provide air barrier continuity.

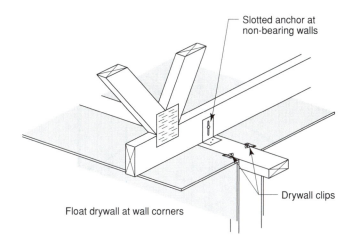

Slotted anchor at non-bearing walls

Drywall clips

Float drywall at wall corners

Figure 17.5
Drywall Clips and Slotted Anchor on Non-Bearing Wall

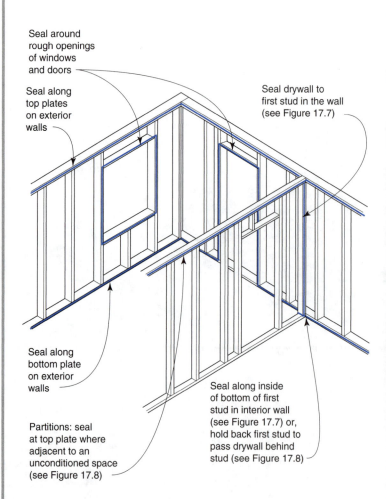

Seal around rough openings of windows and doors

Seal along top plates on exterior walls

Seal drywall to first stud in the wall (see Figure 17.7)

Seal along bottom plate on exterior walls

Partitions: seal at top plate where adjacent to an unconditioned space (see Figure 17.8)

Seal along inside of bottom of first stud in interior wall (see Figure 17.7) or, hold back first stud to pass drywall behind stud (see Figure 17.8)

Figure 17.6
Interior Air Barrier Details at Walls and Ceilings
- Use caution when installing drywall at gaskets or over polyethylene so that gaskets are not moved out of position or polyethylene is not cut or torn.
- Gasket can be used in place of sealants or adhesives

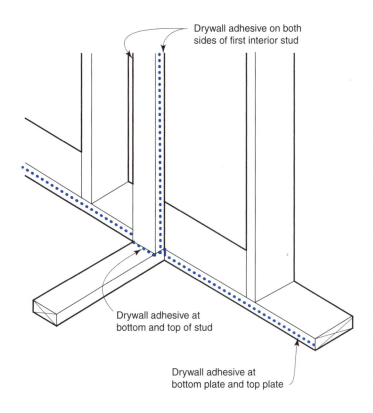

Drywall adhesive on both
sides of first interior stud

Drywall adhesive at
bottom and top of stud

Drywall adhesive at
bottom plate and top plate

Figure 17.7
Interior/Exterior Walls

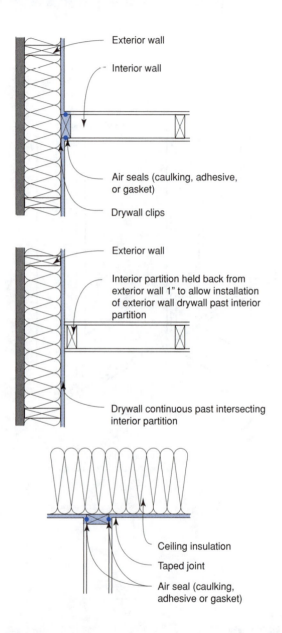

Exterior wall

Interior wall

Air seals (caulking, adhesive, or gasket)

Drywall clips

Exterior wall

Interior partition held back from exterior wall 1" to allow installation of exterior wall drywall past interior partition

Drywall continuous past intersecting interior partition

Ceiling insulation

Taped joint

Air seal (caulking, adhesive or gasket)

Figure 17.8
Intersection Interior Partitions

Painting

Paint has to:

- keep rain out of substrates

- breathe when it's on the outside or on the inside

- protect substrates from getting a sunburn

- look nice

Concerns

Exposure to sunlight (ultraviolet radiation), moisture and heat are the major factors affecting the durability of paint coatings and the durability of substrates (wood, plastics, etc.). Recall that UV, water and heat are "damage functions." Although, each factor can independently lead to deterioration, the effect of the combination of the three factors is much more severe than each factor separately. Ultraviolet radiation, moisture and heat can each lead to the breakdown of the resin in painted surfaces which binds (holds) the pigment to the substrate surface. When the resin breaks down, pigment is lost (washed away from the surface) and fading occurs. In some instances, rubbing the surface with a cloth or a hand will remove a white powder from the paint surface (chalking).

Paints vs. Stains

A paint coating's resistance to ultraviolet radiation and moisture is dependent on the ratio of resin to pigment in the paint. The more resin available to completely coat a pigment particle, the more forcefully the particle is bound to a surface. Premium paints have a high ratio of resin to pigment. A low cost paint typically has a high pigment content relative to resin content as pigment is less expensive than resin. Although a high pigment content paint has an excellent "hiding" ability, high pigment content paints with low resin contents are unable to resist exposure to sunlight and moisture. Gloss paints have more resin than semi-gloss paints, and semi-gloss paints have more resin than flat paints. Gloss paints have the most resistance to ultraviolet radiation and moisture; flat paints have the least.

Stains are not as hydrophobic or resistant to ultraviolet light as paints but are more vapor permeable. Since stains break down more rapidly due to ultraviolet light than do paints, re-coating more frequently with stains will be likely. Solid body stains are thin paints and should not be used. Do not use solid body stains.

Primers

The function of a primer is to prepare the substrate for the paint layer. A primer blocks the migration of extractives (tannins, wood sugars) from wood and the migration of salts from stuccos.

Primers also help the adhesion of paints to surfaces such as concrete, metal, wood and plaster. They also provide a layer of "uniform" porosity that prevents the "over-absorption" of paint on non-uniform surfaces.

Oil-based primers are typically used on wood substrates as tannins and wood sugars, which are water soluble, are not soluble in oil.

Latex or water-based primers are typically reserved for plaster and drywall surfaces — or concrete and stucco.

Wood Substrates

Ultraviolet radiation, moisture and heat have a similar effect on wood as they do on the paint coating itself. Wood breaks down under exposure to ultraviolet radiation and heat and wood surface erosion is increased with exposure to ultraviolet radiation coupled with heat and rain.

The ideal coating system for wood is a system that is hydrophobic (sheds water), vapor permeable (breathes), resistant to ultraviolet light (sunlight) and has good adhesion (sticks to wood) and cohesion (stretching) properties.

Acrylic latex top coats coupled with premium latex primers are recommended when they are applied over stable substrates (dry, dimensionally stable and able to hold paint) as they are more vapor permeable than other paint finishes while providing similar hydrophobic, ultraviolet resistance, adhesion and cohesion properties. Two coats of all-acrylic latex paint over a premium latex primer are recommended.

The optimum thickness for the total dry paint coat (primer and two top-coats) is 3.5 to 5 mils.

Oil-based prime coats coupled with latex top coats do not provide as permeable a system as a latex prime coat-based system. However, oil-based prime coats provide superior adhesion and stain blocking characteristics for difficult substrates. For woods with water-soluble extractives, such as redwood and cedar, oil-based prime coats are recommended. Do not use latex-based prime coats with these type of substrates.

Exposure of unprotected wood to sunlight can adversely affect the adhesion of paint to wood within as little as 3 to 4 week exposures. Wood surfaces should be painted as soon as possible, weather permitting. All exterior wood (except decking materials*) should be back primed or prime coated on all six surfaces. Ideally, wood should be pre-primed on all surfaces prior to arrival at the job site. Field cut edges should be sealed with primer during installation. Top coats should be applied within 2 weeks of field exposure of the prime coat. The sooner, the better. Some prime coats weather by forming a soap-like film that can interfere with adhesion of top coats. Washing aged prime coats (exposed to sunlight) is recommended prior to top coat application. Re-priming may be necessary if prime coats have excessively weathered. Ideally, the temperature should not drop below 50°F for at least 24 hours after paint application. Winter, late fall or early spring topcoat application is not recommended.

Pre-primed material should be utilized during winter construction and not topcoated (finished) until weather permits.

Stucco Substrates

Ideally, in cold climates, stucco claddings/renderings should not be topcoated with paints. Color pigments should be integrated into the topcoat/finish layer of the stucco itself. Unfortunately, this practice is no longer common, and stucco now is typically painted.

When painting stucco, water repellant vapor permeable coating paint systems are recommended. Recall that water vapor flow occurs from both a higher concentration to a lower concentration and from the warm side of an assembly to the cold side of an assembly. A rain wetted stucco cladding that is heated by solar radiation will be warmer and wetter than both the interior and the exterior air. Drying will be to the outside and to the inside. A heavy coat of impermeable paint over exterior stucco will blister under such conditions. The more vapor perme-

---

* Decking materials should be stained on all six surfaces, not painted.

able the exterior paint coating, the better under such circumstances. Acrylic latex paints formulated for exterior use are recommended for almost all stucco applications.

Acrylic latex paints generally outperform "elastomeric" paints over stucco renderings due to the lower permeability of the elastomeric paints. Elastomeric paints have excellent crack spanning characteristics, but give up a degree of permeability to achieve the crack spanning ability. If water enters the stucco rendering at a joint or reveal or flashing or flaw, elastomeric paints have been known to blister. Elastomeric paints should be reserved for special conditions where substrates are severely cracked and crack spanning coatings are necessary and no other coating approaches are practical.

It is important that the stucco has cured sufficiently to reduce its pH[*] prior to painting neutral pH (not acidic, not basic, just right — the "Goldilocks" pH is 7) is ideal for paint application. To determine the pH of stucco prior to painting, use "litmus paper" or just wait — 28 days is often recommended.

If waiting is not practical, apply a masonry primer containing alkali prior to painting.

Deck Substrates

Deck materials should never be painted as even vapor permeable paint coatings serve to inhibit drying of absorbed moisture beyond acceptable levels for such a hostile moisture environment (horizontal, exposed to rain and sun). However, deck materials can be coated with penetrating water repellents or stains. Both of these serve to reduce water absorption without reducing drying ability. In fact, it is critical to "back prime" or "back coat" or "back stain" decking materials to prevent cupping.

Untreated deck materials will deteriorate from both water absorption and exposure to ultraviolet light. Even preservative treated deck materials deteriorate from exposure to ultraviolet light. Stains provide satisfactory protection against ultraviolet exposure (sunlight). Straight water repellents do not. Stains act as a type of "suntan lotion" for the wood. Like most typical suntan lotions, stains must be regularly reapplied. Superior performance may be achieved when stains are applied over preservative treated decking materials.

[*] The negative logarithm of the concentration of hydrogen. Knowing this definition is important only when confronted by a Jeopardy-watching crazed stucco contractor who wants to know why some idiot screwed up his finish coat by painting it.

Peeling Paint on Siding and Trim

Many paint problems with wood siding and wood trim are not paint problems at all, but are due to a fundamental defect associated with the underlying housewrap.

The problem with some housewraps (not all housewraps) is a loss of water repellency. The primary function of a housewrap is rain penetration control. Water repellency is a key element of rain penetration control. Once the water repellency of a housewrap is compromised, water can be absorbed by sheathings such as plywood leading to decay, mold and loss of strength. Additionally, water that is absorbed by sheathings is by definition water that is not drained out of the wall assembly. This absorbed water leads to paint problems with wood claddings and trim materials and in some cases decay of wood claddings and trim. This absorbed water migrates by capillarity and diffusion into wood claddings and trim out of the plywood sheathing through the hydrophobic degraded housewrap.

Contaminants referred to as surfactants ("surface-active contaminants") either raise the surface energy of the housewrap or building paper or lower the surface energy of the water allowing the "wetting" of the housewrap or building paper surface by water. Once wetting of the housewrap or building paper surface occurs, material pores in the housewrap or building paper become filled allowing transport of liquid phase water across the housewrap or building paper via capillarity or hydrostatic pressure (gravity).

Figure 18.01 graphically illustrates some of the terms just used.

Water soluble extractives in wood such as tannins and wood sugars in redwood and cedar are a type of surfactant that contaminate the surface of housewraps and building papers raising their surface energy. Detergents and soaps are another type of surfactant that contaminate the surface of water lowing its surface energy. Both result in the liquid (in this instance "water") being able to "wet" the surface (in this case the building paper or housewrap). The surface energies of either the liquid or surface or both are altered so that the surface energy of the surface becomes greater than the surface energy of the liquid.

Back priming or back coating wood clapboards and trim helps to isolate the surfactants in the wood from the housewrap or building paper surface. Similarly, providing an airspace between wood trim and clapboards using furring or some other spacer ("cedar-breather") reduces the quantity and time liquid phase water is trapped in the exterior of the wall assembly thereby reducing the potential of surfactant movement. Where wood is concerned, both back-priming and an airspace are highly recommended.

Stucco and Housewrap

Where stucco is concerned, stucco should never be installed in direct contact with any of the plastic based housewraps. Stucco can "bond" or adhere to the housewrap surface altering its surface energy thereby allowing housewrap pores to become "wetted" and subsequently establish capillary flow. Another issue with stuccos is many stuccos have additives that improve workability and freeze-thaw resistance. These additives are typically surfactants.

A drainage space between stucco and building papers or housewraps is essential to control liquid phase water penetration. Bonding typically does not happen between stucco and building papers. However, with most stucco applications over building papers insufficient drainage results. It is recommended to use at least two layers of building paper under stucco in order to allow some drainage between the two layers. Even better is to provide a spacer between the two layers of building paper by using a textured building paper or a building paper with granules or cork adhered to its surface thereby creating a space.

Another analogy that might be useful in understanding some of the concepts involved is one of being inside of a tent during a rainstorm and pushing your finger against the inside of the tent surface creating a leak by both compressing and tensioning the fabric altering its surface energy.

Dirt and dust can also affect housewrap performance. Think of a fabric that is "Scotchgard™-ed" to be water repellent that subsequently becomes dirty. The fabric must be cleaned to remove the contaminants and retreated to reestablish water repellency. Don't let you housewraps become muddy or dirty.

Interior Surfaces

Paint coatings installed on interior surfaces can be either permeable or impermeable depending on the design of the wall assembly. On wall and basement assemblies which are designed to dry to the interior, only vapor permeable paint systems (latex paint) should be used.

On wall and roof assemblies which require a surface applied interior vapor retarder, a vapor impermeable paint system should be used. These coatings are available in both latex-based and oil-based systems.

Paints with low or no emissions of volatile organic compounds (VOC's) should be selected for interior applications to reduce concentrations of interior contaminants.

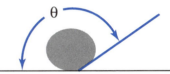
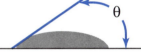

- "non-wettable" surface
- water repellant surface
- hygrophobic surface
- water more attracted to itself than to surface
- surface energy of water greater than surface energy of surface
- water "beads up"
- "greasy" surface
- high contact angle "θ"

- "wettable" surface
- non-water repellant surface
- hygroscopic surface
- water more attracted to surface than to itself
- surface energy of surface greater than surface energy of water
- water "spreads out"
- "non-greasy" surface
- low contact angle "θ"

Figure 18.1
Surface Active Contaminant Properties

Part III

&

19
20
21

Insulating Concrete Forms (ICF)

Insulating concrete forms (ICFs) are hollow building elements made of plastic foam that are assembled, often like building blocks, into the shape of a building's exterior walls. The ICFs are filled with reinforced concrete to create structural walls. Unlike traditional forms, the ICFs are left in place to provide insulation and a surface for finishes.

Concerns

The major concern with ICFs is their ability to control rain entry. Additional concerns relate to insects and groundwater.

An ICF exterior wall can be classified as a face-sealed barrier system with the ability for the assembly to act as a storage reservoir for limited quantities of rain penetrating the exterior face. The plastic foam ICF material and reinforced concrete are not generally adversely affected by penetrating rain water. However, most interior finish materials will be adversely affected. As a result, permeable interior finishes are recommended to facilitate drying to the interior of any rain water that has penetrated the exterior face. Gypsum board can be adhered directly to the interior of the ICFs. Similarly, interior plaster finishes or renderings can be directly applied to the interior of the ICFs.

As a face-sealed barrier system, window and door detailing is important. Details similar to historically successfully performing details developed for face-sealed masonry stucco systems should be used. With appropriate window and door detailing it is not necessary to provide a drainage plane (such as felt building paper or a housewrap) immediately behind exterior cladding systems.

Where wood siding is installed over furring strips, a ventilated rain screen control strategy can be employed. The drainage and ventilated air space make it unnecessary to require an additional drainage plane material such as a building paper over the exterior face of the plastic foam material under the furring strips.

Insect control and groundwater control principles developed for rigid exterior foam insulations used in below grade applications or ground contact applications should be used.

Typical ICF Flat Wall

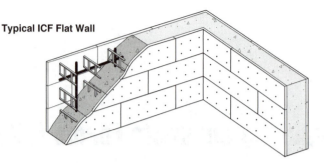

Typical ICF Grid Wall

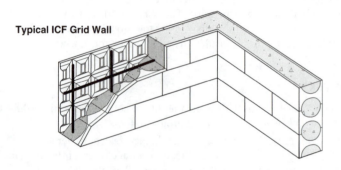

Typical ICF Post-and-Beam Wall

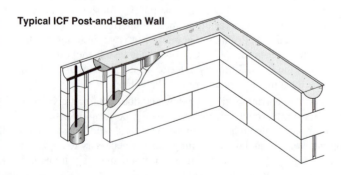

Figure 19.1
Types of ICF Walls

- ICF's are forms for cast concrete walls that stay in place as a part of the wall after the concrete is poured
- Flat systems are filled with concrete like conventionally cast walls
- Grid systems have a waffle pattern where the concrete varies in thickness
- Post and beam systems have discrete columns of horizontal and vertical concrete completely surrounded by foam

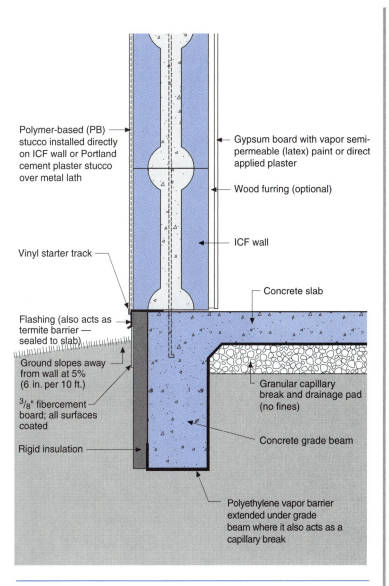

Polymer-based (PB) stucco installed directly on ICF wall or Portland cement plaster stucco over metal lath

Gypsum board with vapor semi-permeable (latex) paint or direct applied plaster

Wood furring (optional)

ICF wall

Vinyl starter track

Concrete slab

Flashing (also acts as termite barrier — sealed to slab)

Ground slopes away from wall at 5% (6 in. per 10 ft.)

Granular capillary break and drainage pad (no fines)

$3/8$" fibercement board; all surfaces coated

Rigid insulation

Concrete grade beam

Polyethylene vapor barrier extended under grade beam where it also acts as a capillary break

 19

Figure 19.2
Monolithic Slab — ICF Above Grade Wall

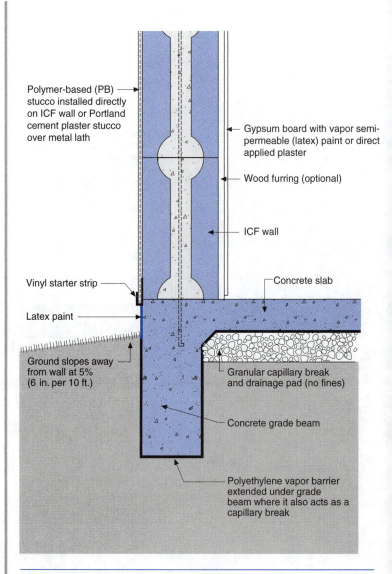

Polymer-based (PB) stucco installed directly on ICF wall or Portland cement plaster stucco over metal lath

Gypsum board with vapor semi-permeable (latex) paint or direct applied plaster

Wood furring (optional)

ICF wall

Vinyl starter strip

Concrete slab

Latex paint

Ground slopes away from wall at 5% (6 in. per 10 ft.)

Granular capillary break and drainage pad (no fines)

Concrete grade beam

Polyethylene vapor barrier extended under grade beam where it also acts as a capillary break

**Figure 19.3
Monolithic Slab — ICF Above Grade Wall and Stucco**

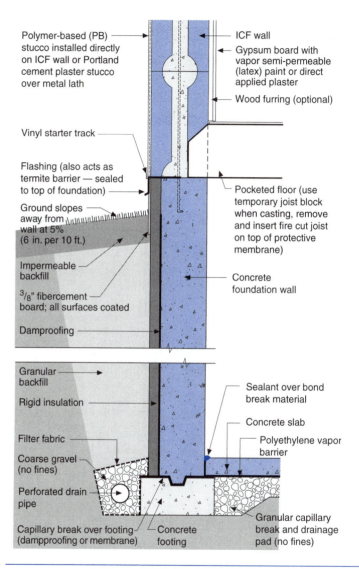

Polymer-based (PB) stucco installed directly on ICF wall or Portland cement plaster stucco over metal lath

ICF wall

Gypsum board with vapor semi-permeable (latex) paint or direct applied plaster

Wood furring (optional)

Vinyl starter track

Flashing (also acts as termite barrier — sealed to top of foundation)

Ground slopes away from wall at 5% (6 in. per 10 ft.)

Impermeable backfill

$^3/_8$" fibercement board; all surfaces coated

Damproofing

Pocketed floor (use temporary joist block when casting, remove and insert fire cut joist on top of protective membrane)

Concrete foundation wall

Granular backfill

Rigid insulation

Filter fabric

Coarse gravel (no fines)

Perforated drain pipe

Capillary break over footing (dampproofing or membrane)

Concrete footing

Sealant over bond break material

Concrete slab

Polyethylene vapor barrier

Granular capillary break and drainage pad (no fines)

 19

Figure 19.4
Externally Insulated Concrete Basement — ICF Above Grade Wall

• Concrete wall warm, can dry to the interior; extremely low likelihood of mold
• Basement floor slab can dry to the interior

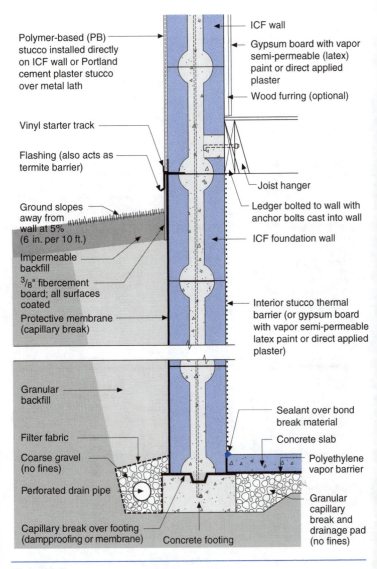

Polymer-based (PB) stucco installed directly on ICF wall or Portland cement plaster stucco over metal lath

ICF wall

Gypsum board with vapor semi-permeable (latex) paint or direct applied plaster

Wood furring (optional)

Vinyl starter track

Flashing (also acts as termite barrier)

Joist hanger

Ledger bolted to wall with anchor bolts cast into wall

Ground slopes away from wall at 5% (6 in. per 10 ft.)

ICF foundation wall

Impermeable backfill

$3/8$" fibercement board; all surfaces coated

Protective membrane (capillary break)

Interior stucco thermal barrier (or gypsum board with vapor semi-permeable latex paint or direct applied plaster)

Granular backfill

Sealant over bond break material

Filter fabric

Concrete slab

Coarse gravel (no fines)

Polyethylene vapor barrier

Perforated drain pipe

Granular capillary break and drainage pad (no fines)

Capillary break over footing (dampproofing or membrane)

Concrete footing

19

Figure 19.5
ICF Basement — ICF Above Grade Wall

- Foundation wall warm, can dry to the interior; extremely low likelihood of mold
- Basement floor slab can dry to the interior
- Protective membrane can be adhesive-backed roll roofing, below grade sheet waterproofing or an ice-dam protection membrane

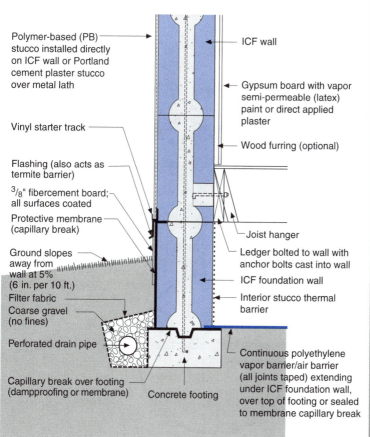

Polymer-based (PB) stucco installed directly on ICF wall or Portland cement plaster stucco over metal lath

ICF wall

Gypsum board with vapor semi-permeable (latex) paint or direct applied plaster

Vinyl starter track

Wood furring (optional)

Flashing (also acts as termite barrier)

$3/8$" fibercement board; all surfaces coated

Protective membrane (capillary break)

Joist hanger

Ledger bolted to wall with anchor bolts cast into wall

Ground slopes away from wall at 5% (6 in. per 10 ft.)

ICF foundation wall

Interior stucco thermal barrier

Filter fabric

Coarse gravel (no fines)

Perforated drain pipe

Continuous polyethylene vapor barrier/air barrier (all joints taped) extending under ICF foundation wall, over top of footing or sealed to membrane capillary break

Capillary break over footing (dampproofing or membrane)

Concrete footing

Figure 19.6
ICF Crawlspace — ICF Above Grade Wall

- Foundation wall warm, can dry to the interior; extremely low likelihood of mold
- Protective membrane can be adhesive-backed roll roofing, below grade sheet waterproofing or an ice-dam protection membrane

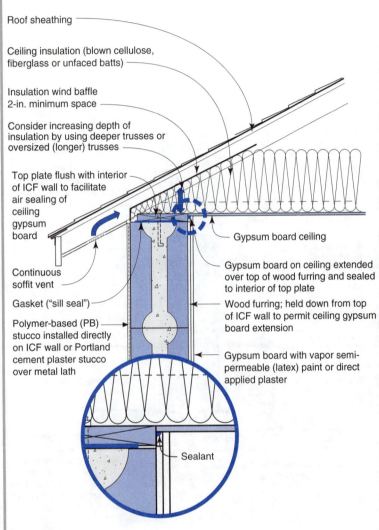

Roof sheathing

Ceiling insulation (blown cellulose, fiberglass or unfaced batts)

Insulation wind baffle
2-in. minimum space

Consider increasing depth of insulation by using deeper trusses or oversized (longer) trusses

Top plate flush with interior of ICF wall to facilitate air sealing of ceiling gypsum board

Gypsum board ceiling

Continuous soffit vent

Gasket ("sill seal")

Gypsum board on ceiling extended over top of wood furring and sealed to interior of top plate

Wood furring; held down from top of ICF wall to permit ceiling gypsum board extension

Polymer-based (PB) stucco installed directly on ICF wall or Portland cement plaster stucco over metal lath

Gypsum board with vapor semi-permeable (latex) paint or direct applied plaster

Sealant

Figure 19.7
Roof Framing — Top Plate
• Wall can dry to interior; extremely low likelihood of mold

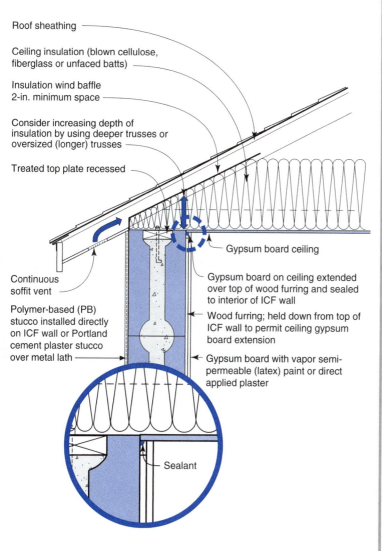

Roof sheathing

Ceiling insulation (blown cellulose, fiberglass or unfaced batts)

Insulation wind baffle 2-in. minimum space

Consider increasing depth of insulation by using deeper trusses or oversized (longer) trusses

Treated top plate recessed

Continuous soffit vent

Polymer-based (PB) stucco installed directly on ICF wall or Portland cement plaster stucco over metal lath

Gypsum board ceiling

Gypsum board on ceiling extended over top of wood furring and sealed to interior of ICF wall

Wood furring; held down from top of ICF wall to permit ceiling gypsum board extension

Gypsum board with vapor semi-permeable (latex) paint or direct applied plaster

Sealant

19

Figure 19.8
Roof Framing — Recessed Plate
• Wall can dry to interior; extremely low likelihood of mold

417

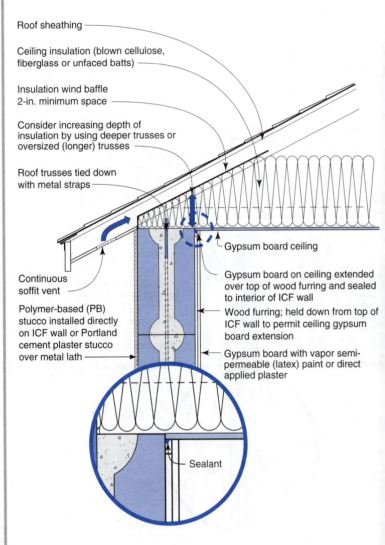

Roof sheathing

Ceiling insulation (blown cellulose, fiberglass or unfaced batts)

Insulation wind baffle 2-in. minimum space

Consider increasing depth of insulation by using deeper trusses or oversized (longer) trusses

Roof trusses tied down with metal straps

Continuous soffit vent

Polymer-based (PB) stucco installed directly on ICF wall or Portland cement plaster stucco over metal lath

Gypsum board ceiling

Gypsum board on ceiling extended over top of wood furring and sealed to interior of ICF wall

Wood furring; held down from top of ICF wall to permit ceiling gypsum board extension

Gypsum board with vapor semi-permeable (latex) paint or direct applied plaster

Sealant

19

Figure 19.9
Roof Framing — Metal Straps
• Wall can dry to interior; extremely low likelihood of mold

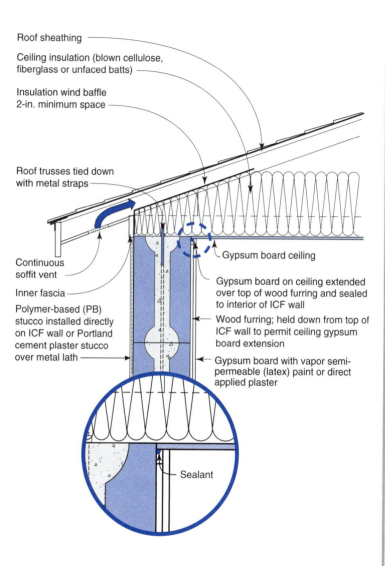

Roof sheathing

Ceiling insulation (blown cellulose, fiberglass or unfaced batts)

Insulation wind baffle 2-in. minimum space

Roof trusses tied down with metal straps

Continuous soffit vent

Inner fascia

Polymer-based (PB) stucco installed directly on ICF wall or Portland cement plaster stucco over metal lath

Gypsum board ceiling

Gypsum board on ceiling extended over top of wood furring and sealed to interior of ICF wall

Wood furring; held down from top of ICF wall to permit ceiling gypsum board extension

Gypsum board with vapor semi-permeable (latex) paint or direct applied plaster

Sealant

 19

Figure 19.10
Alternative Roof Truss to Increase Depth of Ceiling Insulation at Perimeter
- Wall can dry to interior; extremely low likelihood of mold

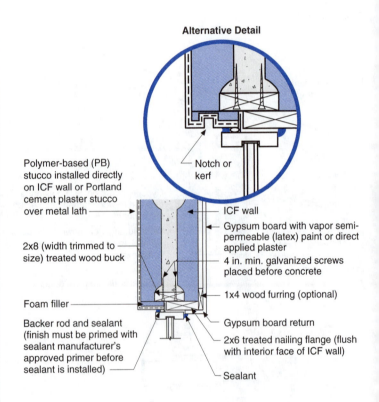

Alternative Detail

Notch or kerf

Polymer-based (PB) stucco installed directly on ICF wall or Portland cement plaster stucco over metal lath

ICF wall

Gypsum board with vapor semi-permeable (latex) paint or direct applied plaster

2x8 (width trimmed to size) treated wood buck

4 in. min. galvanized screws placed before concrete

Foam filler

1x4 wood furring (optional)

Backer rod and sealant (finish must be primed with sealant manufacturer's approved primer before sealant is installed)

Gypsum board return

2x6 treated nailing flange (flush with interior face of ICF wall)

Sealant

Figure 19.11
ICF Recessed Window Head — Stucco
   • Windows installed before stucco

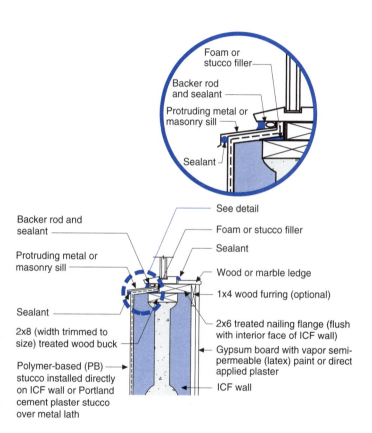

Foam or
stucco filler

Backer rod
and sealant

Protruding metal or
masonry sill

Sealant

See detail

Backer rod and
sealant

Foam or stucco filler

Sealant

Protruding metal or
masonry sill

Wood or marble ledge

1x4 wood furring (optional)

Sealant

2x8 (width trimmed to
size) treated wood buck

2x6 treated nailing flange (flush
with interior face of ICF wall)

Gypsum board with vapor semi-
permeable (latex) paint or direct
applied plaster

Polymer-based (PB)
stucco installed directly
on ICF wall or Portland
cement plaster stucco
over metal lath

ICF wall

**Figure 19.12
ICF Recessed Window Sill — Stucco**
• Windows installed before stucco

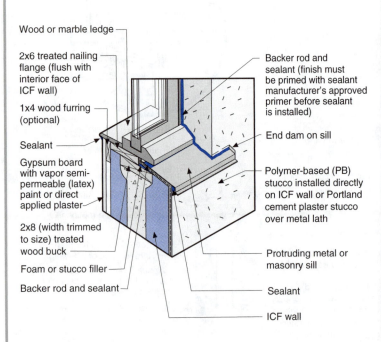

Wood or marble ledge

2x6 treated nailing flange (flush with interior face of ICF wall)

1x4 wood furring (optional)

Sealant

Gypsum board with vapor semi-permeable (latex) paint or direct applied plaster

2x8 (width trimmed to size) treated wood buck

Foam or stucco filler

Backer rod and sealant

Backer rod and sealant (finish must be primed with sealant manufacturer's approved primer before sealant is installed)

End dam on sill

Polymer-based (PB) stucco installed directly on ICF wall or Portland cement plaster stucco over metal lath

Protruding metal or masonry sill

Sealant

ICF wall

19

Figure 19.13
ICF Recessed Window Sill — Stucco
• Windows installed before stucco

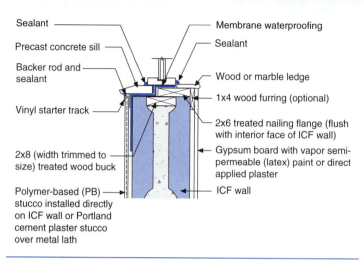

Sealant

Precast concrete sill

Backer rod and sealant

Vinyl starter track

2x8 (width trimmed to size) treated wood buck

Polymer-based (PB) stucco installed directly on ICF wall or Portland cement plaster stucco over metal lath

Membrane waterproofing

Sealant

Wood or marble ledge

1x4 wood furring (optional)

2x6 treated nailing flange (flush with interior face of ICF wall)

Gypsum board with vapor semi-permeable (latex) paint or direct applied plaster

ICF wall

Figure 19.14
Alternative ICF Recessed Window Sill — Stucco
• Windows installed before stucco

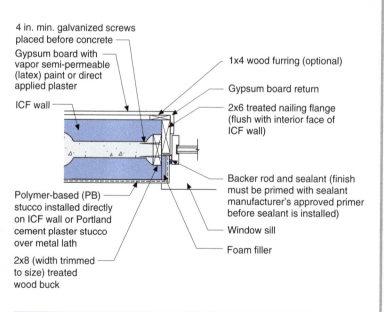

4 in. min. galvanized screws placed before concrete

Gypsum board with vapor semi-permeable (latex) paint or direct applied plaster

ICF wall

1x4 wood furring (optional)

Gypsum board return

2x6 treated nailing flange (flush with interior face of ICF wall)

Backer rod and sealant (finish must be primed with sealant manufacturer's approved primer before sealant is installed)

Window sill

Foam filler

Polymer-based (PB) stucco installed directly on ICF wall or Portland cement plaster stucco over metal lath

2x8 (width trimmed to size) treated wood buck

 19

Figure 19.15
ICF Recessed Window Jamb — Stucco
• Windows installed before stucco

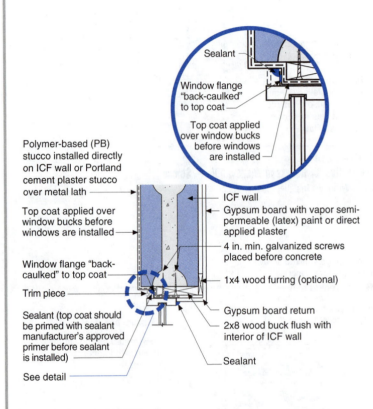

Sealant

Window flange "back-caulked" to top coat

Top coat applied over window bucks before windows are installed

Polymer-based (PB) stucco installed directly on ICF wall or Portland cement plaster stucco over metal lath

Top coat applied over window bucks before windows are installed

Window flange "back-caulked" to top coat

Trim piece

Sealant (top coat should be primed with sealant manufacturer's approved primer before sealant is installed)

See detail

ICF wall

Gypsum board with vapor semi-permeable (latex) paint or direct applied plaster

4 in. min. galvanized screws placed before concrete

1x4 wood furring (optional)

Gypsum board return

2x8 wood buck flush with interior of ICF wall

Sealant

Figure 19.16
Alternative ICF Recessed Window Head — Stucco
• Windows installed after stucco

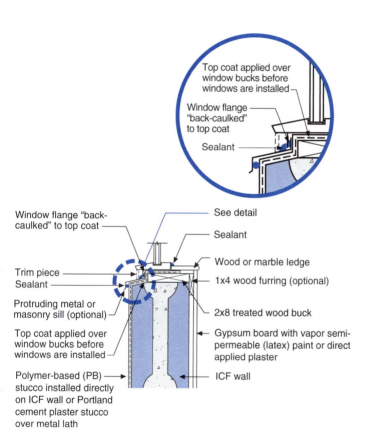

Top coat applied over window bucks before windows are installed

Window flange "back-caulked" to top coat

Sealant

Window flange "back-caulked" to top coat

See detail

Sealant

Trim piece
Sealant

Wood or marble ledge

1x4 wood furring (optional)

Protruding metal or masonry sill (optional)

2x8 treated wood buck

Top coat applied over window bucks before windows are installed

Gypsum board with vapor semi-permeable (latex) paint or direct applied plaster

Polymer-based (PB) stucco installed directly on ICF wall or Portland cement plaster stucco over metal lath

ICF wall

Figure 19.17
Alternative ICF Recessed Window Sill — Stucco
- Windows installed after stucco

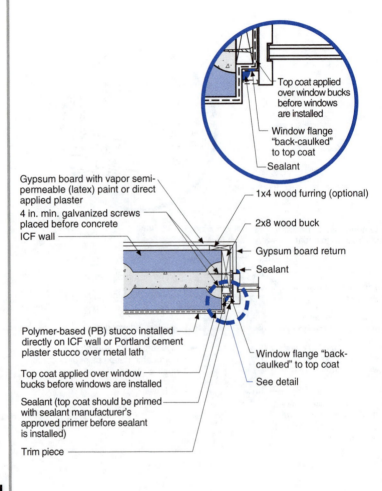

Top coat applied over window bucks before windows are installed

Window flange "back-caulked" to top coat

Sealant

Gypsum board with vapor semi-permeable (latex) paint or direct applied plaster

1x4 wood furring (optional)

4 in. min. galvanized screws placed before concrete

2x8 wood buck

ICF wall

Gypsum board return

Sealant

Polymer-based (PB) stucco installed directly on ICF wall or Portland cement plaster stucco over metal lath

Window flange "back-caulked" to top coat

Top coat applied over window bucks before windows are installed

See detail

Sealant (top coat should be primed with sealant manufacturer's approved primer before sealant is installed)

Trim piece

Figure 19.18
Alternative ICF Recessed Window Jamb — Stucco
- Windows installed after stucco

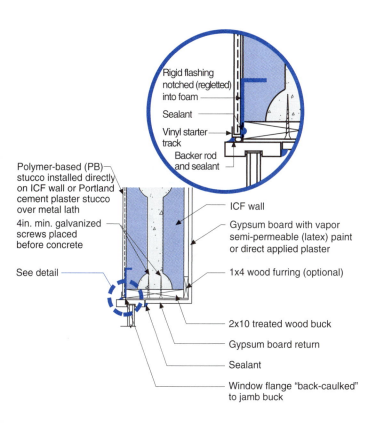

Rigid flashing notched (regletted) into foam

Sealant

Vinyl starter track

Backer rod and sealant

Polymer-based (PB) stucco installed directly on ICF wall or Portland cement plaster stucco over metal lath

4in. min. galvanized screws placed before concrete

See detail

ICF wall

Gypsum board with vapor semi-permeable (latex) paint or direct applied plaster

1x4 wood furring (optional)

2x10 treated wood buck

Gypsum board return

Sealant

Window flange "back-caulked" to jamb buck

Figure 19.19
ICF Flush Window Head — Stucco
• Windows installed before stucco

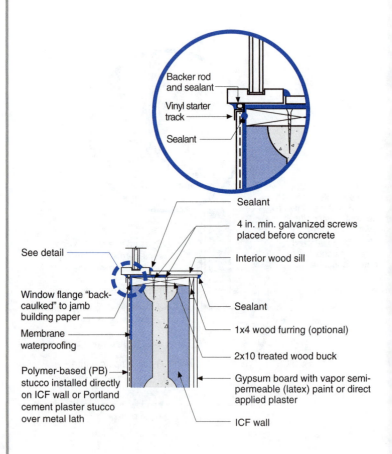

Backer rod and sealant

Vinyl starter track

Sealant

Sealant

4 in. min. galvanized screws placed before concrete

See detail

Interior wood sill

Window flange "back-caulked" to jamb building paper

Sealant

Membrane waterproofing

1x4 wood furring (optional)

2x10 treated wood buck

Polymer-based (PB) stucco installed directly on ICF wall or Portland cement plaster stucco over metal lath

Gypsum board with vapor semi-permeable (latex) paint or direct applied plaster

ICF wall

Figure 19.20
ICF Flush Window Sill — Stucco
- Windows installed before stucco

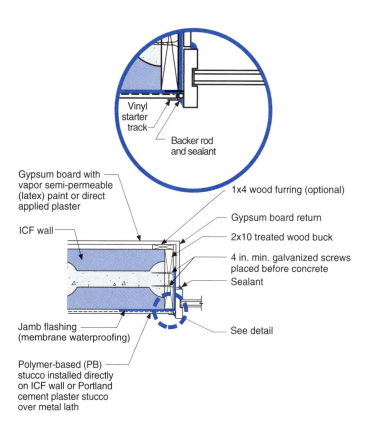

Vinyl starter track

Backer rod and sealant

Gypsum board with vapor semi-permeable (latex) paint or direct applied plaster

1x4 wood furring (optional)

Gypsum board return

ICF wall

2x10 treated wood buck

4 in. min. galvanized screws placed before concrete

Sealant

Jamb flashing (membrane waterproofing)

See detail

Polymer-based (PB) stucco installed directly on ICF wall or Portland cement plaster stucco over metal lath

**Figure 19.21
ICF Flush Window Jamb — Stucco**
• Windows installed before stucco

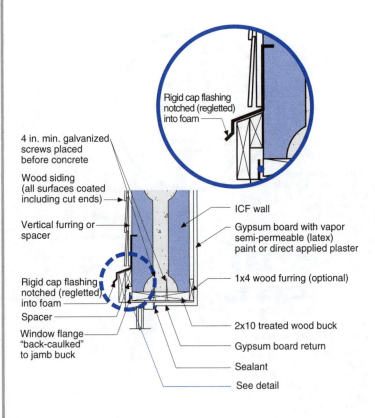

4 in. min. galvanized
screws placed
before concrete

Wood siding
(all surfaces coated
including cut ends)

Vertical furring or
spacer

Rigid cap flashing
notched (regletted)
into foam

Spacer

Window flange
"back-caulked"
to jamb buck

Rigid cap flashing
notched (regletted)
into foam

ICF wall

Gypsum board with vapor
semi-permeable (latex)
paint or direct applied plaster

1x4 wood furring (optional)

2x10 treated wood buck

Gypsum board return

Sealant

See detail

Figure 19.22
ICF Flush Window Head — Wood Siding

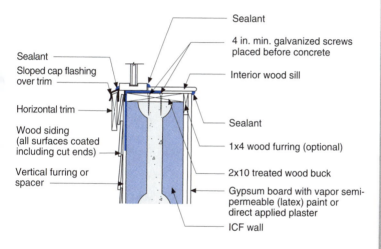

Sealant

4 in. min. galvanized screws placed before concrete

Sealant
Sloped cap flashing over trim

Interior wood sill

Horizontal trim

Sealant

Wood siding (all surfaces coated including cut ends)

1x4 wood furring (optional)

2x10 treated wood buck

Vertical furring or spacer

Gypsum board with vapor semi-permeable (latex) paint or direct applied plaster

ICF wall

19

Figure 19.23
ICF Flush Window Sill — Wood Siding

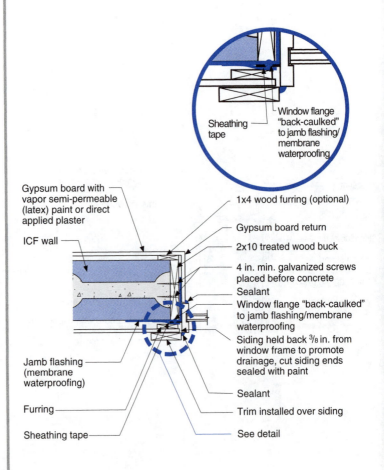

Window flange "back-caulked" to jamb flashing/membrane waterproofing

Sheathing tape

Gypsum board with vapor semi-permeable (latex) paint or direct applied plaster

ICF wall

1x4 wood furring (optional)

Gypsum board return

2x10 treated wood buck

4 in. min. galvanized screws placed before concrete

Sealant

Window flange "back-caulked" to jamb flashing/membrane waterproofing

Siding held back ³/₈ in. from window frame to promote drainage, cut siding ends sealed with paint

Jamb flashing (membrane waterproofing)

Furring

Sheathing tape

Sealant

Trim installed over siding

See detail

Figure 19.24
ICF Flush Window Jamb — Wood Siding

19

Structural Insulated Panel Systems (SIPS)

Structural Insulated Panel Systems (SIPS) are prefabricated building panels consisting of plastic foam cores sandwiched between two skins; typically oriented strand board (OSB) or plywood.

Concerns

As with wood frame construction, the single greatest concern with SIPS is their ability to control rain water entry. In this regard, SIPS should be treated similarly to wood frame construction where OSB or plywood is used as an exterior sheathing. Accordingly, an exterior drainage plane of #30 felt building paper installed shingle fashion and detailed at window openings in a similar manner to wood frame construction is recommended (see Chapter 2 — Rain, Drainage Planes and Flashings).

Since SIPS components are rigid panels made from laminated or adhered elements they are prone to dimensional stability problems due to thermal induced moisture differentials between their wood-based hygroscopic facings. They bow and move differentially with the changing moisture contents of their exterior skins. This can result in the "telegraphing" of panel joints in roof assemblies under asphalt shingles.

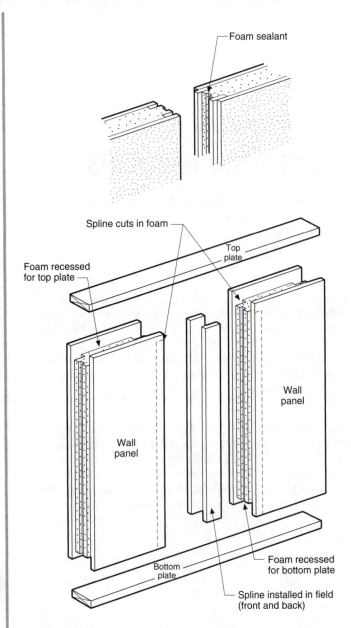

Foam sealant

Spline cuts in foam

Top plate

Foam recessed for top plate

Wall panel

Wall panel

Bottom plate

Foam recessed for bottom plate

Spline installed in field (front and back)

Figure 20.1
SIPS Wall Assembly

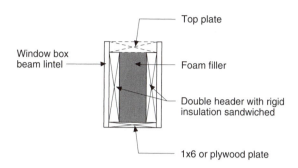

- Top plate
- Window box beam lintel
- Foam filler
- Double header with rigid insulation sandwiched
- 1x6 or plywood plate

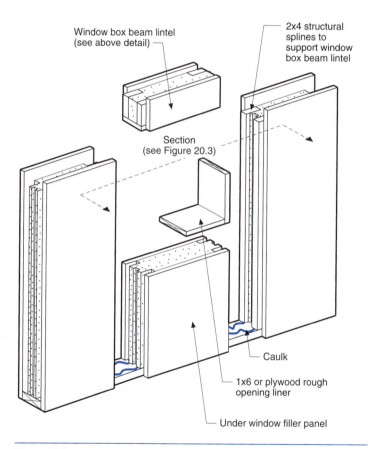

Window box beam lintel (see above detail)

2x4 structural splines to support window box beam lintel

Section (see Figure 20.3)

Caulk

1x6 or plywood rough opening liner

Under window filler panel

**Figure 20.2
SIPS Load Bearing Window/Sliding Door Opening**

20

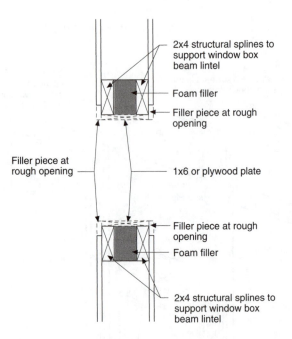

Figure 20.3
SIPS Load Bearing Window Jamb Assembly

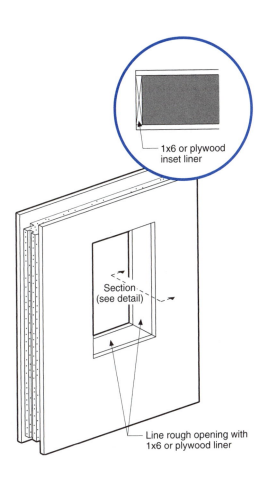

1x6 or plywood
inset liner

Section
(see detail)

Line rough opening with
1x6 or plywood liner

Figure 20.4
SIPS Non-Load Bearing Window Opening

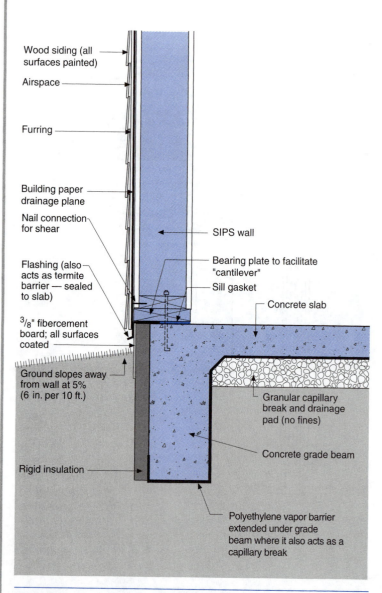

Wood siding (all surfaces painted)

Airspace

Furring

Building paper drainage plane

Nail connection for shear

Flashing (also acts as termite barrier — sealed to slab)

$3/8$" fibercement board; all surfaces coated

Ground slopes away from wall at 5% (6 in. per 10 ft.)

Rigid insulation

SIPS wall

Bearing plate to facilitate "cantilever"

Sill gasket

Concrete slab

Granular capillary break and drainage pad (no fines)

Concrete grade beam

Polyethylene vapor barrier extended under grade beam where it also acts as a capillary break

Figure 20.5
Monolithic Slab — SIPS Above Grade Wall
• Cantilever SIPS over foundation slab

20

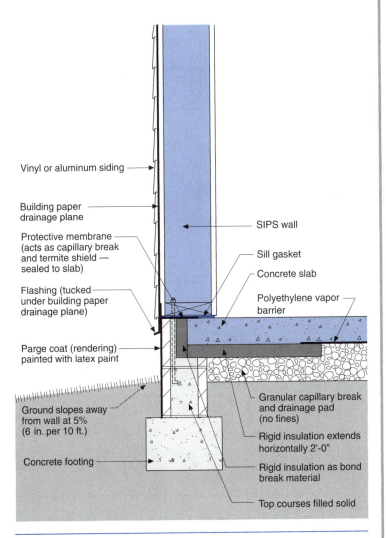

Vinyl or aluminum siding

Building paper drainage plane

Protective membrane (acts as capillary break and termite shield — sealed to slab)

Flashing (tucked under building paper drainage plane)

Parge coat (rendering) painted with latex paint

Ground slopes away from wall at 5% (6 in. per 10 ft.)

Concrete footing

SIPS wall

Sill gasket

Concrete slab

Polyethylene vapor barrier

Granular capillary break and drainage pad (no fines)

Rigid insulation extends horizontally 2'-0"

Rigid insulation as bond break material

Top courses filled solid

Figure 20.6
Slab with Masonry Perimeter — SIPS Above Grade Wall with Vinyl or Aluminum Siding

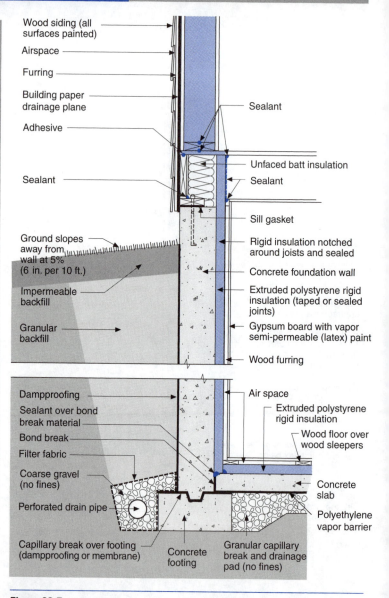

Wood siding (all surfaces painted)

Airspace

Furring

Building paper drainage plane

Adhesive

Sealant

Sealant

Ground slopes away from wall at 5% (6 in. per 10 ft.)

Impermeable backfill

Granular backfill

Dampproofing

Sealant over bond break material

Bond break

Filter fabric

Coarse gravel (no fines)

Perforated drain pipe

Capillary break over footing (dampproofing or membrane)

Concrete footing

Granular capillary break and drainage pad (no fines)

Sealant

Unfaced batt insulation

Sealant

Sill gasket

Rigid insulation notched around joists and sealed

Concrete foundation wall

Extruded polystyrene rigid insulation (taped or sealed joints)

Gypsum board with vapor semi-permeable (latex) paint

Wood furring

Air space

Extruded polystyrene rigid insulation

Wood floor over wood sleepers

Concrete slab

Polyethylene vapor barrier

Figure 20.7
Internally Insulated Concrete Basement — SIPS Above Grade Wall with Wood Siding

20

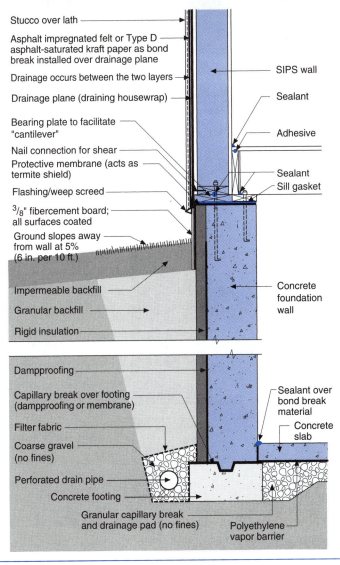

Stucco over lath

Asphalt impregnated felt or Type D asphalt-saturated kraft paper as bond break installed over drainage plane

Drainage occurs between the two layers

Drainage plane (draining housewrap)

Bearing plate to facilitate "cantilever"

Nail connection for shear

Protective membrane (acts as termite shield)

Flashing/weep screed

$^3/_8$" fibercement board; all surfaces coated

Ground slopes away from wall at 5% (6 in. per 10 ft.)

Impermeable backfill

Granular backfill

Rigid insulation

SIPS wall

Sealant

Adhesive

Sealant

Sill gasket

Concrete foundation wall

Dampproofing

Capillary break over footing (dampproofing or membrane)

Filter fabric

Coarse gravel (no fines)

Perforated drain pipe

Concrete footing

Granular capillary break and drainage pad (no fines)

Sealant over bond break material

Concrete slab

Polyethylene vapor barrier

Figure 20.8
Externally Insulated Concrete Basement — SIPS Above Grade Wall with Stucco

- Cantilever SIPS over foundation wall
- Stucco applied over water sensitive materials must be uncoupled from the water sensitive materials. This can be accomplished via drainage and a capillary break or vapor retarder such as foam or both.
- For drainage to occur both a drainage plane and a drainage space are required. In some assemblies a ventilated air space is used. The stucco must not bond continuously to the drainage plane in order for drainage to occur; a bond break is necessary.

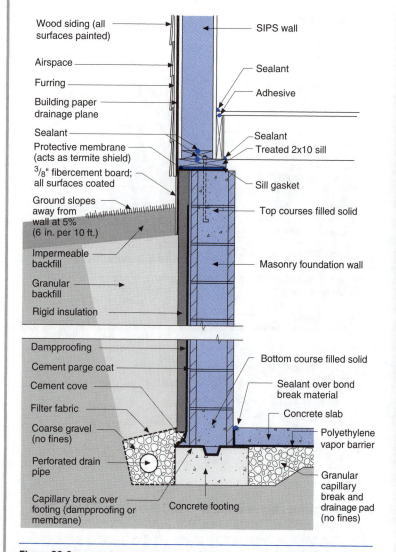

Wood siding (all surfaces painted)

Airspace

Furring

Building paper drainage plane

Sealant

Protective membrane (acts as termite shield)

$^3/_8$" fibercement board; all surfaces coated

Ground slopes away from wall at 5% (6 in. per 10 ft.)

Impermeable backfill

Granular backfill

Rigid insulation

Dampproofing

Cement parge coat

Cement cove

Filter fabric

Coarse gravel (no fines)

Perforated drain pipe

Capillary break over footing (dampproofing or membrane)

SIPS wall

Sealant

Adhesive

Sealant

Treated 2x10 sill

Sill gasket

Top courses filled solid

Masonry foundation wall

Bottom course filled solid

Sealant over bond break material

Concrete slab

Polyethylene vapor barrier

Granular capillary break and drainage pad (no fines)

Concrete footing

Concrete slab

Figure 20.9
Externally Insulated Masonry Basement — SIPS Above Grade Wall with Wood Siding

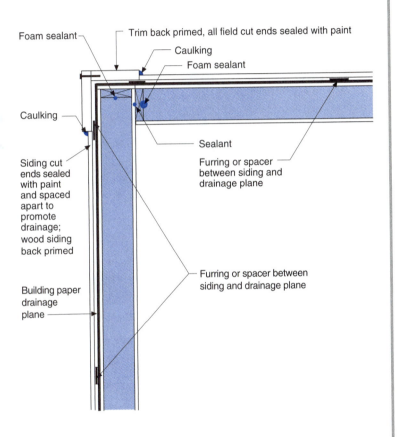

Foam sealant

Trim back primed, all field cut ends sealed with paint

Caulking

Foam sealant

Caulking

Sealant

Siding cut ends sealed with paint and spaced apart to promote drainage; wood siding back primed

Furring or spacer between siding and drainage plane

Building paper drainage plane

Furring or spacer between siding and drainage plane

20

Figure 20.10
Exterior Corner Trim Detail for Rigid Insulation — Wood Siding

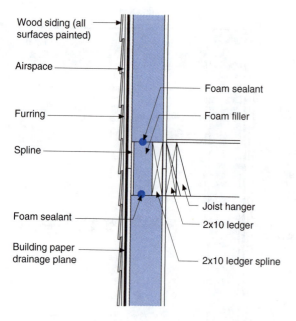

Wood siding (all surfaces painted)

Airspace

Furring

Spline

Foam sealant

Building paper drainage plane

Foam sealant

Foam filler

Joist hanger

2x10 ledger

2x10 ledger spline

Figure 20.11
SIPS Above Grade Wall Spline

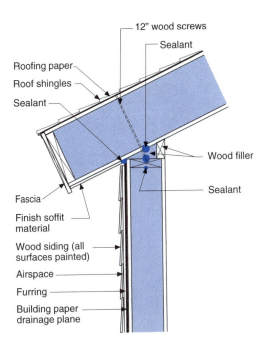

12" wood screws
Sealant
Roofing paper
Roof shingles
Sealant
Wood filler
Sealant
Fascia
Finish soffit material
Wood siding (all surfaces painted)
Airspace
Furring
Building paper drainage plane

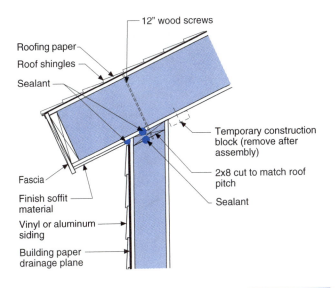

12" wood screws
Roofing paper
Roof shingles
Sealant
Temporary construction block (remove after assembly)
2x8 cut to match roof pitch
Fascia
Sealant
Finish soffit material
Vinyl or aluminum siding
Building paper drainage plane

20

Figure 20.12
SIPS Roof Framing

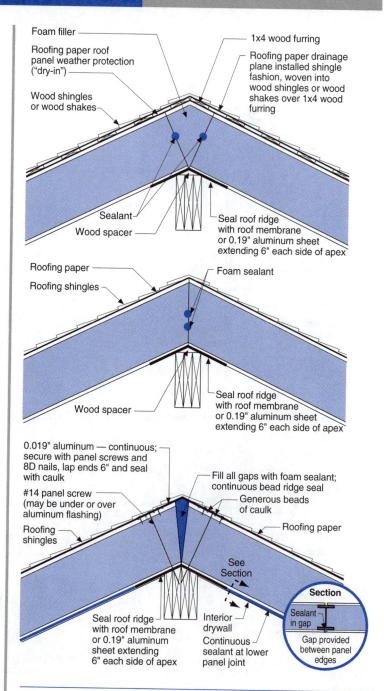

Foam filler

Roofing paper roof panel weather protection ("dry-in")

Wood shingles or wood shakes

1x4 wood furring

Roofing paper drainage plane installed shingle fashion, woven into wood shingles or wood shakes over 1x4 wood furring

Sealant

Wood spacer

Seal roof ridge with roof membrane or 0.19" aluminum sheet extending 6" each side of apex

Roofing paper

Roofing shingles

Foam sealant

Wood spacer

Seal roof ridge with roof membrane or 0.19" aluminum sheet extending 6" each side of apex

0.019" aluminum — continuous; secure with panel screws and 8D nails, lap ends 6" and seal with caulk

#14 panel screw (may be under or over aluminum flashing)

Roofing shingles

Fill all gaps with foam sealant; continuous bead ridge seal

Generous beads of caulk

Roofing paper

See Section

Seal roof ridge with roof membrane or 0.19" aluminum sheet extending 6" each side of apex

Interior drywall

Continuous sealant at lower panel joint

Section

Sealant in gap

Gap provided between panel edges

Figure 20.13
SIPS Roof Ridge

20

Wood siding (all surfaces painted)

Airspace

Furring

Building paper drainage plane

Nail connection for shear

Flashing (also acts as termite barrier — sealed to slab

$3/8$" fibercement board; all surfaces coated

Ground slopes away from wall at 5% (6 in. per 10 ft.)

Rigid insulation

Bearing plate to facilitate "cantilever"

Sill gasket

$3/4$" pine spacer

Baseboard trim

Electrical service

$3/4$" pine spacer

Granular capillary break and drainage pad (no fines)

Concrete slab

Concrete grade beam

Polyethylene vapor barrier extended under grade beam where it also acts as a capillary break

Figure 20.14
SIPS Service Chase

- Cantilever for SIPS over foundation wall depends on shear connection between panel and bottom plate

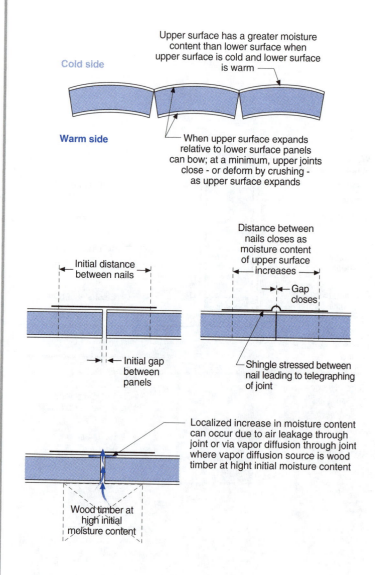

Cold side

Upper surface has a greater moisture content than lower surface when upper surface is cold and lower surface is warm

Warm side

When upper surface expands relative to lower surface panels can bow; at a minimum, upper joints close - or deform by crushing - as upper surface expands

Initial distance between nails

Initial gap between panels

Distance between nails closes as moisture content of upper surface increases

Gap closes

Shingle stressed between nail leading to telegraphing of joint

Localized increase in moisture content can occur due to air leakage through joint or via vapor diffusion through joint where vapor diffusion source is wood timber at hight initial moisture content

Wood timber at high initial moisture content

Figure 20.15
Telegraphing of Panel Joints
- Telegraphing can occur due to panel movement
- Telegraphing can also occur due to localized increase in moisture content

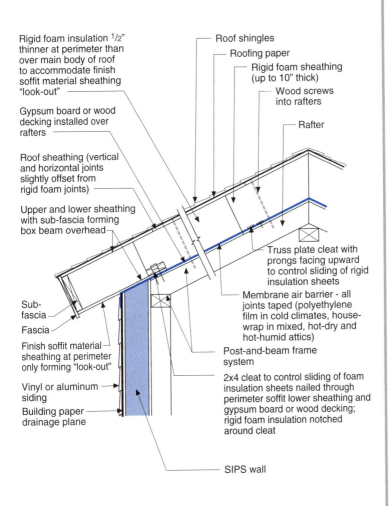

Rigid foam insulation ¹/₂"
thinner at perimeter than
over main body of roof
to accommodate finish
soffit material sheathing
"look-out"

Gypsum board or wood
decking installed over
rafters

Roof sheathing (vertical
and horizontal joints
slightly offset from
rigid foam joints)

Upper and lower sheathing
with sub-fascia forming
box beam overhead

Sub-
fascia

Fascia

Finish soffit material
sheathing at perimeter
only forming "look-out"

Vinyl or aluminum
siding

Building paper
drainage plane

Roof shingles

Roofing paper

Rigid foam sheathing
(up to 10" thick)

Wood screws
into rafters

Rafter

Truss plate cleat with
prongs facing upward
to control sliding of rigid
insulation sheets

Membrane air barrier - all
joints taped (polyethylene
film in cold climates, housewrap in mixed, hot-dry and
hot-humid attics)

Post-and-beam frame
system

2x4 cleat to control sliding of foam
insulation sheets nailed through
perimeter soffit lower sheathing and
gypsum board or wood decking;
rigid foam insulation notched
around cleat

SIPS wall

Figure 20.16
Post-and-Beam Construction
• Site-built foam roof in place of SIPS roof panels to eliminate telegraphing of
shingles

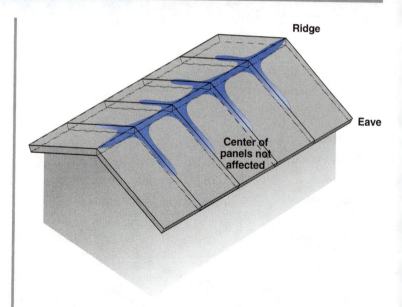

Figure 20.17
SIP "Ridge Rot"

 • Air transported moisture leads to decay when interior surface of panel joints is not sealed

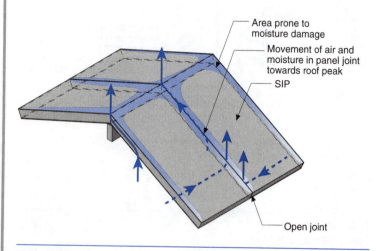

Figure 20.18
Air Leakage Pathway

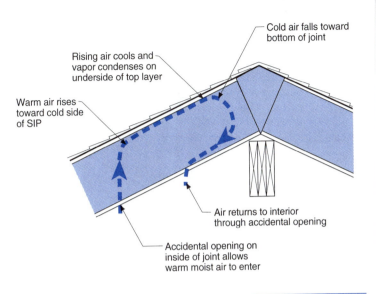

Cold air falls toward bottom of joint

Rising air cools and vapor condenses on underside of top layer

Warm air rises toward cold side of SIP

Air returns to interior through accidental opening

Accidental opening on inside of joint allows warm moist air to enter

Figure 20.19
Air Leakage Pathway

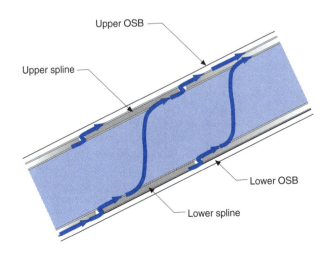

Upper OSB

Upper spline

Lower OSB

Lower spline

Figure 20.20
Air Leakage Pathway

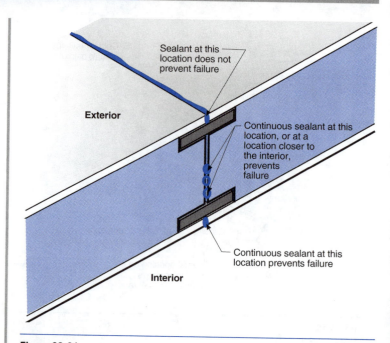

Sealant at this location does not prevent failure

Exterior

Continuous sealant at this location, or at a location closer to the interior, prevents failure

Continuous sealant at this location prevents failure

Interior

Figure 20.21
Seal Locations

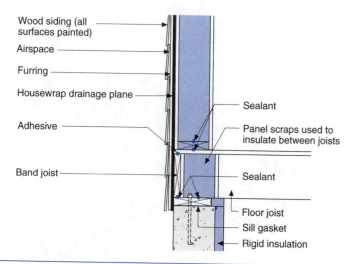

Wood siding (all surfaces painted)

Airspace

Furring

Housewrap drainage plane

Sealant

Adhesive

Panel scraps used to insulate between joists

Band joist

Sealant

Floor joist

Sill gasket

Rigid insulation

20

Figure 20.22
Panel Resting on Floor Deck — Detail

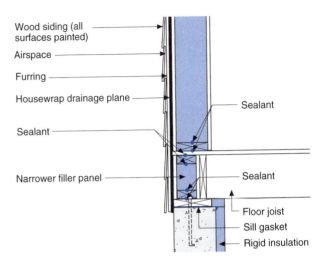

Wood siding (all surfaces painted)

Airspace

Furring

Housewrap drainage plane

Sealant

Narrower filler panel

Sealant

Sealant

Floor joist

Sill gasket

Rigid insulation

**Figure 20.23
Filler Panel at Band Joist — Detail**

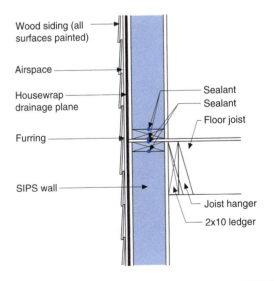

Wood siding (all surfaces painted)

Airspace

Housewrap drainage plane

Sealant

Sealant

Floor joist

Furring

SIPS wall

Joist hanger

2x10 ledger

20

**Figure 20.24
Floor/Wall Detail**

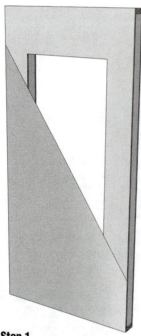

Step 1
SIPS panel with housewrap

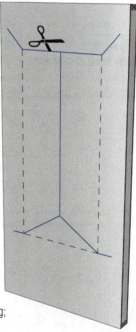

Step 2
Cut housewrap at rough opening;
note cutting pattern

Figure 20.25
Installing Window in a SIPS Wall in Twelve Steps

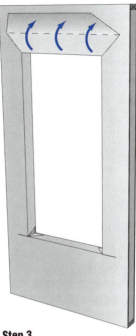

Step 3
Fold housewrap in at jambs and sill and secure tightly; head flap folded up and outward

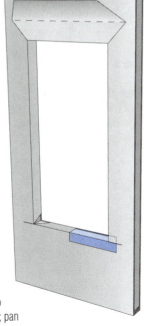

Step 4
Slip first piece of sill pan into horizontal slit in housewraps; pan must fit tightly; mechanically fasten at exterior vertical face only

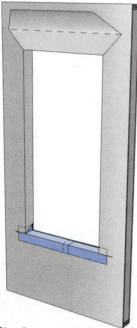

Step 5
Slip second piece of sill pan into horizontal slit in housewrap; pan must fit tightly; mechanically fasten at exterior vertical face only; ensure a minimum 3" overlap at sill

Step 6
Tape or flash sill pan at jambs and sill joint

Step 7

Back-caulk window; apply sealant at jambs and head; do not apply to sill; alternatively, caulk can be applied to window nailing flange prior to installation

Step 8

Install window plumb, level and square per manufacturer's instructions

20

Step 9

Install jamb flashing; extend minimum 1" above head nailing flange and a minimum of 3" below sill nailing flange

Step 10

Install head flashing; extend minimum 2" past edge of jamb flashing

Step 11
Fold housewrap down at head;
ensure head flap has not been
damaged during installation
process

Step 12
Apply corner patches at head;
extend minimum 1" beyond cut in
housewrap; air seal window
around entire perimeter on the
interior with sealant or non-
expanding foam

Precast Autoclaved Aerated Concrete (PAAC)

Precast autoclaved aerated concrete (PAAC) is a building material manufactured from sand, cement, lime, gypsum, an expanding agent and water. The materials are chemically expanded and steam cured under pressure in an autoclave to form building elements that are assembled, often like building blocks, into the shape of a building's exterior walls. When PAAC is kept dry its significant advantage over other materials is its thermal performance. PAAC is also insect resistant and fire resistant.

Concerns

The single greatest concern with PAAC is its water sensitivity and the ability of PAAC assemblies to control rain entry. Additional concerns relate to contact with groundwater.

A PAAC exterior wall can be classed as a face-sealed barrier system with some ability for the assembly to act as a storage reservoir for limited quantities of rain penetrating the exterior face. The durability of the PAAC material is not generally adversely affected by penetrating rain water. However, thermal performance is significantly affected. PAAC is a hygroscopic material that must be isolated from contact with groundwater sources and protected from exterior rain. Provision for drying towards interior conditioned spaces through interior finishes is recommended.

Accordingly, vapor permeable interior finishes are recommended to facilitate drying to the interior of any rain water that has penetrated the exterior face. Interior plaster finishes or renderings can be directly applied to the interior of the PAAC elements, components or blocks. The lack of a paper-facing associated with interior plaster or renderings significantly enhances moisture resistance.

As a face-sealed barrier system, window and door detailing is important. Details similar to historically successfully performing details developed for face-sealed masonry stucco systems should be used. Due to the water sensitivity of PAAC to thermal performance, high polymer content polymer-based (PB) synthetic stucco topcoats should be used as exterior renderings rather than traditional stucco coatings.

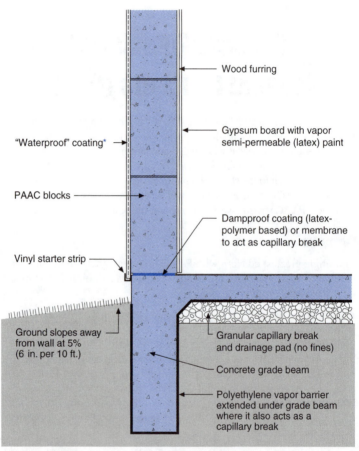

Wood furring

Gypsum board with vapor
semi-permeable (latex) paint

"Waterproof" coating*

PAAC blocks

Dampproof coating (latex-
polymer based) or membrane
to act as capillary break

Vinyl starter strip

Ground slopes away
from wall at 5%
(6 in. per 10 ft.)

Granular capillary break
and drainage pad (no fines)

Concrete grade beam

Polyethylene vapor barrier
extended under grade beam
where it also acts as a
capillary break

* High polymer content polymer-based (PB) synthetic stucco top coat such as
Dryflex® by Dryvit, Flexyl® by Sto or equal. Waterproof in this instance does
not mean vapor proof. These proprietary coatings pass water vapor; they are
not vapor barriers.

**Figure 21.1
Monolithic Slab — PAAC Above Grade Wall and Stucco**

- PAAC is a water sensitive material and must be protected from ground moisture
  sources and rain

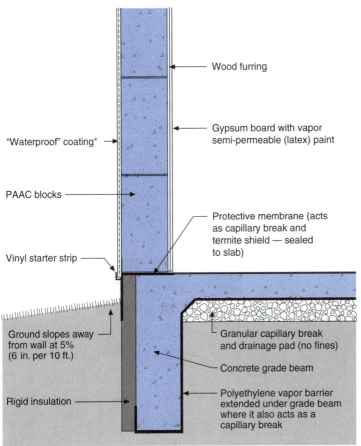

Wood furring

Gypsum board with vapor
semi-permeable (latex) paint

"Waterproof" coating*

PAAC blocks

Protective membrane (acts
as capillary break and
termite shield — sealed
to slab)

Vinyl starter strip

Ground slopes away
from wall at 5%
(6 in. per 10 ft.)

Granular capillary break
and drainage pad (no fines)

Concrete grade beam

Rigid insulation

Polyethylene vapor barrier
extended under grade beam
where it also acts as a
capillary break

* High polymer content polymer-based (PB) synthetic stucco top coat such as
Dryflex® by Dryvit, Flexyl® by Sto or equal. Waterproof in this instance does
not mean vapor proof. These proprietary coatings pass water vapor; they are
not vapor barriers.

Figure 21.2
Monolithic Slab — PAAC Above Grade Wall and Stucco

- PAAC is a water sensitive material and must be protected from ground moisture
  sources and rain

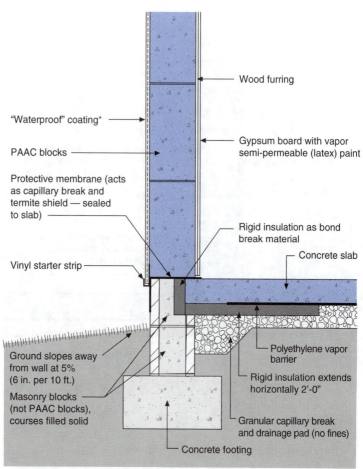

Wood furring

"Waterproof" coating*

PAAC blocks

Protective membrane (acts as capillary break and termite shield — sealed to slab)

Gypsum board with vapor semi-permeable (latex) paint

Rigid insulation as bond break material

Concrete slab

Vinyl starter strip

Ground slopes away from wall at 5% (6 in. per 10 ft.)

Masonry blocks (not PAAC blocks), courses filled solid

Polyethylene vapor barrier

Rigid insulation extends horizontally 2'-0"

Granular capillary break and drainage pad (no fines)

Concrete footing

* High polymer content polymer-based (PB) synthetic stucco top coat such as Dryflex® by Dryvit, Flexyl® by Sto or equal. Waterproof in this instance does not mean vapor proof. These proprietary coatings pass water vapor; they are not vapor barriers.

Figure 21.3
Slab with Masonry Perimeter —PAAC Above Grade Wall and Stucco

- PAAC is a water sensitive material and must be protected from ground moisture sources and rain

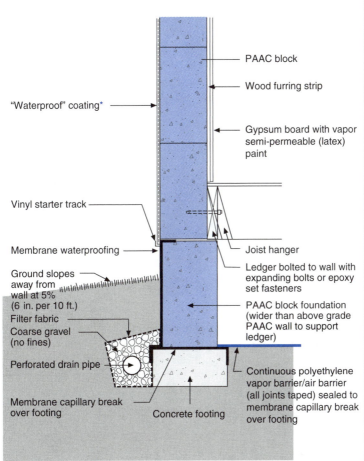

"Waterproof" coating*

PAAC block

Wood furring strip

Gypsum board with vapor semi-permeable (latex) paint

Vinyl starter track

Membrane waterproofing

Ground slopes away from wall at 5% (6 in. per 10 ft.)

Filter fabric

Coarse gravel (no fines)

Perforated drain pipe

Membrane capillary break over footing

Joist hanger

Ledger bolted to wall with expanding bolts or epoxy set fasteners

PAAC block foundation (wider than above grade PAAC wall to support ledger)

Concrete footing

Continuous polyethylene vapor barrier/air barrier (all joints taped) sealed to membrane capillary break over footing

* High polymer content polymer-based (PB) synthetic stucco top coat such as Dryflex® by Dryvit, Flexyl® by Sto or equal. Waterproof in this instance does not mean vapor proof. These proprietary coatings pass water vapor; they are not vapor barriers.

Figure 21.4
PAAC Crawl Space — PAAC Above Grade Wall and Stucco

• PAAC is a water sensitive material and must be protected from ground moisture sources and rain

 21

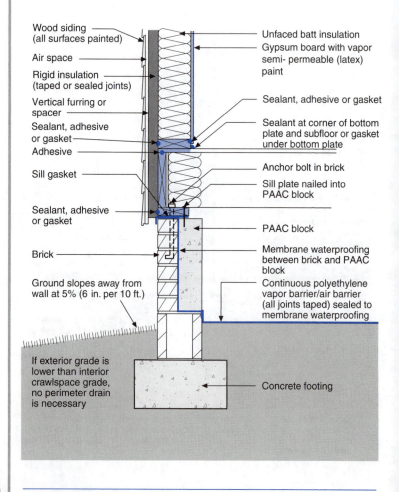

Wood siding
(all surfaces painted)

Air space

Rigid insulation
(taped or sealed joints)

Vertical furring or
spacer

Sealant, adhesive
or gasket

Adhesive

Sill gasket

Sealant, adhesive
or gasket

Brick

Ground slopes away from
wall at 5% (6 in. per 10 ft.)

If exterior grade is
lower than interior
crawlspace grade,
no perimeter drain
is necessary

Unfaced batt insulation

Gypsum board with vapor
semi- permeable (latex)
paint

Sealant, adhesive or gasket

Sealant at corner of bottom
plate and subfloor or gasket
under bottom plate

Anchor bolt in brick

Sill plate nailed into
PAAC block

PAAC block

Membrane waterproofing
between brick and PAAC
block

Continuous polyethylene
vapor barrier/air barrier
(all joints taped) sealed to
membrane waterproofing

Concrete footing

Figure 21.5
PAAC Crawl Space — Wood Frame Wall

- PAAC is a water sensitive material and must be protected from ground moisture sources and rain
- PAAC is used to provide insect resistant crawlspace perimeter insulation

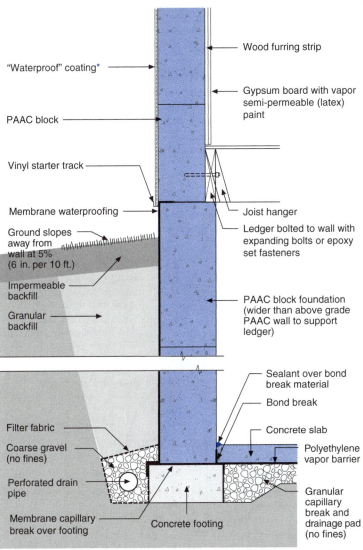

"Waterproof" coating*

Wood furring strip

PAAC block

Gypsum board with vapor semi-permeable (latex) paint

Vinyl starter track

Membrane waterproofing

Joist hanger

Ground slopes away from wall at 5% (6 in. per 10 ft.)

Ledger bolted to wall with expanding bolts or epoxy set fasteners

Impermeable backfill

Granular backfill

PAAC block foundation (wider than above grade PAAC wall to support ledger)

Sealant over bond break material

Bond break

Filter fabric

Concrete slab

Coarse gravel (no fines)

Polyethylene vapor barrier

Perforated drain pipe

Granular capillary break and drainage pad (no fines)

Membrane capillary break over footing

Concrete footing

* High polymer content polymer-based (PB) synthetic stucco top coat such as Dryflex® by Dryvit, Flexyl® by Sto or equal. Waterproof in this instance does not mean vapor proof. These proprietary coatings pass water vapor; they are not vapor barriers.

Figure 21.6
PAAC Basement — Stucco

- PAAC is a water sensitive material and must be protected from ground moisture sources and rain

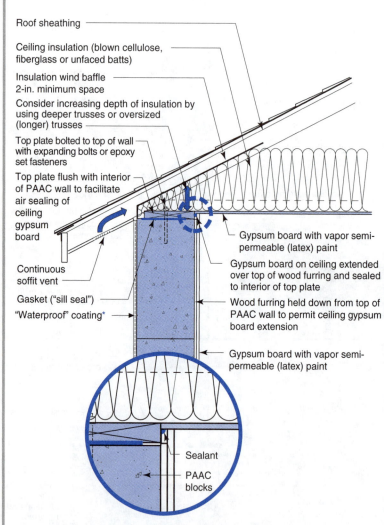

Roof sheathing

Ceiling insulation (blown cellulose, fiberglass or unfaced batts)

Insulation wind baffle
2-in. minimum space

Consider increasing depth of insulation by using deeper trusses or oversized (longer) trusses

Top plate bolted to top of wall with expanding bolts or epoxy set fasteners

Top plate flush with interior of PAAC wall to facilitate air sealing of ceiling gypsum board

Continuous soffit vent

Gasket ("sill seal")

"Waterproof" coating*

Gypsum board with vapor semi-permeable (latex) paint

Gypsum board on ceiling extended over top of wood furring and sealed to interior of top plate

Wood furring held down from top of PAAC wall to permit ceiling gypsum board extension

Gypsum board with vapor semi-permeable (latex) paint

Sealant

PAAC blocks

* High polymer content polymer-based (PB) synthetic stucco top coat such as Dryflex® by Dryvit, Flexyl® by Sto or equal. Waterproof in this instance does not mean vapor proof. These proprietary coatings pass water vapor; they are not vapor barriers.

21

**Figure 21.7
Roof Framing — Top Plate**

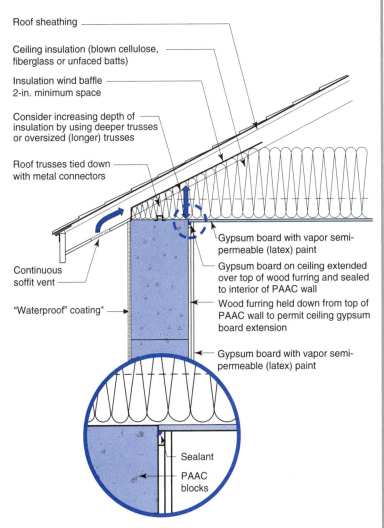

Roof sheathing

Ceiling insulation (blown cellulose, fiberglass or unfaced batts)

Insulation wind baffle
2-in. minimum space

Consider increasing depth of insulation by using deeper trusses or oversized (longer) trusses

Roof trusses tied down with metal connectors

Continuous soffit vent

"Waterproof" coating*

Gypsum board with vapor semi-permeable (latex) paint

Gypsum board on ceiling extended over top of wood furring and sealed to interior of PAAC wall

Wood furring held down from top of PAAC wall to permit ceiling gypsum board extension

Gypsum board with vapor semi-permeable (latex) paint

Sealant

PAAC blocks

* High polymer content polymer-based (PB) synthetic stucco top coat such as Dryflex® by Dryvit, Flexyl® by Sto or equal. Waterproof in this instance does not mean vapor proof. These proprietary coatings pass water vapor; they are not vapor barriers.

Figure 21.8
Roof Framing — Metal Straps

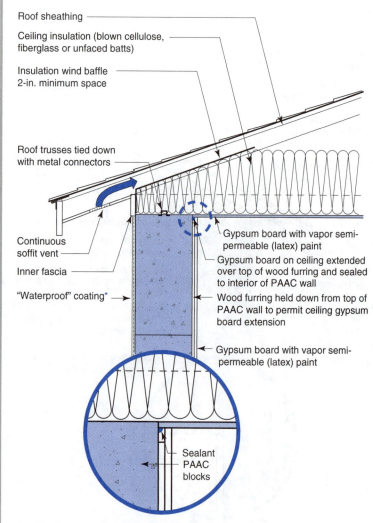

Roof sheathing

Ceiling insulation (blown cellulose, fiberglass or unfaced batts)

Insulation wind baffle 2-in. minimum space

Roof trusses tied down with metal connectors

Continuous soffit vent

Inner fascia

"Waterproof" coating*

Gypsum board with vapor semi-permeable (latex) paint

Gypsum board on ceiling extended over top of wood furring and sealed to interior of PAAC wall

Wood furring held down from top of PAAC wall to permit ceiling gypsum board extension

Gypsum board with vapor semi-permeable (latex) paint

Sealant
PAAC blocks

* High polymer content polymer-based (PB) synthetic stucco top coat such as Dryflex® by Dryvit, Flexyl® by Sto or equal. Waterproof in this instance does not mean vapor proof. These proprietary coatings pass water vapor; they are not vapor barriers.

21

Figure 21.9
Alternative Roof Truss to Increase Depth of Ceiling Insulation at Perimeter

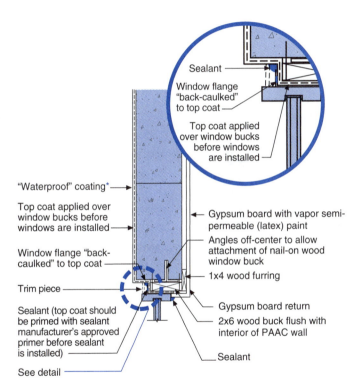

"Waterproof" coating* ⟶

Top coat applied over window bucks before windows are installed ⟶

Window flange "back-caulked" to top coat ⟶

Trim piece ⟶

Sealant (top coat should be primed with sealant manufacturer's approved primer before sealant is installed) ⟶

See detail ⟶

Sealant ⟶

Window flange "back-caulked" to top coat ⟶

Top coat applied over window bucks before windows are installed ⟶

⟵ Gypsum board with vapor semi-permeable (latex) paint

Angles off-center to allow attachment of nail-on wood window buck

1x4 wood furring

Gypsum board return

2x6 wood buck flush with interior of PAAC wall

Sealant

* High polymer content polymer-based (PB) synthetic stucco top coat such as Dryflex® by Dryvit, Flexyl® by Sto or equal. Waterproof in this instance does not mean vapor proof. These proprietary coatings pass water vapor; they are not vapor barriers.

Figure 21.10
PAAC Recessed Window Head — Stucco
• Windows installed after stucco

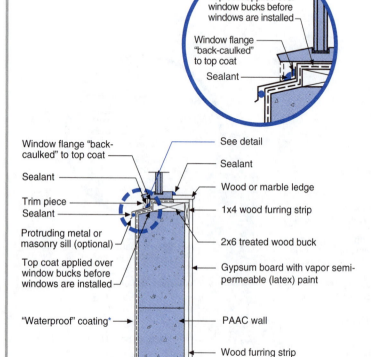

Top coat applied over window bucks before windows are installed

Window flange "back-caulked" to top coat

Sealant

Window flange "back-caulked" to top coat

See detail

Sealant

Sealant

Wood or marble ledge

Trim piece

1x4 wood furring strip

Sealant

Protruding metal or masonry sill (optional)

2x6 treated wood buck

Top coat applied over window bucks before windows are installed

Gypsum board with vapor semi-permeable (latex) paint

"Waterproof" coating*

PAAC wall

Wood furring strip

* High polymer content polymer-based (PB) synthetic stucco top coat such as Dryflex® by Dryvit, Flexyl® by Sto or equal. Waterproof in this instance does not mean vapor proof. These proprietary coatings pass water vapor; they are not vapor barriers.

Figure 21.11
PAAC Recessed Window Sill — Stucco
- Windows installed after stucco

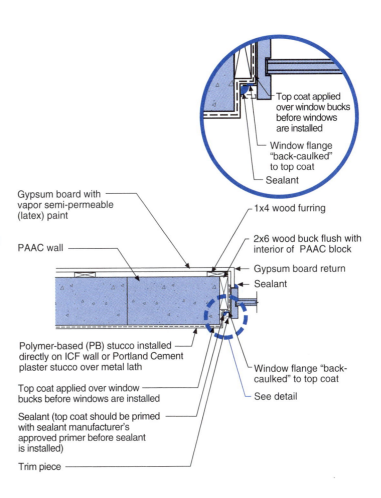

Top coat applied
over window bucks
before windows
are installed

Window flange
"back-caulked"
to top coat

Sealant

Gypsum board with
vapor semi-permeable
(latex) paint

1x4 wood furring

PAAC wall

2x6 wood buck flush with
interior of PAAC block

Gypsum board return

Sealant

Polymer-based (PB) stucco installed
directly on ICF wall or Portland Cement
plaster stucco over metal lath

Top coat applied over window
bucks before windows are installed

Sealant (top coat should be primed
with sealant manufacturer's
approved primer before sealant
is installed)

Trim piece

Window flange "back-
caulked" to top coat

See detail

 21

Figure 21.12
PAAC Recessed Window Jamb — Stucco
  • Windows installed after stucco

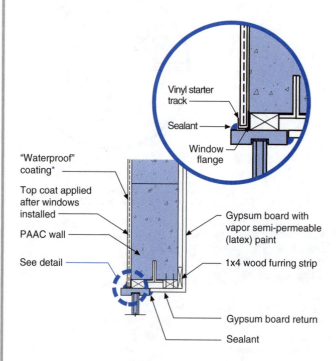

"Waterproof" coating*

Top coat applied after windows installed

PAAC wall

See detail

Vinyl starter track

Sealant

Window flange

Gypsum board with vapor semi-permeable (latex) paint

1x4 wood furring strip

Gypsum board return

Sealant

* High polymer content polymer-based (PB) synthetic stucco top coat such as Dryflex® by Dryvit, Flexyl® by Sto or equal. Waterproof in this instance does not mean vapor proof. These proprietary coatings pass water vapor; they are not vapor barriers.

**Figure 21.13
PAAC Flush Window Head — Stucco**
- Windows installed before stucco

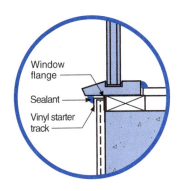

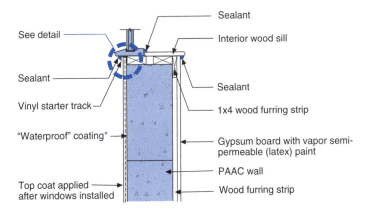

See detail

Sealant

Vinyl starter track

"Waterproof" coating*

Top coat applied
after windows installed

Sealant

Interior wood sill

Sealant

1x4 wood furring strip

Gypsum board with vapor semi-
permeable (latex) paint

PAAC wall

Wood furring strip

* High polymer content polymer-based (PB) synthetic stucco top coat such as
Dryflex® by Dryvit, Flexyl® by Sto or equal. Waterproof in this instance does
not mean vapor proof. These proprietary coatings pass water vapor; they are
not vapor barriers.

Figure 21.14
PAAC Flush Window Sill — Stucco
- Windows installed before stucco

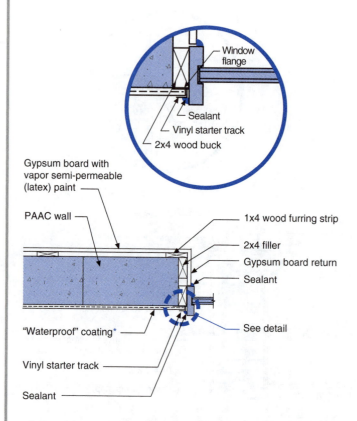

Window flange

Sealant

Vinyl starter track

2x4 wood buck

Gypsum board with vapor semi-permeable (latex) paint

PAAC wall

1x4 wood furring strip

2x4 filler

Gypsum board return

Sealant

See detail

"Waterproof" coating*

Vinyl starter track

Sealant

* High polymer content polymer-based (PB) synthetic stucco top coat such as Dryflex® by Dryvit, Flexyl® by Sto or equal. Waterproof in this instance does not mean vapor proof. These proprietary coatings pass water vapor; they are not vapor barriers.

Figure 21.15
PAAC Flush Window Jamb — Stucco
- Windows installed before stucco

Appendix I

Design Data

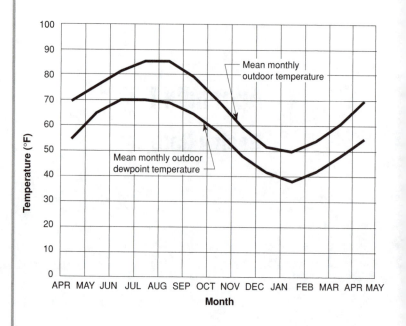

	Apr	May	Jun	Jul	Aug	Sep	Oct	Nov	Dec	Jan	Feb	Mar	Apr
Mean outdoor temp. (°F)	69	75	81	85	85	79	70	59	52	50	54	61	69
Mean dew point temp. (°F)	55	65	70	70	69	65	56	48	42	38	42	48	55

Figure I.1
Austin, Texas

- Average annual precipitation 31.9 inches
- 1,688 heating degree days; 3,016 cooling degree days
- Winter design temperature 28°F
- Summer design temperature 98°F dry bulb, 74°F wet bulb
- Average deep ground temperature 71°F
- Latitude 30.17°

I

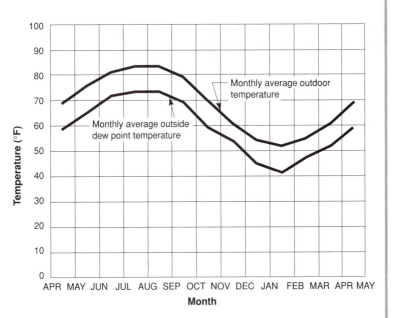

	Apr	May	Jun	Jul	Aug	Sep	Oct	Nov	Dec	Jan	Feb	Mar	Apr
Mean outdoor temp. (°F)	69	76	81	83	83	79	70	61	54	52	55	61	69
Mean dew point temp. (°F)	58	66	72	73	73	58	59	54	45	42	47	52	58

Figure I.2
Biloxi, Mississippi
- Average annual precipitation 62 inches
- 1,171 heating degree days; 3,201 cooling degree days
- Winter design temperature 31°F
- Summer design temperature 92°F dry bulb, 81°F wet bulb
- Average deep ground temperature 69°F
- Latitude 30.23°

I

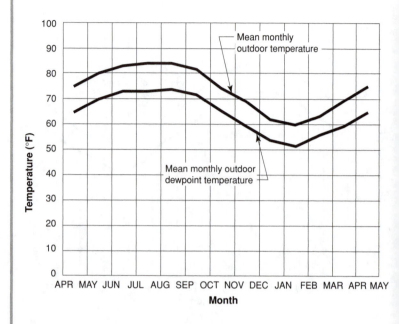

	Apr	May	Jun	Jul	Aug	Sep	Oct	Nov	Dec	Jan	Feb	Mar	Apr
Mean outdoor temp. (°F)	75	80	83	84	84	82	75	69	62	60	63	69	75
Mean dew point temp. (°F)	65	70	73	73	74	72	66	59	54	52	55	59	65

Figure I.3
Brownsville, Texas

- Average annual precipitation 26.6 inches
- 635 heating degree days; 3,888 cooling degree days
- Winter design temperature 39°F
- Summer design temperature 93°F dry bulb, 77°F wet bulb
- Average deep ground temperature 74°F
- Latitude 25.54°

I

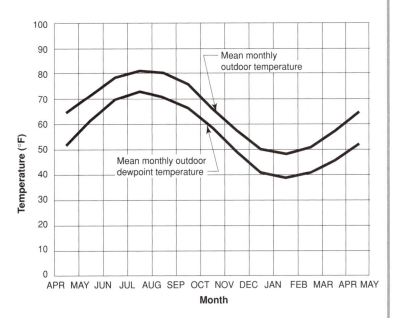

	Apr	May	Jun	Jul	Aug	Sep	Oct	Nov	Dec	Jan	Feb	Mar	Apr
Mean outdoor temp. (°F)	65	72	78	81	80	76	66	57	50	48	51	57	65
Mean dew point temp. (°F)	52	62	69	73	72	68	58	49	41	38	41	46	52

Figure I.4
Charleston, South Carolina
- Average annual precipitation 51.5 inches
- 2,013 heating degree days; 2,266 cooling degree days
- Winter design temperature 27°F
- Summer design temperature 91°F dry bulb, 78°F wet bulb
- Average deep ground temperature 68°F
- Latitude 32.53°

I

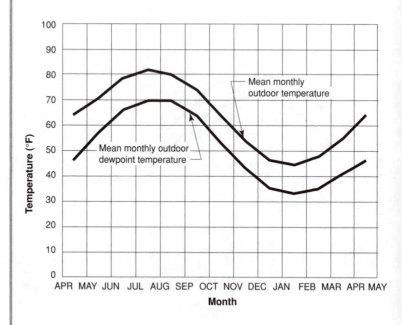

	Apr	May	Jun	Jul	Aug	Sep	Oct	Nov	Dec	Jan	Feb	Mar	Apr
Mean outdoor temp. (°F)	63	71	78	82	80	74	64	54	47	45	48	55	63
Mean dew point temp. (°F)	47	57	66	70	70	64	53	44	36	33	36	41	47

Figure I.5
Columbia, South Carolina

- Average annual precipitation 50 inches
- 2,649 heating degree days; 1,966 cooling degree days
- Winter design temperature 24°F
- Summer design temperature 95°F dry bulb, 75°F wet bulb
- Average deep ground temperature 64°F
- Latitude 33.56°

I

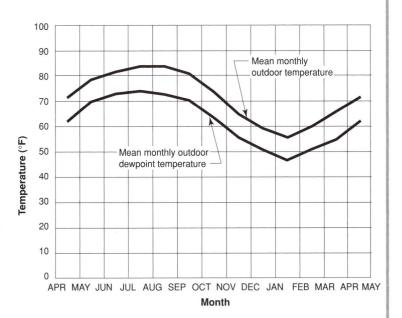

	Apr	May	Jun	Jul	Aug	Sep	Oct	Nov	Dec	Jan	Feb	Mar	Apr
Mean outdoor temp. (°F)	72	78	82	84	84	81	74	65	59	56	60	66	72
Mean dew point temp. (°F)	62	70	73	74	73	71	64	56	51	47	51	55	62

Figure I.6
Corpus Christi, Texas
- Average annual precipitation 30.1 inches
- 1,016 heating degree days; 3,439 cooling degree days
- Winter design temperature 35°F
- Summer design temperature 94°F dry bulb, 78°F wet bulb
- Average deep ground temperature 72°F
- Latitude 27.46°

I

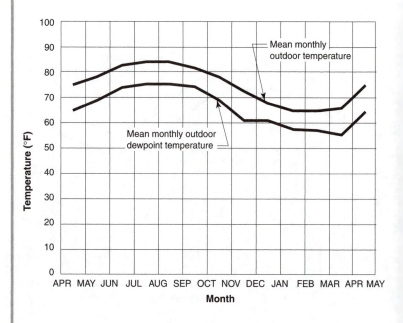

	Apr	May	Jun	Jul	Aug	Sep	Oct	Nov	Dec	Jan	Feb	Mar	Apr
Mean outdoor temp. (°F)	75	78	82	83	83	81	78	72	68	66	66	67	75
Mean dew point temp. (°F)	65	69	73	75	75	74	69	61	61	57	57	55	65

Figure I.7
Fort Myers, Florida

- Average annual precipitation 53 inches
- 225 heating degree days, 3,702 cooling degree days
- Winter design temperature 41°F
- Summer design temperature 93°F dry bulb, 80°F wet bulb
- Average deep ground temperature 77°F
- Latitude 26.35°

I

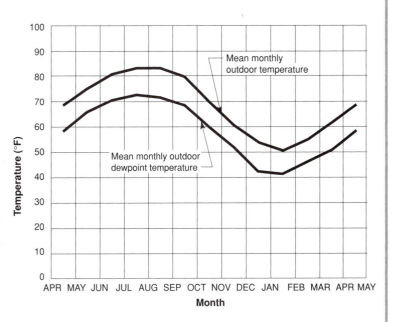

	Apr	May	Jun	Jul	Aug	Sep	Oct	Nov	Dec	Jan	Feb	Mar	Apr
Mean outdoor temp. (°F)	68	75	81	83	83	79	70	61	54	51	55	62	68
Mean dew point temp. (°F)	58	66	71	73	72	68	60	52	43	42	46	51	58

Figure I.8
Houston, Texas

- Average annual precipitation 46.1 inches
- 1,599 heating degree days; 2,700 cooling degree days
- Winter design temperature 32°F
- Summer design temperature 94°F dry bulb, 77°F wet bulb
- Average deep ground temperature 72°F
- Latitude 29.59°

I

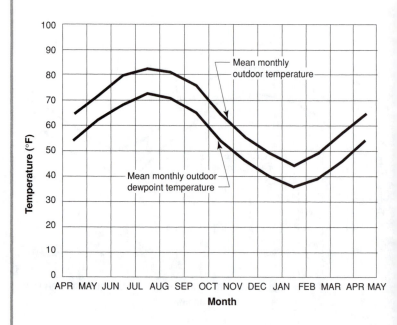

	Apr	May	Jun	Jul	Aug	Sep	Oct	Nov	Dec	Jan	Feb	Mar	Apr
Mean outdoor temp. (°F)	65	72	79	82	81	76	65	56	49	45	49	57	65
Mean dew point temp. (°F)	54	62	68	72	71	66	54	47	40	37	39	46	54

Figure I.9
Jackson, Mississippi

- Average annual precipitation 55.4 inches
- 2,467 heating degree days; 2,215 cooling degree days
- Winter design temperature 25°F
- Summer design temperature 95°F dry bulb, 76°F wet bulb
- Average deep ground temperature 64°F
- Latitude 32.19°

I

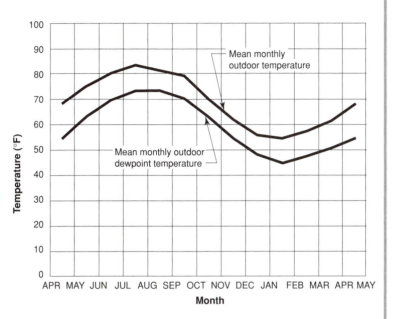

	Apr	May	Jun	Jul	Aug	Sep	Oct	Nov	Dec	Jan	Feb	Mar	Apr
Mean outdoor temp. (°F)	68	75	80	83	82	79	70	62	56	54	57	62	68
Mean dew point temp. (°F)	55	63	70	73	73	71	63	55	48	45	47	51	55

Figure I.10
Jacksonville, Florida
- Average annual precipitation 51.3 inches
- 1,434 heating degree days; 2,551 cooling degree days
- Winter design temperature 32°F
- Summer design temperature 94°F dry bulb, 77°F wet bulb
- Average deep ground temperature 70°F
- Latitude 30.29°

I

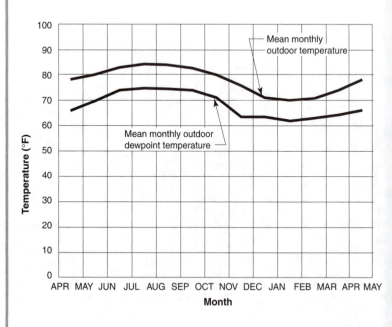

	Apr	May	Jun	Jul	Aug	Sep	Oct	Nov	Dec	Jan	Feb	Mar	Apr
Mean outdoor temp. (°F)	77	80	83	84	84	83	80	76	71	70	71	74	77
Mean dew point temp. (°F)	66	70	74	75	75	75	72	64	64	62	63	64	66

Figure I.11
Key West, Florida
- Average annual precipitation 40 inches
- 100 heating degree days; 4,798 cooling degree days
- Winter design temperature 57°F
- Summer design temperature 90°F dry bulb, 78°F wet bulb
- Average deep ground temperature 78°F
- Latitude 24.33°

I

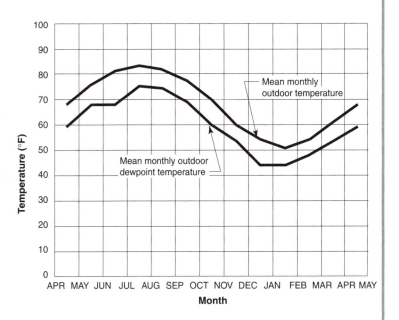

	Apr	May	Jun	Jul	Aug	Sep	Oct	Nov	Dec	Jan	Feb	Mar	Apr
Mean outdoor temp. (°F)	68	75	81	83	82	78	70	60	54	51	54	61	68
Mean dew point temp. (°F)	59	68	68	75	74	69	60	53	44	44	48	53	59

Figure I.12
Lake Charles, Louisiana
- Average annual precipitation 55 inches
- 1,209 heating degree days; 2,922 cooling degree days
- Winter design temperature 27°F
- Summer design temperature 95°F dry bulb, 80°F wet bulb
- Average deep ground temperature 69°F
- Latitude 30.07°

I

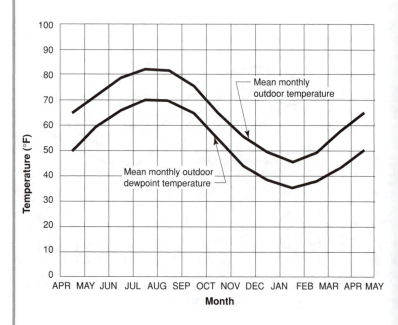

	Apr	May	Jun	Jul	Aug	Sep	Oct	Nov	Dec	Jan	Feb	Mar	Apr
Mean outdoor temp. (°F)	64	72	78	82	81	76	65	56	49	46	49	57	64
Mean dew point temp. (°F)	50	59	66	70	70	65	54	44	38	36	38	43	50

Figure I.13
Macon, Georgia

- Average annual precipitation 44.6 inches
- 2,334 heating degree days; 2,125 cooling degree days
- Winter design temperature 25°F
- Summer design temperature 93°F dry bulb, 76°F wet bulb
- Average deep ground temperature 65°F
- Latitude 32.41°

I

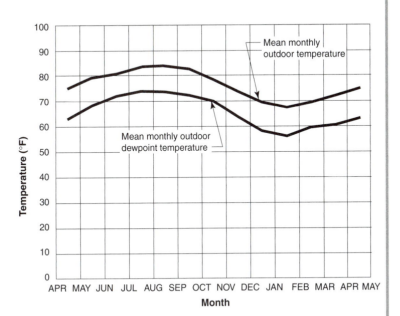

	Apr	May	Jun	Jul	Aug	Sep	Oct	Nov	Dec	Jan	Feb	Mar	Apr
Mean outdoor temp. (°F)	75	79	81	83	83	82	78	73	69	68	69	72	75
Mean dew point temp. (°F)	63	68	72	74	74	73	70	63	58	57	59	61	63

Figure I.14
Miami, Florida

- Average annual precipitation 55.9 inches
- 200 heating degree days; 4,198 cooling degree days
- Winter design temperature 47°F
- Summer design temperature 90°F dry bulb, 77°F wet bulb
- Average deep ground temperature 77°F
- Latitude 25.49°

I

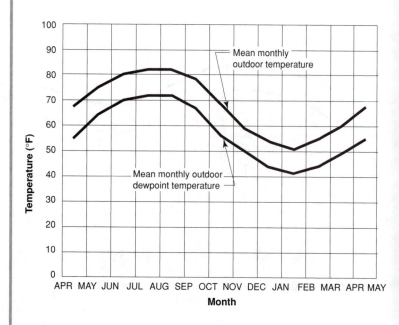

	Apr	May	Jun	Jul	Aug	Sep	Oct	Nov	Dec	Jan	Feb	Mar	Apr
Mean outdoor temp. (°F)	67	75	80	82	82	78	68	59	53	51	54	60	67
Mean dew point temp. (°F)	55	64	70	72	72	67	57	50	44	41	44	49	55

Figure I.15
Mobile, Alabama

- Average annual precipitation 64 inches
- 1,702 heating degree days; 2,627 cooling degree days
- Winter design temperature 29°F
- Summer design temperature 93°F dry bulb, 77°F wet bulb
- Average deep ground temperature 68°F
- Latitude 30.41°

I

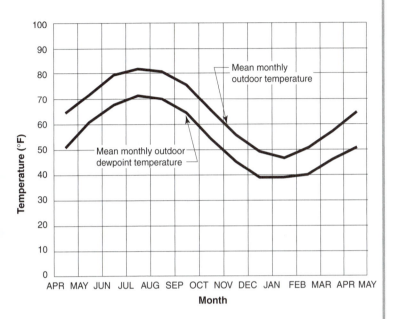

	Apr	May	Jun	Jul	Aug	Sep	Oct	Nov	Dec	Jan	Feb	Mar	Apr
Mean outdoor temp. (°F)	65	72	79	82	81	76	66	56	49	47	51	57	65
Mean dew point temp. (°F)	51	61	68	71	70	65	55	46	39	39	40	45	51

Figure I.16
Montgomery, Alabama
- Average annual precipitation 53.4 inches
- 2,224 heating degree days; 2,212 cooling degree days
- Winter design temperature 25°F
- Summer design temperature 95°F dry bulb, 76°F wet bulb
- Average deep ground temperature 65°F
- Latitude 32.18` °

I

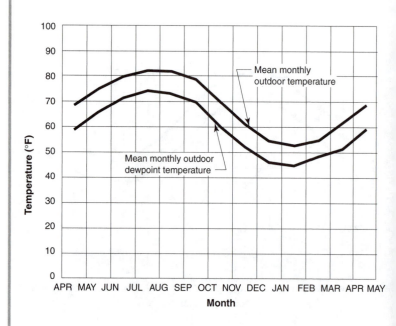

	Apr	May	Jun	Jul	Aug	Sep	Oct	Nov	Dec	Jan	Feb	Mar	Apr
Mean outdoor temp. (°F)	68	75	80	82	82	78	70	61	55	53	55	62	68
Mean dew point temp. (°F)	58	66	72	74	73	70	60	52	47	46	48	52	58

Figure I.17
New Orleans, Louisiana

- Average annual precipitation 61.9 inches
- 1,513 heating degree days; 2,655 cooling degree days
- Winter design temperature 33°F
- Summer design temperature 92°F dry bulb, 78°F wet bulb
- Average deep ground temperature 70°F
- Latitude 29.59°

I

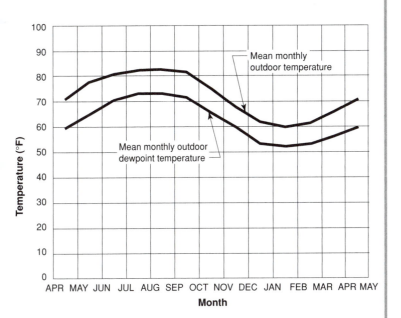

	Apr	May	Jun	Jul	Aug	Sep	Oct	Nov	Dec	Jan	Feb	Mar	Apr
Mean outdoor temp. (°F)	71	77	81	82	82	81	75	68	62	60	62	66	71
Mean dew point temp. (°F)	59	65	71	73	73	72	66	60	54	52	53	57	59

Figure I.18
Orlando, Florida
- Average annual precipitation 48.1 inches
- 686 heating degree days; 3,381 cooling degree days
- Winter design temperature 38°F
- Summer design temperature 93°F dry bulb, 76°F wet bulb
- Average deep ground temperature 72°F
- Latitude 28.26°

I

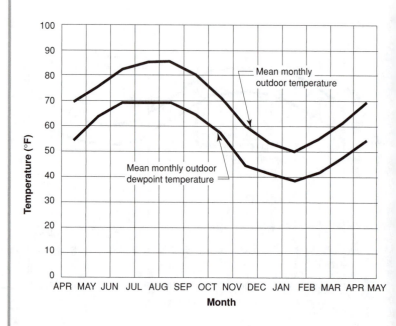

	Apr	May	Jun	Jul	Aug	Sep	Oct	Nov	Dec	Jan	Feb	Mar	Apr
Mean outdoor temp. (°F)	69	76	82	85	85	80	71	60	53	50	55	62	69
Mean dew point temp. (°F)	54	64	69	69	69	65	58	45	42	38	42	48	54

Figure I.19
San Antonio, Texas
- Average annual precipitation 31 inches
- 1,644 heating degree days; 2,996 cooling degree days
- Winter design temperature 30°F
- Summer design temperature 97°F dry bulb, 73°F wet bulb
- Average deep ground temperature 71°F
- Latitude 29.31°

I

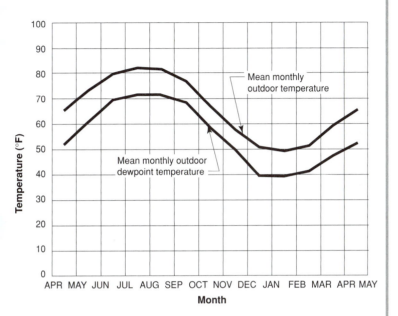

	Apr	May	Jun	Jul	Aug	Sep	Oct	Nov	Dec	Jan	Feb	Mar	Apr
Mean outdoor temp. (°F)	66	73	79	82	81	77	67	58	51	49	52	59	66
Mean dew point temp. (°F)	52	61	69	72	72	68	58	50	39	39	42	47	52

Figure I.20
Savannah, Georgia
- Average annual precipitation 49.2 inches
- 1,847 heating degree days; 2,365 cooling degree days
- Winter design temperature 27°F
- Summer design temperature 93°F dry bulb, 77°F wet bulb
- Average deep ground temperature 70°F
- Latitude 32.07°

I

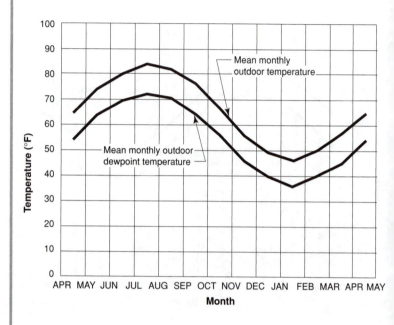

	Apr	May	Jun	Jul	Aug	Sep	Oct	Nov	Dec	Jan	Feb	Mar	Apr
Mean outdoor temp. (°F)	65	73	80	83	82	77	67	56	49	46	50	57	65
Mean dew point temp. (°F)	53	63	69	72	71	65	56	47	40	36	40	45	53

Figure I.21
Shreveport, Louisiana
- Average annual precipitation 46 inches
- 2,264 heating degree days; 2,368 cooling degree days
- Winter design temperature 25°F
- Summer design temperature 96°F dry bulb, 76°F wet bulb
- Average deep ground temperature 65°F
- Latitude 32.26°

I

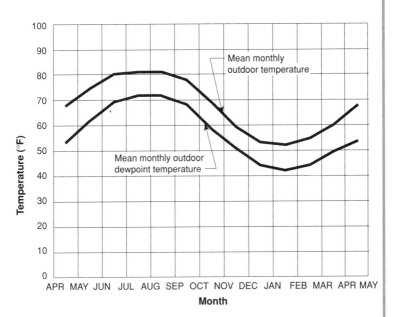

	Apr	May	Jun	Jul	Aug	Sep	Oct	Nov	Dec	Jan	Feb	Mar	Apr
Mean outdoor temp. (°F)	67	74	80	81	81	78	69	59	53	52	55	60	67
Mean dew point temp. (°F)	53	62	69	72	72	68	58	51	44	42	44	49	53

Figure I.22
Tallahassee, Florida
- Average annual precipitation 65.7 inches
- 1,705 heating degree days; 2,518 cooling degree days
- Winter design temperature 30°F
- Summer design temperature 92°F dry bulb, 76°F wet bulb
- Average deep ground temperature 70°F
- Latitude 30.23°

I

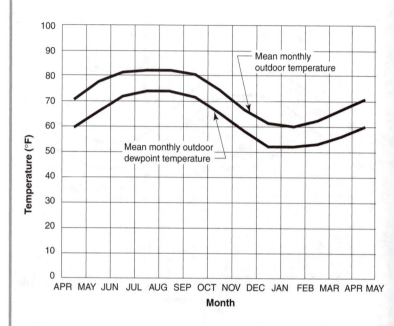

	Apr	May	Jun	Jul	Aug	Sep	Oct	Nov	Dec	Jan	Feb	Mar	Apr
Mean outdoor temp. (°F)	71	77	81	82	82	81	75	67	62	60	62	67	71
Mean dew point temp. (°F)	60	66	72	74	74	72	66	58	52	52	53	56	60

Figure I.23
Tampa, Florida

- Average annual precipitation 43.9 inches
- 725 heating degree days; 3,427 cooling degree days
- Winter design temperature 40°F
- Summer design temperature 91°F dry bulb, 77°F wet bulb
- Average deep ground temperature 74°F
- Latitude 27.57°

I

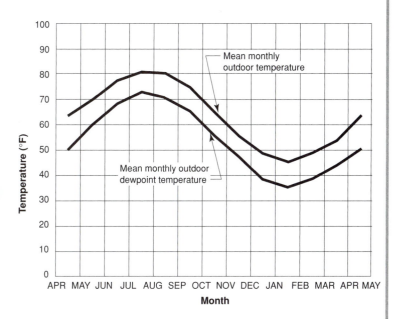

	Apr	May	Jun	Jul	Aug	Sep	Oct	Nov	Dec	Jan	Feb	Mar	Apr
Mean outdoor temp. (°F)	63	70	77	81	80	75	65	56	48	46	48	54	63
Mean dew point temp. (°F)	50	60	68	73	71	66	56	47	39	36	38	44	50

Figure I.24
Wilmington, North Carolina
- Average annual precipitation 54.3 inches
- 2,470 heating degree days; 1,926 cooling degree days
- Winter design temperature 26°F
- Summer design temperature 91°F dry bulb, 78°F wet bulb
- Average deep ground temperature 64°F
- Latitude 34.16°

I

I

Appendix II
Air Leakage Testing, Pressure Balancing, and Combustion Safety

Air leakage testing is a method for determining the total leakage area of a building enclosure or the leakage of air distribution systems (ductwork leakage in ducted forced air space conditioning systems). Air leakage testing is not a method for determining the actual air leakage or air change which occurs through the building enclosure under the influence of air pressure differences created by wind, stack action (the buoyancy of heated air) and mechanical systems (duct leakage and unbalanced forced air systems).

Air leakage testing for both building enclosures and ductwork is based on the fundamental properties of air flow through openings. The amount of air flow through an opening is determined by two principle factors:

- the area/size/geometry of the opening; and

- the air pressure difference across the opening.

The three parameters — air flow, area and air pressure difference — can be related to each other by applying a simple mathematical relationship. Measuring two of the three parameters and applying the mathematical relationship can determine the third parameter. For example, if the air flow through an opening is measured, as well as the air pressure difference across the opening when air flow is occurring, the area of the opening can be calculated by applying the mathematical relationship. Applying this relationship to a flow rate of 1,000 cfm through an opening with an air pressure differential of 50 Pascals across the opening obtains a mathematically calculated area of approximately 1 square foot. In other words 1,000 cfm air flow occurs through a 1 square foot opening as a result of an air pressure difference of 50 Pascals.

II

Air Leakage Testing of Building Enclosures

Air leakage testing of building enclosures involves placing a large calibrated fan in an exterior door and creating an air flow through the fan. The calibrated fan is often referred to as a "blower door". Exhausting air depressurizes the building. Supplying air pressurizes the building. When air is exhausted from a building through a blower door, air leaks into the building through openings to replace the air exhausted. If sufficient air is exhausted to overcome any naturally occurring pressures, the quantity of air exhausted will equal the quantity of air supplied. The quantity of air exhausted through the blower door can be readily measured. This exhaust quantity can be equated to the air leaking into the building enclosure through all of the openings in the building enclosure. If the air pressure difference between the interior of the building enclosure and the exterior is also measured, this can be used to approximate the air pressure difference across all of the openings in the building enclosure.

By determining the quantity of air exhausted and the air pressure difference across the building enclosure, the combined air flow through all of the openings in the building enclosure as well as the air pressure difference across all of the openings is known. Applying a mathematical relationship converts the combined air flow and air pressure difference to the combined leakage area of all of the openings in the building enclosure. In this manner a blower door can determine the combined area of all of the openings in a building enclosure, including random cracks, flaws, openings built into the building as a result of the building process, without actually determining where the leaks and openings are.

The procedure can also be applied to building enclosures pressurized by calibrated fans. Either depressurization or pressurization can be employed. Depressurization is more common in building enclosure leakage testing as a result of tradition rather than accuracy (the procedures were initially popularized in cold climates where pressurization during the heating season typically created comfort problems during testing).

Air leakage test results are expressed several ways. Test results can be presented as a flow rate at 50 Pascals air pressure difference (CFM50). In other words, the volume flow rate of air extracted out of the building enclosure necessary to depressurize the building enclosure 50 Pascals relative to the exterior is measured and reported. The combined or equivalent leakage area of the building (EqLA) in square inches can be determined from the CFM50 measurement by the application of the mathematical relationship.

The combined leakage area of a building enclosure can also be compared to the total surface area of the building enclosure using a parameter called a leakage ratio. Leakage ratios are typically expressed as square inches of leakage for every 100 square feet of building enclosure area (or $cm^2/100m^2$ SI). In this manner the measured EqLA is related to the measured surface area of the building enclosure.

Test results can also be expressed in the form of air changes per hour at a pressure difference of 50 Pascals (ach @ 50 Pa). In this approach, the volume flow rate extracted out of the building is related to the volume of the building enclosure. For example, consider a blower door extracting air out of a building enclosure at a rate of 1,000 cubic feet per minute establishing an air pressure difference of 50 Pa. This is equivalent to an air extraction rate of 60,000 cubic feet per hour. If the volume of the conditioned space is 10,000 cubic feet, this would result in 6 air changes per hour at a 50 Pascal air pressure difference (60,000/ 10,000 = 6). A flow rate of 1,000 cfm through an opening with a 50 Pascals air pressure difference was previously determined to require a 1 foot square opening. In other words, for this particular volume of building, 6 ach @ 50 Pa is equivalent to 1 foot square of leakage or an EqLA of 1 foot square or 144 square inches. If the surface area of the building enclosure was approximately 3,000 square feet, the leakage ratio would be approximately 4.8 square inches of leakage for every 100 square feet of building enclosure surface area (3,000/100 = 30 and 144/30 = 4.8).

The following information all describes the same building. It came from the blower door test conducted on the building described in the previous example.

$$CFM50 = 1,000$$

$$EqLA = 144 \ in^2$$

6 ach @ 50 Pa (where building volume is $10,000 \ ft^3$)

Leakage Ratio = $4.8 \ in^2/100 \ ft^2$ (where building enclosure surface area is $3,000 \ ft^2$)

Not all of these values are always determined or recorded. All can be related to each other and the building enclosure tested. The leakage ratio value is the most descriptive as it can be used to compare buildings of differing volumes and surface areas to each other. Since the other values are related to specific buildings with given building volumes and building surface areas, comparisons between buildings of diverse construction are less meaningful. However, where buildings are of approximately the same floor area and volume, all of the values provide reasonably comparative information.

II

Air leakage testing of building enclosures is typically conducted as a method of quality control to ensure that control of air flow occurs in constructed buildings. It is also used to identify leakage areas which may have been missed during construction thereby facilitating repairs and remediation.

Air Leakage Testing of Air Distribution Systems

Air leakage testing of air distribution systems is similar to air leakage testing of building enclosures in that both procedures involve using a calibrated fan to create an air pressure difference. In addition, both procedures require that air flows through the calibrated fans and pressure differentials be determined.

Air leakage testing of air distribution systems involves sealing the supply and return registers and depressurizing the system using a calibrated fan. In this approach, the ductwork and the air handler is considered a closed system. The calibrated fan, sometimes referred to as a "duct blaster" is typically attached to the air handler. The quantity of air moved by the duct blaster to pressurize or depressurize the system is directly related to the leakage area of the system. By determining the quantity of air supplied by the calibrated fan and the air pressure difference between the ductwork and the conditioned space, the combined air flow through all of the leakage openings in the air distribution system is determined.

Air leakage test results for air distribution systems are often presented as the flow rate through the calibrated fan required to pressurize or depressurize the duct system to a specific pressure differential. For example CFM25 values are typical. A reading of CFM25 = 60 cfm translates to: an air flow rate of 60 cfm through the calibrated fan (duct blaster) depressurized the air distribution system (ductwork and air handler) to a negative of 25 Pascals (0.1" w.c.) relative to the conditioned space.

Air leakage testing of air distribution systems is typically conducted as a method of quality control to ensure control of air flow occurs in air distribution systems. It is also used to identify leakage areas thereby facilitating repairs and remediation.

Pressure Balancing

Air pressure differentials between conditioned spaces and the surroundings, as well as between rooms and between building assemblies and rooms, affect the health, safety and durability of the building enclosure.

II

Infiltration of humid air during cooling periods is a concern as is the exfiltration of interior heated, moist air during heating periods. Infiltration of soil gas (moisture, radon, pesticides, other) below grade or from crawlspaces and slabs is a concern throughout the year. High interior negative air pressures can lead to the spillage and backdrafting of combustion appliances such as fireplaces, wood stoves, combustion water heaters and furnaces. Finally, significant depressurization can lead to flame roll out and fire in some combustion appliances such as furnaces and water heaters.

With respect to combustion appliances, installation of appliances which are not sensitive to negative air pressures as well as limiting the negative air pressures which can occur within building enclosures is an appropriate strategy for control.

With respect to the infiltration of exterior pollutants such as soil gas, pesticides, radon and below grade moisture, limiting the negative air pressure which can occur, providing positive air pressurization of building enclosures (or portions of building enclosures) as well as providing sub-slab or crawlspace depressurization are appropriate strategies for control.

With respect to the exfiltration of interior moisture during heating periods, constructing building assemblies which are forgiving and/or tolerant of interior moisture, as well as limiting the positive air pressure which can occur during heating periods (and/or providing negative pressures during heating periods) are appropriate strategies for control.

In all cases the maximum pressurization or depressurization relative to exterior (ambient) conditions should be limited to less than 3 Pascals.

Ducted forced air distribution systems are traditionally viewed as interior circulation systems which move air from place to place within a conditioned space, with a neutral effect on the pressure differences between the interior and exterior. However, as a result of installation practices and design/sizing faults ducted forced air distribution systems can have significant effects on air pressure relationships.

Duct leakage can result in either pressurization or depressurization of entire conditioned spaces or specific rooms. Duct leakage can significantly increase space conditioning energy requirements. Incorrect duct sizing, distribution layout or lack of adequate returns can lead to pressurization and depressurization of rooms and interstitial spaces. These effects should be limited to less than 3 Pascals positive or negative relative to the exterior or between rooms and/or interstitial (between two surfaces) cavities within building enclosures.

Exhaust fans and appliances such as whole house fans, attic ventilation

II

fans, indoor grills, clothes dryers, kitchen exhaust range hoods can also significantly alter air pressure relationships. These effects should also be limited to less than 3 Pascals positive or negative relative to the exterior or between rooms and/or interstitial cavities within building enclosures (except in the case of whole house fans and indoor grills). Where whole house fans and/or indoor grills are operating, no other combustion appliances or heating and/or mechanical cooling systems should be in simultaneous use. Furthermore, windows and/or doors should be opened when whole house fans and/or indoor grills are operating.

Testing Pressure Differentials and Commissioning

Air pressure relationships between conditioned spaces and the exterior, as well as between rooms and between rooms and interstitial spaces should be measured under all operating conditions. Equipment should be cycled on and off at all speed settings. Interior doors should be both opened and closed during all testing. Measurements are typically taken with a digital micromanometer. Where any air pressure differential greater than 3 Pascals is measured, remediation work and/or adjustments to equipment, or the building enclosure will be necessary.

Combustion Safety

If combustion appliances are selected, they should not interact aerodynamically with the building. In other words, changing interior air pressure differentials should not be able to influence the operation of combustion appliances. In order to meet this requirement only sealed combustion, power vented, induced draft or direct vented combustion appliances should be used for space conditioning and domestic hot water.

Gas cook tops and ovens should be only installed in conjunction with direct vented (to the exterior) exhaust range hoods. Recirculating range hoods should be avoided even in the absence of combustion appliances as they become breeding grounds for biological growth and a source of odors.

Fireplaces and wood stoves should only be installed with their own correctly sized air supply from the exterior. Fireplaces should also be provided with tight-fitting glass doors.

High interior negative air pressures can lead to the spillage and backdrafting of combustion appliances and significant depressurization can lead to flame roll out and fire. In all enclosures containing combustion appliances, the maximum depressurization relative to the exterior should be limited to less than 3 Pascals.

II

All combustion appliances (except fireplaces and wood stoves) should be tested for carbon monoxide production prior to occupancy and on a yearly basis thereafter. Carbon monoxide production of any appliance should not exceed 50 ppm. All measurements should be taken in the draft hood or vent before exhaust gases are mixed with dilution air. The installation of household carbon monoxide detectors is recommended.

Air Leakage — Determining Leakage Ratios and Leakage Coefficients

Using a blower door, measure the flow rate necessary to depressurize the building 50 Pascals. This flow rate is defines as CFM50. Alternatively, determine the Equivalent Leakage Area (EqLA) in square inches at 10 Pascals using the procedure outlined by the Canadian General Standards Board (or, alternatively, ASTM calculated at 4 Pascals). When determining these values, intentional openings (design openings) should be closed or blocked. These openings include fireplace dampers and fireplace glass doors, dryer vents, bathroom fans, exhaust fans, heat recovery ventilators (HRV's), wood stove flues, water heater flues, furnaces flues and combustion air openings.

Calculate the leakage ratio or the leakage coefficient using the entire surface area of the building enclosure. When determining the surface area of the building enclosure, below grade surface areas such as basement perimeter walls and basement floor slabs are included.

For example, a 2,550 square foot house constructed in Grayslake, Ill., has a building enclosure surface areas of 6,732 square feet and a conditioned space volume of 33,750 cubic fee (including the basement). The measure Equivalent Leakage Area (EqLA) using a blower door is 128 square inches. This also correspond to a blower door measure CFM50 value of 1,320 cfm and a blower door measure 2.3 airchanges per hour at 50 Pascals.

Surface Area	EqLA	CFM50	ach@50 Pa	Volume
6,723 ft^2	128in^2	1,320 cfm	2.3	33,750 ft^3

To determine the Leakage Ratio, divide the surface area of the building enclosure by 100 square feet and take this interim value and divide it into the EqLA.

$$6,732 \text{ ft}^2 \div 100 \text{ ft}^2 = 67.32$$
$$128 \text{ in}^2 \div 67.32 = 1.9 \text{ in}^2/100 \text{ ft}^2$$
(Leakage Ratio)

II

To determine the Leakage Coefficient, divide the CFM50 value by the surface area of the building enclosure.

$$1,320 \text{ CFM50} \div 6,732 \text{ ft}^2 = 0.20 \text{ cfm/ft}^2$$
(Leakage Coefficient)

Many airtightness measurements are recorded as air changes per hour at a pressure differential of 50 Pascals (ach @ 50 Pa). To convert ach @ 50 Pa to CFM50, multiply the volume of the building enclosure (including the basement) by the ach @ 50 Pa and divide by 60 min/hour.

For example, 2.3 ach @ 50 Pa across the building enclosure of volume 33,750 ft^3 is equivalent to a CFM50 value of 1,320 cfm.

$$33,750 \text{ ft}^3 \times 2.3 \text{ ach @ } 50 \text{ Pa} \div 60 \text{ min/hour} = 1,320 \text{ CFM50}$$

Ductwork Leakage

To determine the allowable limit for ductwork leakage, determine the rated air flow rate of the air handler, furnace, air conditioner, etc. at high speed from the manufacturer's literature. For example, a Carrier or Lennox or York heat pump system may have a high speed flow rate of 1,200 cfm across the blower according to the literature supplied with the unit. Ten percent of this value is 120 cfm. This 10 percent value becomes the ductwork leakage limit when the total air handling system is depressurized to 25 Pascals with a duct blaster.

II

Appendix III
Best Practices for High Performance Homes in Hot-Humid Climates

The primary consideration for high performance Building America homes in hot-humid climates is maintaining moisture control both inside the home and within building assemblies, particularly as energy conservation shifts the relationship between sensible and latent loads. Reducing solar gain, using energy conserving appliances and compact fluorescent lighting, reduces the sensible load. This affects the ability of the air conditioner to remove moisture or dehumidify the air.

The following Best Practices are based on our Building America Performance Targets (http://www.buildingscience.com/buildingamerica/targets.htm).

1. Process

Building Design, Systems Engineering, and Commissioning

- *Design for Energy Performance*
  Energy performance 40% better than the 1995 Model Energy Code base case house (i.e. equal to 10% better than Energy Star performance requirements).

- *Systems Engineering*
  - Design structure using advanced framing methods (see Chapter 11: Wood Frame Construction).

- Design structure to accommodate the most efficient duct distribution system that places all ducts and air handling equipment within conditioned space (see Part I: Design and Chapter 13: HVAC).

- Design and detail structure for durability, in terms of wall and roof assembly drying potential, continuous drainage plane, and continuous thermal barrier that clearly defines the conditioned space.

- *Commissioning – Performance Testing*
  - Air leakage (determined by blower door depressurization testing) should be less than 2.5 square inches/100 square feet surface area leakage ratio (CGSB, calculated at a 10 Pa pressure differential); or 1.25 square inches/100 square feet leakage ratio (ASTM, calculated at a 4 Pa pressure differential); or 0.25 CFM/square foot of building enclosure surface area at a 50 Pa air pressure differential. If the house is divided into multiple conditioned zones, such as a conditioned attic or conditioned crawl space, the blower door requirement must be met with the access to the space open, connecting the zones.

  - Ductwork leakage to the exterior for ducts distributing conditioned air should be limited to 5.0 percent of the total air handling system rated air flow at high speed (nominal 400 CFM per ton) determined by pressurization testing at 25 Pa. Two acceptable compliance mechanisms are (1) test duct leakage to outside at finish stage, or (2) test total duct leakage at duct rough-in stage.

  - Forced air systems that distribute air for heating and cooling should be designed to supply airflow to all conditioned spaces and zones (bedrooms, hallways, basements) as well as to provide a return path from all conditioned spaces or zones. Interzonal air pressure differences, when doors are closed, should be limited to 3 Pa. This is typically achieved by installing properly sized transfer grilles or jump ducts (see and Chapter 13: HVAC).

  - Mechanical ventilation system airflow should be tested during commissioning of the building.

  - Testing of the house should be completed as part of the commissioning process. The SNAPSHOT form is available for download as a convenient way to record the testing information (Snapshot Form – see www.buildingscience.com/buildingamerica/snapshot_form.pdf). Instructions for completing the form are also available (Snapshot Instructions — see www.buildingscience.com/buildingamerica/snapshot_instructions.pdf). Unique or custom house plans should each be tested. In a production setting, each model type (i.e., floor plan) should be tested until two consecutive houses of this model type meet testing requirements. At

III

this point, testing on this model type can be reduced to a sampling rate of 1 in 7 (i.e., 1 test, with 6 "referenced" houses). Small additions to a floor plan (e.g., bay window, conversion of den to bedroom) should be considered the same model type; major changes (e.g., bonus room over the garage, conversion of garage into a hobby room, etc) should be considered a separate model type.

2. Site

Drainage, Pest Control, and Landscaping

- *Drainage* – Grading and landscaping shall be planned for movement of building run-off away from the home and its foundation, with roof drainage directed at least 3 feet beyond the building, and a surface grade of at least 5% maintained for at least 10 feet around and away from the entire structure.

- *Pest Control* - Based on local code and Termite Infestation Probability (TIP) maps, use environmentally-appropriate termite treatments, bait systems, and treated building materials that are near or have ground contact (see www.uky.edu/Agriculture/Entomology/entfacts/).

- *Landscaping* – Plantings should be held back as much as 3 feet and no less than 18 inches from the finished structure, with any supporting irrigation directed away from the finished structure. Decorative ground cover—mulch or pea stone, for example—should be thinned to no more than 2 inches for the first 18 inches from the finished structure.

3. Foundation

Moisture Control and Energy Performance

- *Moisture Control* - The building foundation shall be designed and constructed to prevent the entry of moisture and other soil gases.
  - Slabs require a 6-mil polyethylene vapor barrier directly beneath the concrete or an equivalent approach (such as rigid insulation) that accomplishes vapor and capillary control for the slab. The vapor barrier must continuously wrap the slab as well as the grade beam.
  - Sub-slab drainage shall consist of a granular capillary break directly beneath the slab vapor barrier.
  - Perimeter drainage should be used (see Chapter 7: Foundation Design)

III

- Use radon resistant construction practices as referenced in the ASTM Standard "Radon Resistant Design and Construction of New Low Rise Residential Buildings.

4. Enclosure

Moisture Control and Energy Performance

- *Moisture Control*
  - Water management - Roof and wall assemblies must contain elements that provide drainage in a continuous manner over the entire surface area of the building enclosure, including lapped flashing systems at all penetrations. See the *EEBA Water Management Guide* for specific details for various wall assemblies.

  - Vapor management – Roof and wall assemblies must contain elements that, individually and in combination, permit drying of interstitial spaces.

- *Energy Performance*
  - Air leakage - Exterior air barrier—foam sheathing; interior air barrier—gypsum board is sealed to the slab and frame walls (Airtight Drywall Approach).

  - Windows - Recommend one of two approaches:
    - U-factor 0.35 or lower and SHGC (solar heat gain coefficient) 0.35 or less, regardless of climate.

    - Climate-specific glazing properties if passive solar orientation and design can be employed by the builder and occupants employ proper window treatments and their use.

5. Mechanicals/Electrical/Plumbing

Systems Engineering, Energy Performance, Occupant Health and Safety, and Enclosure/Mechanicals Management

- *Systems Engineering*
  - HVAC system design, both equipment and duct, should be done as an integral part of the architectural design process.

  - HVAC system sizing should follow ACCA Manual J and duct sizing should follow Manual D (see Chapter 13: HVAC).

  - Mechanical ventilation should be an integral part of the HVAC system design; see Chapter 13: HVAC).

III

- A whole house dehumidification system integral to the air handling system is recommended for this climate (see www.building science.com/resources/mechanical/conditioning_air.pdf).

- *Energy Performance*

  - Air conditioner or heat pump should be installed according to best practices; see www.buildingscience.com/resources/mechanical/air_conditioning_equipment_efficiency.pdf.

  - Energy Star rated appliances should be selected. See www.eren.doe.gov/consumerinfo/energy_savers/appliances.html.

- *Occupant health and safety*
  - Base rate ventilation: controlled mechanical ventilation at a minimum base rate of 15 CFM per master bedroom and 7.5 CFM for each additional bedroom should be provided when the building is occupied.

  - Spot ventilation: intermittent spot ventilation of 100 CFM should be provided for each kitchen; all kitchen range hoods should be vented to the outside (no recirculating hoods). Intermittent spot ventilation of 50 CFM, or continuous ventilation of 20 CFM when the building is occupied, should be provided for each washroom/bathroom.

  - All combustion appliances in the conditioned space should be sealed combustion or power vented. Specifically, any furnace inside conditioned space should be a sealed-combustion 90%+ unit. Any water heater inside conditioned space should be power vented or power-direct vented. Designs that incorporate passive combustion air supply openings or outdoor supply air ducts not directly connected to the appliance should be avoided.

  - Provide filtration systems for forced air systems that provide a minimum atmospheric dust spot efficiency of 30% or MERV of 10 or higher.

  - Indoor humidity should be maintained in the range of 25 to 60% by controlled mechanical ventilation, mechanical cooling, or dehumidification. See www.buildingscience.com/resources/moisture/relative_humidity_0402.pdf.

  - Carbon monoxide detectors (hard-wired units) should be installed (at one per every approximate 1000 square feet) in any house containing combustion appliances and/or an attached garage.

  - Information relating to the safe, healthy, comfortable operation and maintenance of the building and systems that provide control

III

over space conditioning, hot water, or lighting energy use should be provided to occupants.

- *Enclosure/Mechanicals Management*
  - Plumbing - No plumbing in exterior walls. Air seal around plumbing penetrations in pressure boundary (air barrier) such as rim (band) joist or ceiling.

  - Electrical - Seal around wires penetrating air barrier or pressure boundary.

Appendix IV

Transfer Grille Sizing Charts

- Goal is to prevent pressurization of individual bedrooms when door is closed

- Maximum pressurization of 3 Pa (0.012 WIC) is recommended

- Transfer grille size is based on supply flow to room

- At master bedroom suite (or any other multi-room suite), supply flows must be totaled, and transfer grille sized based on that total

Given:

Door width:	32 inch
Door undercut:	0.5 inch
Transfer grille width:	10, 12, or 14 inches
Net free area:	0.75 fraction of total grille area

Find: How high does the return air transfer grille have to be to meet the 3 Pa criteria?

Based on $Q = 1.07 \times A \times P^{1/2}$ for square edged orifice flow

where: Q = room supply flow (CFM)

1.07 = constant, including unit conversions

A = free area (in^2)

P = pressure difference between room and central area (Pa)

IV

Transfer Grille Sizing Table

Room supply air flow	Net free area required	Area required after door undercut	Transfer grille height required for listed width in inches			Jump duct diameter required
			10	12	14	
(CFM)	(in²)	(in²)	(in)	(in)	(in)	(in)
50	27.0	11.0	1.5	1.2	1.0	3.7
75	40.5	24.5	3.3	2.7	2.3	5.6
100	54	38	5.1	4.2	3.6	7.0
125	67.4	51.4	6.9	5.7	4.9	8.1
150	80.9	64.9	8.7	7.2	6.2	9.1
175	94.4	78.4	10.5	8.7	7.5	10.0
200	107.9	91.9	12.3	10.2	8.8	10.8
225	121.4	105.4	14.1	11.7	10.0	11.6
250	134.9	118.9	15.9	13.2	11.3	12.3
275	148.4	132.4	17.7	14.7	12.6	13.0
300	161.9	145.9	19.4	16.2	13.9	13.6
325	175.4	159.4	21.2	17.7	15.2	14.2
350	188.9	172.9	23.0	19.2	16.5	14.8

IV

Typical Free Area Measurements of Return Air Grilles

- Ameri-Flow Return Air Grilles
- Data from Grainger Catalog (2001)

Width	Nominal Height	Width	Overall size Height	Free area (in²)
10	6	11.5	7.5	51
12	6	13.5	7.5	61
12	12	13.5	13.5	121
14	6	15.5	7.5	71
14	14	15.5	15.5	163
20	20	21.5	21.5	336
30	6	31.5	7.5	153

IV

IV

Appendix V
Additional Resources

Organizations

Advanced Energy Corporation
      909 Capability Drive
      Suite 2100
      Raleigh, NC 27606-3870
      (919) 857-9000
      www.advancedenergy.org

American Council for an Energy-Efficient Economy
      1001 Connecticut Avenue, NW
      Suite 801
      Washington, DC 20036
      Research and Conferences (202) 429-8873
      Publications (202) 429-0063
      aceee.org

American Society of Heating, Refrigeration and Air Conditioning
Engineers
      1792 Tullie Circle, NE
      Atlanta, GA 30329-2305
      (404) 636-8400
      www.ashrae.org

Energy and Environmental Building Association
      10740 Lyndale Avenue South, Suite 10W
      Bloomington, MN 55420
      (952) 881-1098
      www.eeba.org

Florida Solar Energy Center
A Research Institute of the University of Central Florida
1679 Clearlake Road
Cocoa, FL 32992
(321) 638-1000
www.fsec.ucf.edu

Rocky Mountain Institute
1739 Snowmass Creek Road
Snowmass, CO 81654-9199
(303) 927-3851
www.rmi.org
www.natcap.org

Southface Energy Institute
241 Pine Street
Atlanta, GA 30308
(404) 872-3549
www.southface.org

U.S. EPA ENERGY STAR Buildings Program
U.S. EPA Atmospheric Pollution Prevention Division
401 M Street, SW (6202J)
Washington, DC 20460
(888) STAR-YES
www.epa.gov/energystar.html

V

Publications - Books

Building Air Quality
> U.S. Environmental Protection Agency
> Indoor Air Division
> Office of Air and Radiation
> Washington, DC 20460
> (202) 564-7400

Canadian Home Builders' Association Builders Manual
> Canadian Home Builders' Association
> 150 Laurier Avenue West
> Suite 500
> Ottawa, Ontario, Canada K1P 5J4
> (613) 230-3060

Energy Source Directory: A Guide to Products Used in Energy Efficient Construction
> Iris Communications, Inc.
> P.O. Box 5920
> Eugene, OR 97405
> (541) 484-9353

The Healthy House
> Bower, J.
> The Healthy House Institute
> 430 N. Sewell Road
> Bloomington, IN 47408
> (812) 332-5073

Understanding Ventilation: How to Design, Select and Install Residential Ventilation Systems
> Bower, J.
> The Healthy House Institute
> 430 N. Sewell Road
> Bloomington, IN 47408
> (812) 332-5073

Publications - Periodicals and Catalogs

Energy Design Update
> Cutter Information Corporation
> 37 Broadway
> Arlington, MA 02174
> (800) 964-5118

Environmental Building News
R.R. 1
Box 161
Brattleboro, VT 05301
(802) 257-7300

Fine Homebuilding
The Taunton Press
63 S. Main Street, P.O. Box 5506
Newtown, CT 04670-5506
(203) 426-8171

Home Energy Magazine
2124 Kittredge Street
No. 95
Berkeley, CA 94704
(510) 524-5405

Journal of Light Construction
R.R. 2
Box 146
Richmond, VT 05477
(800) 375-5981

Solplan Review
Box 86627
North Vancouver, British Columbia, Canada V7L 4L2
(604) 689-1841

V

Appendix VI
Glossary

Air Barrier Air barriers are systems of materials designed and constructed to control airflow between a conditioned space and an unconditioned space. The air barrier system is the primary air enclosure boundary that separates indoor (conditioned) air and outdoor (unconditioned) air. In multi-unit/townhouse/apartment construction the air barrier system also separates the conditioned air from any given unit and adjacent units. Air barrier systems also typically define the location of the pressure boundary of the building enclosure. In multi-unit/townhouse/apartment construction the air barrier system is also the fire barrier and smoke barrier in inter-unit separations. In such assemblies the air barrier system must also meet the specific fire resistance rating requirement for the given separation.

Air Barrier; Performance Requirements Air barrier systems typically are assembled from materials incorporated in assemblies that are interconnected to create enclosures. Each of these three elements has measurable resistance to airflow. The recommended minimum resistances or air permeances for the three components are listed as follows:

- Material $0.02\ l/(s\text{-}m^2)@75$ Pa
- Assembly $0.20\ l/(s\text{-}m^2)@75$ Pa
- Enclosure $2.00\ l/(s\text{-}m^2)@75$ Pa

Air Retarder Materials and assemblies that do not meet the performance requirements of air barrier materials and air barrier assemblies, but are nevertheless designed and constructed to control airflow are said to be air retarders.

Building Enclosure A building enclosure is an environmental separator. It separates the interior environment from the exterior environment. A building enclosure controls heat flow, air flow, water vapor flow, rain, groundwater, light and solar radiation, noise and vibrations, contaminants, environmental hazards and odors, insects, rodents and vermin, and fire. A building enclosure provides strength and rigidity and must be durable, aesthetically pleasing and economical.

Conditioned Space A conditioned space is the part of the building that is designed to be thermally conditioned for the comfort of occupants or for other occupancies or for other reasons.

Diffusion The movement of individual molecules through a material. The movement occurs because of concentration gradients and thermal gradients, independent of airflow.

Drainage Plane Drainage planes are water repellent materials (building paper, housewrap, foam insulation, etc.) which are typically located behind the cladding and are designed and constructed to drain water that passes through the cladding. They are interconnected with flashings, window and door openings, and other penetrations of the building enclosure to provide drainage of water to the exterior of the building. The materials that form the drainage plane overlap each other shingle fashion or are sealed so that water drains down and out of the assembly. The drainage plane is also referred to as the "water resistant barrier" or WRB.

Equivalent Leakage Area of a building (EqLA or ELA) Quantitative expression of the airtightness of a building enclosure. EqLA is the method set by the Canadian General Standards Board in which a blower door depressurizes the building enclosure to 10 Pascals and the leakiness of the enclosure is expressed as a summary hole in square inches. ELA is set by the ASTM equivalent procedure at a pressure differential of 4 Pascals. *See also* Builder's Guides Appendix II: Air Leakage Testing, Pressure Balancing, and Combustion Safety for a complete discussion of these and other expressions of air leakage.

Foundation, Water-managed Systems for at or below-grade enclosure assemblies where gravity (drainage) is used to move liquid water away from the structure, relieving hydrostatic water forces. *See* Chapter 7: Foundations — Water Managed Foundations.

Grade Beam A foundation wall that is cast at or just below the grade of the earth, most often associated with the deepened perimeter concrete section in slab-on-grade foundations.

Housewrap Any of the numerous spun-fiber polyolefin rolled sheet goods, or perforated plastic films designed to function as drainage planes.

VI

Indoor Air Air in a conditioned space.

Insulating Sheathing Non-structural insulating board products with varying R-values and a wide variation in vapor permeability and drainage characteristics. Materials include expanded polystyrene (EPS), extruded polystyrene (XPS), polyisocyanurate (most often foil-faced), rigid fiberglass, and mineral wool. *See also* Chapter 4: Insulations, Sheathings and Vapor Retarders.

Jump Duct A flexible, short, U-shaped duct (typically 10-inch diameter) that connects a room to a common space as a pressure balancing mechanism. Jump ducts serve the same function as transfer grilles. Used when return ducts are not located in every room. *See also* Chapter 13: HVAC.

Kiln-dried Lumber Any lumber placed in a heated chamber or "shed" to reduce its moisture content to a specified range or average under controlled conditions. For softwood framing lumber, the moisture content of KD lumber is somewhat based on regional conventions but is most often an average of 12% by weight. In comparison, the moisture content of thoroughly air-dried softwood framing lumber is 15% to 20%. *See also* Chapter 17: Drywall — Truss Uplift.

Low-E Most often used in reference to a coating for high-performance windows, the "e" stands for emissivity or re-radiated heat flow. The thin metallic oxide coating increases the U-value of the window by reducing heat flow from a warm(er) air space to a cold(er) glazing surface. The best location for the coating is based on whether the primary heat flow you want to control is from the inside out (heating climates) or the outside in (cooling climates).

Mechanical Ventilation Controlled, purposeful introduction of outdoor air to the conditioned space. *See also* Chapter 13: HVAC.

Outdoor Air Air outside the building. It can enter the conditioned space via the ventilation system, or by infiltration through holes in the pressure boundary or designed ventilation openings.

Ozone O_3 instead of O_2. This 3-atom molecule is an even more active oxidizing agent than its more common 2-atom relative. At ground level, ozone is a pollutant and in the upper atmosphere it is a solar shield (location, location, location). Touted for its ability to "clean" air in room or household ozone generators, this application actually does more harm than good-ozone's highly reactive nature tends to accelerate the breakdown of synthetic materials in homes such as paints, plastics, and ever-available volatile organic compounds, often with less-than-desirable results. All told, we look to protect ozone in the heavens and shun it here at home, inside and out.

VI

Appendices

Permeance The physical property that defines the ease at which water molecules diffuse through a material. It is to vapor diffusion what conductance is to heat transfer. The unit of measurement is typically the "perm."

Pressure Boundary Air barriers define the location of the pressure boundary of the building enclosure. *See also* Air Barrier.

R-value Quantitative measure of resistance to heat flow or conductivity, the reciprocal of U-factor. The units for R-value are ft2 °F hr/Btu (English) or m2 °K hr/W (SI or metric). While many in the building community consider R-value to be the primary or paramount indicator of energy efficiency, it only deals conduction, one of three modes of heat flow, (the other two being convection and radiation). As an example of the context in to which R-value should be placed, 25% to 40% of a typical home's energy use can be attributed to air infiltration.

Thermal Boundary The layer in a building enclosure that controls the transfer of energy (heat) between the interior and the exterior. It is a component of the building enclosure and it may, but does not have to align with the pressure boundary.

U-factor Quantitative measure of heat flow or conductivity, the reciprocal of R-value. While building scientists will use R-values for measures of the resistance to heat flow for individual building materials, U-factor is always used as a summary measure for the conductive energy measure of building enclosures.

Vapor Barrier A vapor barrier is a Class I vapor retarder. Vapor barriers are materials that are vapor impermeable.

Vapor Impermeable Materials with a permeance of 0.1 perm or less (rubber membranes, polyethylene film, glass, aluminum foil)

Vapor Permeable Materials with a permeance of greater than 10 perms (housewraps, building papers)

Vapor Retarder A vapor retarder is the element that is designed and installed in an assembly to retard the movement of water by vapor diffusion. There are several classes of vapor retarders:

Class I vapor retarder	0.1 perm or less
Class II vapor retarder	1.0 perm or less and greater than 0.1 perm
Class III vapor retarder	10 perms or less and greater than 1.0 perm

VI

The test procedure for classifying vapor retarders is ASTM E-96 Test Method A — the desiccant or dry cup method. *See also* Chapter 4: Insulations, Sheathings and Vapor Retarders.

Vapor Semi-Impermeable Materials with a permeance of 1.0 perm or less and greater than 0.1 perm (oil-based paints, most vinyl coverings)

Vapor Semi-Permeable Materials with a permeance of 10 perms or less and greater than 1.0 perm (plywood, OSB, most latex-based paints)

Water Resistant Barrier A water resistant barrier (WRB) is also referred to as a drainage plane.

Wind-Washing The phenomenon of air movement that occurs due to wind entering building enclosures typically at the outside corners and roof eaves of buildings. Wind-washing can have significant impact on thermal and moisture movement and hence thermal and moisture performance of exterior wall assemblies.

Xeriscaping Climate-tuned landscaping that minimizes outdoor water use while maintaining soil integrity and building aesthetics. Typically includes emphasis on native plantings, mulching, and no or limited drip/subsurface irrigation.

Zero Energy House Any house that averages out to net zero energy consumption. A zero energy home can supply more than its needs during peak demand, typically using one or more solar energy strategies, energy storage and/or net metering. In a zero energy home, efficiencies in the building enclosure and HVAC are great enough that plug loads tend to dominate and so these homes must have the added focus of high efficiency appliances and lighting.

VI

VI

Index

A

adobe. *See* rammed earth
air barrier(s) 101-111, 115, 118-120, 155-157, 162, 165, 173, 182, 187, 204-206, 210, 211, 245, 272, 273, 289, 381, 388, 390, 394, 396, 415, 465, 466, 516, 525, 528
air leakage 125, 341, 371, 375, 379, 514
  testing 503, 504, 506, 517
air-to-air heat exchanger 337
air-to-air heat pump 357. *See also* heat pump: air-to-air
air retarder 35, 41, 97, 266, 281, 295, 449, 525. *See also* polyethylene
Airtight Drywall Approach (ADA) 109, 110
ants 224
asthma trigger 223, 225
attic
  access
    removable cover 286
    scuttlehole 286
  ventilation fan

B

backdraft 4, 5, 42-50, 53, 56, 323, 327, 328, 507, 508
baffle 156, 308, 384, 385, 387, 416, 417, 468-470. *See also* insulation: baffle
balancing damper 339
barbecue 349
blower door 504, 505, 509, 526
break points xviii, 35
building enclosure xii, xviii, 2, 4, 5, 8, 32, 40, 43, 46, 47, 61, 62, 101, 102, 113, 119-121, 158, 245, 277, 278, 304, 323, 324, 360, 375, 503, 525, 526, 528, 529

C

calibrated fan 504, 506. *See also* blower door; duct blaster; fan
capillarity 64, 465, 176, 235, 403
capillary break 64, 69, 81, 177, 178, 180, 185, 192, 194-201, 203-207, 212-219, 221, 239, 320, 411, 413-415, 438-442, 447, 462-465, 467, 513
carbon dioxide 3, 14, 391, 392
carbon monoxide 12, 61, 509, 515

carpet
  dust marking xix, 42
chimney 4, 57, 277, 278, 392
chlorofluorocarbons (CFCs) 14
clip support 248, 249, 393. *See also* drywall: clips
clothes dryer 349, 508
cockroaches 223. *See also* roaches
combustion safety 503, 508
commissioning 61, 62, 508
concrete
  control joints 176, 234, 235, 237, 238, 240,
  expansion joint 236, 237
  seat 191-200, 203, 235, 304-307, 320
  step down 190
contaminants 404, 526
control joint 310, 390. *See also* concrete: control joints; flashing
crawlspace 3, 5, 32, 41, 42, 49, 57, 115, 175, 177, 179, 182, 185, 186, 188,
    190, 204, 207-212, 324, 342, 343, 346, 388, 466, 507
cross bracing 230, 254

D

dampproofing 59, 150, 151, 176, 177, 193-195, 204, 207, 211, 413-415,
    440-442
daylighting 17, 19, 21, 24, 25, 376
decommissioning
dehumidification 57, 114, 120, 122, 132-134, 322, 326-328, 331, 337, 341
desiccant 528
dewpoint 124, 157, 223,
diffusion 114, 115, 122, 239, 242, 403, 526
dilution 4, 121, 130, 133, 134, 327, 328, 509
draftstop 119, 120, 244, 245, 270, 272-274, 276, 281-284, 287, 308, 363,
    365, 367, 369, 371, 375, 380, 390
drain pipe 182, 211, 320, 413-415, 440, 442, 465, 467. See also perimeter
    drain
drywall
  clips 109, 395. *See also* clip support
  floating corners 390, 392-394
duct
  leakage 40, 338, 341, 507
duct blaster 506, 510
dust mites 4, 184, 223, 224
dustmarking 349. See also carpet; dustmarking

E

earthquake 243
EER 27, 31

EIFS 66, 70, 211, 214
electrical
    boxes 319, 377. *See also* outlet boxes
    panel 285, 375, 376
    sealing 378
    wiring 383. *See also* insulation: installation
Energy Star 515
entropy 114
Equivalent Leakage Area (EqLA or ELA) 504, 505, 509, 526
expanded polystyrene (EPS). *See* insulation
exfiltration 101, 120, 507
extruded polystyrene (XPS). *See* insulation

F

fan
    attic 181-183, 508
    calibrated 504, 506
    exhaust 4, 60, 176, 327, 334, 349, 507
    recirculating 329
    supply 334
    whole house 329, 340, 349, 507, 508, 508
fan-coil 44-50, 323, 360. *See* water-to-air heat exchanger
fenestration. *See* windows
fire 5, 13, 101, 156, 277, 278, 335-337, 346, 381, 526
fireplace 4, 42, 53, 57, 60, 61, 277, 278, 327, 328, 340, 349, 507-509
    sealed combustion, direct vent 277, 278
firestop 244, 245, 277, 279, 280, 390
flashing 212, 216-219, 271, 284, 290, 294, 297-301, 305, 310, 374, 441.
      See also windows: sill pan
    kick-out 72, 94
    step 78, 94, 98, 99
flue 277-280, 358
fly ash 240

G

gas barrier 101, 104, 187
glazing 3, 24, 25, 35, 37. *See also* window(s)
ground source heat pump 357. *See also* heat pump: ground source
groundwater 35, 175, 176, 179, 180, 185, 188, 239, 241, 409, 461, 526

H

heat flow 162, 522, 528
heat gain 21, 24-26, 39, 155, 322, 340, 341
heat loss 21, 24, 155, 186, 340, 341
heat pump 4, 43-48, 50, 60, 341, 515

heat recovery ventilators (HRV's) 337, 509
herbicides 176, 346
housewrap(s) 3, 35, 67, 73, 74, 76, 103, 106, 108, 115, 116, 118, 245, 295,
    403, 404, 449, 454-456, 459, 526, 528
humidity 136, 321-324, 327, 331, 334
hydrophobic 116, 400, 403
hydrostatic (water forces) 176, 403, 526
hygric buffer 158
hygroscopic 115, 116, 128, 329, 433, 461

I

ice-dam 51, 155, 162, 165, 211, 295, 297, 387
infiltration 5, 12, 52, 103, 120, 122, 131, 133, 134, 322, 332, 333, 507, 527,
    528
insect(s) 1, 36, 409, 526
isocyanurates. *See* insulation
insulating concrete forms (ICFs) 32, 409-414, 416
    basement 413, 414
    crawlspace 415
    slab 411
    systems 410
    windows 420-432
insulating sheathing 32, 45-47, 117, 122-125, 142, 158, 161, 168, 205, 245,
    290, 527
insulation(s)
    baffle 288, 291
    batt 119, 120, 159, 160, 169, 171, 184, 213-219, 195, 204, 265, 286,
        375, 384, 391, 416-419, 440, 466, 468-470
    blown/dry spray cellulose 188, 193, 204, 205, 207, 289, 304, 381, 387,
        468-470
    damp spray cellulose 113, 126, 150, 169, 193, 381, 387
    expanded polystyrene (EPS) 94, 117, 132, 189, 196, 197, 199, 212, 214,
        221, 289, 304, 381, 527
    extruded polystyrene (XPS) 93, 93, 117, 132, 184, 187, 189, 197, 199,
        212, 214, 221, 289, 304, 307, 381, 440, 527
    fiberglass 42, 63, 113, 118, 120, 122-124, 144, 145, 147, 148, 198-203,
        304, 381, 382, 418, 419, 469, 470, 527
    isocyanurate 117
    installation 382, 383
    mineral/rock/slag wool 113, 381, 527
    spray foam 103, 157, 162, 166-168, 212, 266, 268, 272, 273, 320, 381,
        388, 390

J

joist
    hanger 465, 467

    set-back rim 265
jump duct 43, 53, 347, 361, 518, 527

K

kerf 64, 83, 420
kiln-dried lumber 527

L

leakage coefficient 509, 510
leakage ratio 505, 509
lighting 9, 376, 516, 529
    fixtures
        insulated cover (IC) rated 375, 379
        recessed 375, 379
        sealing 380
Low-E 527. *Seealso* window(s): Low-E

M

Manual D 341, 514
Manual J 341, 514
massing 17, 32
mice 223, 224
mold 4, 36, 157, 162, 177, 184, 207, 211, 213, 232, 343, 347, 348, 403,
        413-415

N

nail pops xix, 389
noise 35, 321, 368, 369, 526

O

odor 1, 8, 13, 321, 325, 329, 343, 371, 508, 526
outlet boxes 375
overhang 2, 20, 32, 64, 65, 69, 72, 180, 290
ozone 13, 14, 527

P

perimeter drain 180, 211, 466
pest 185, 186, 188, 223
pesticide 5, 186, 223, 224, 346, 507
pollutant 3, 4, 20, 42, 321, 324, 325, 328, 343, 507, 527
polyethylene 32, 91, 92, 94, 97, 103, 1-6, 108, 115-117, 119, 130, 150,
        151, 165, 176-178, 182, 185, 192, 194, 195, 197-204, 206, 207, 210-

220, 239-242, 245, 290, 381, 387, 390, 396, 411, 413-415, 438-442, 447, 449, 462-467, 513. *See also* air barrier; vapor barrier
precast autoclaved aerated concrete (PAAC)
   crawlspace 465, 466
   slab 462-464
   window 471-476
pressure balancing 503, 506, 527
pressure boundary 185, 516, 527, 528
pressure relief grille 338. *See also* transfer grille

R

R-value 528
radiant barrier 39
radiation 528
radon 3, 5, 12, 176, 186, 327, 340, 346, 507, 514
rain control 66, 192, 245, 303, 409, 433, 461
rammed earth 25, 26
rats 223, 224
roaches 224. *See also* cockroaches
rodent 224, 526
roof
   framing 288, 416-418, 468, 469
   hot 295
   knee wall 291
   set-back 272
   shed 95, 271
   truss 247, 419, 470
   ventilation 292, 297
      hip 293
      shed 294
   wood shingle or wood shake 297

S

SEER 27
seismic 37
   conditions 17, 35
   zones xiv
shear 254, 294
   forces 37
   loads 35
   panels 38
sheathings
   insulating. *See* insulation: expanded polystyrene (EPS); insulation: extruded polystyrene (XPS)
SHGC (solar heat gain coefficient) 24, 514

siding
    aluminum 96, 131, 133, 134, 136-138, 141, 147, 159, 161, 169, 201, 202,
        205, 211, 213, 216, 217, 219, 257, 260, 267, 295, 384, 385, 439, 449
    fiber cement 69, 141, 144, 147, 201, 202, 308, 311
    vinyl 96, 131, 133, 134, 136-138, 141, 147, 159, 161, 169, 171, 201, 202,
        205, 211, 213, 216, 217, 219, 267, 295, 384, 385, 439, 449
    wood 67, 69, 144, 145, 147, 201, 202, 204, 245, 258, 259, 296, 403, 409,
        430, 431, 432, 438, 440, 443-445, 447, 466
siting 17
skylight 25, 61, 292, 302
soffit 33, 41, 167, 245, 297, 308, 309, 339, 386, 387, 445, 449
    interior 281, 282, 375, 380
    vent 384-387, 416-419, 468-470
soil gas(es) 5, 175, 176, 186, 210, 320, 327, 340, 346, 507, 513. *See also*
        radon
solar
    gain 17, 19, 21, 24, 25, 39, 321
    radiation xiv, 304, 401, 526
solar water heating/solar assisted water heating 60, 61
sound 338, 361, 362
soundproofing 346
stack effect 60, 119, 181-183, 503
stack framing 32, 34, 246, 247
stain
    solid body 400
structural insulated panel systems (SIPS) 32, 433, 44
    basement 440-442
    door 435
    roof
        framing 445
        ridge 446
    service chase 447
    slab 438, 439
    telegraphing 433, 448-450
    wall assembly 434
    window 436, 437
sump (pump) 180, 187, 188
sunspaces 2
surfactant 403, 404

T

termite 36, 185, 188, 207, 211, 213, 215-218, 224, 411, 414, 415, 438, 439,
    447, 462-464, 513
termiticides 176, 221
thermal barrier 41, 162, 166-168, 173, 185, 212
thermal bridging 149-151, 153, 191-194, 197
thermal mass 21, 25, 27, 29-31, 39, 57

thermostat
   setback 341
transfer grilles 43, 45-47, 49, 53, 54, 186, 208, 209, 347, 360-362, 517, 518, 527
truss uplift 390, 394

U

U-value (factor) 24, 528
ultraviolet radiation xv, 8, 399, 400

V

vapor barrier 36, 80, 103, 118, 124, 150, 151, 162, 177, 178, 181-183, 186, 192-200, 202-204, 206, 207, 210, 211, 213, 220-222, 241, 242, 305, 307, 320, 411, 413-415, 462-467, 513, 528. *See also* polyethylene
vapor diffusion 32, 113-115, 118-120, 124, 127, 155, 385, 448, 450, 528
vapor retarder 36, 115, 117-119, 123-125, 140, 141, 143, 146, 148, 150, 151, 155, 157, 158, 161, 163, 164, 193-195, 239, 245, 295, 387, 390, 404, 438-442, 447, 527, 528
vegetation xiv, 17, 19, 24, 341
vent stack 56, 181-183, 371, 374
ventilation (system) 9, 22, 24, 25, 28, 39, 43, 57, 120-122, 176, 181-183, 188, 286, 292, 514, 527
   attic 39, 155, 387, 507
   balanced 321, 326, 327, 330, 336, 337
   exhaust 321, 326, 327, 330, 332, 333
   ridge 384-386
   supply 321, 326, 327, 330, 331, 334-336, 342
   whole house 27, 31
vermin 1, 520, 526
vibration 1, 13, 321, 338, 343, 347, 526
visible light transmittance 24
volatile organic compound(s) (VOC) 3, 404, 527

W

wall coverings
   vinyl 304
wallpaper. *See* wall coverings: vinyl
water resistant barrier (WRB) 67, 222, 526, 529
waterproofing 96, 153, 175, 197, 203, 206, 211-218, 274, 311-313, 391, 414, 415, 423, 428, 429, 432, 462-471, 474-476
whole house fan *See* fan: whole house
wind-washing 101, 103, 387, 529
window(s)
   head 76, 77, 83, 93, 311-313, 420, 424, 427, 430, 455, 457-459, 471, 474
   jamb 74, 76, 93, 110, 111, 253, 423, 426, 429, 432, 456, 457, 473, 476

low E/low emmitance 25, 45-47, 322
pan flashing 456, 305
sill 64, 74, 75, 82, 93, 310-313, 421-423, 425, 428, 431, 456, 457, 472, 475
spectrally selective 25, 45-47, 57, 322. *See also* glazing
wood
  shingle/shake 158, 297, 446
  stove 60, 327-329, 507-509

X

xeriscaping 20, 529

Z

zero energy house 529